Lecture Notes in Artificial Intelligence   4045

Edited by J. G. Carbonell and J. Siekmann

Subseries of Lecture Notes in Computer Science

T0239125

Dave Barker-Plummer   Richard Cox
Nik Swoboda (Eds.)

Diagrammatic Representation and Inference

4th International Conference, Diagrams 2006
Stanford, CA, USA, June 28-30, 2006
Proceedings

 Springer

Series Editors

Jaime G. Carbonell, Carnegie Mellon University, Pittsburgh, PA, USA
Jörg Siekmann, University of Saarland, Saarbrücken, Germany

Volume Editors

Dave Barker-Plummer
Stanford University, CSLI
Cordura Hall, Stanford, CA 94305-4101, USA
E-mail: dbp@csli.stanford.edu

Richard Cox
University of Sussex, School of Science and Technology
Department of Informatics, Falmer, Brighton BN1 9QH, UK
E-mail: richc@sussex.ac.uk

Nik Swoboda
Universidad Politécnica de Madrid, Departamento de Inteligencia Artificial
Campus de Montegancedo, 28660 Boadilla del Monte, Madrid, Spain
E-mail: nswoboda@clip.dia.fi.upm.es

Library of Congress Control Number: 2006927703

CR Subject Classification (1998): I.2, D.1.7, G.2, H.5, J.4, J.5

LNCS Sublibrary: SL 7 – Artificial Intelligence

ISSN   0302-9743
ISBN-10   3-540-35623-1 Springer Berlin Heidelberg New York
ISBN-13   978-3-540-35623-3 Springer Berlin Heidelberg New York

Springer is a part of Springer Science+Business Media

springer.com

© Springer-Verlag Berlin Heidelberg 2006
Printed in Germany

Typesetting: Camera-ready by author, data conversion by Scientific Publishing Services, Chennai, India
Printed on acid-free paper   SPIN: 11783183   06/3142   5 4 3 2 1 0

Preface

The Diagrams conference series is concerned with the study of all aspects of diagram use. While diagrammatic representations have been a corner-stone of communication throughout human history, modern media are much more graphical than those of the past. Recent advances in technologies for information presentation - on computers, PDAs and cell-phone screens - have immersed us in a reality that is richly pervaded by diagrammatic representations.

The diagrams research community tends to define the term "diagram" very broadly. Those studied include familiar notations such as bar charts and line graphs, formally defined notations such as Venn diagrams and Peirce's existential graphs, and complex representations such as the graphical user interfaces of most modern computer systems. Thus, the conference series attracts researchers from virtually all academic fields that are studying the nature of diagrammatic representations. These include disciplines that study human communication, human cognition, computational systems, HCI, information visualization, pure and applied sciences, education and philosophy, etc.

Diagrams 2006 is the fourth event in this conference series, which was launched in Edinburgh in September 2000, and which has emerged as the major international conference on this topic. For Diagrams 2006 we especially solicited papers in the area of diagrams and education—a topic that includes uses of diagrams in all subject areas and phases of education from primary schools to professional development, and which also concerns education for diagrammatic literacy.

The call for papers for Diagrams 2006 solicited contributions of full papers, extended abstracts and tutorial proposals. Submissions were received from authors representing both academia and industry, 23 countries, and disciplines including: aerospace, computer and information science, education, engineering, geographical and spacial information science linguistics, mathematics, philosophy, psychology and telecommunications.

Submissions were reviewed by a distinguished international Program Committee with members from ten countries. Each submission was peer-reviewed by three members of the Program Committee or by reviewers that they nominated. This labor-intensive process was designed to ensure that only the submissions with the greatest technical merit, clearest communication, and widest interest were included in the conference. By this process we sought to maintain the quality and multidisciplinary balance characteristic of the conference series. Keeping to the high standards of previous Diagrams conferences, the acceptance rate for full papers was about 30% (13 papers), and that for extended abstracts 60% (22 abstracts). Nine submissions chosen from the best of the extended abstracts were selected for short paper presentation.

Besides technical paper and poster sessions, Diagrams 2006 included three invited talks and two tutorial sessions. The first invited talk was by Daniel Schwartz, a professor of Education at Stanford University. The second was by Felice Frankel of the Massachusetts Institute of Technology. Laura Novick of Vanderbilt University presented the third invited address. Mary Hegarty of The University of California Santa Barbara presented a tutorial on the collection and use of data from eye-tracking studies, while Aaron Marcus of Aaron Marcus and Associates presented a tutorial on issues in the design of cross-cultural user interfaces.

Diagrams 2006 would not have been possible without the efforts of many people. The Program Committee performed the difficult and time-consuming task of reviewing submissions. Their work made the program what it is. We value the continued support of our publishers, Springer, and particular thanks are due to Ursula Barth. Stanford's Office of Conference Services and the staff at the conference location in Wallenberg Hall ensured that the logistics of the conference ran smoothly. We appreciate the continued financial support of the Cognitive Science Society, who again provided funding for best paper prizes. The conference received funding from the office of the Provost at Stanford University, The Center of the Study of Language and Information, and CSLI Publications. Apple Computer Inc. provided iPods as best paper prizes, as well as equipment and reprographic support. We are grateful to Google Inc. and to Microsoft Corporation for providing funds to assist student travel and registration. Richard Cox would like to acknowledge the support of the Royal Society, Nik Swoboda the support of the Universidad Politécnica de Madrid and Dave Barker-Plummer that of Stanford University.

June 2006   Dave Barker-Plummer
                                                   Richard Cox
                                                   Nik Swoboda

Conference Organization

General Chair

Dave Barker-Plummer   Stanford University, USA

Program Chairs

Richard Cox   University of Sussex, UK
Nik Swoboda   Universidad Politécnica de Madrid, Spain

Sponsoring Institutions

Apple Computer, Inc.
Cognitive Science Society
Google, Inc.
Microsoft Corporation
The Royal Society
Stanford University
Stanford University Center for the Study of Language and Information
Center for the Study of Language and Information Publications

Program Committee

Gerard Allwein   Naval Research Laboratory, USA
Michael Anderson   University of Hartford, USA
Alan Blackwell   Cambridge University, UK
Dorothea Blostein   Queen's University, Canada
Paolo Bottoni   University of Rome - "La Sapienza", Italy
B. Chandrasekaran   Ohio State University, USA
Peter Cheng   University of Sussex, UK
Max J. Egenhofer   University of Maine, USA
Sara Fabrikant   Universität Zürich, Switzerland
George W. Furnas   University of Michigan, USA
Corin Gurr   University of Reading, UK
Volker Haarslev   Concordia Univesity, Canada
Patrick G. T. Healey   Queen Mary, University of London, UK
Mary Hegarty   University of California, Santa Barbara, USA
John Howse   University of Brighton, UK
Roland Hübscher   Bentley College, USA

Table of Contents

Keynote Presentations

Tutorials

Diagram Comprehension by Humans and Machines

Notations: History, Design and Formalization

Diagrams and Education

Reasoning with Diagrams by Humans and Machines

Psychological Issues in Comprehension, Production and Communication

The Importance of Both Diagrammatic Conventions and Domain-Specific Knowledge for Diagram Literacy in Science: The Hierarchy as an Illustrative Case

Laura R. Novick

Dept. of Psychology & Human Development
Peabody College #512
230 Appleton Place
Vanderbilt University
Nashville, TN 37203
Laura.Novick@vanderbilt.edu

"Visual displays are distinctively involved in scientific communication and in the very 'construction' of scientific facts... They are essential to how scientific objects and orderly relationships are revealed and made analyzable" (Lynch, 1990, pp. 153-154).

As noted so eloquently by Lynch (1990), diagrams are critically important in science. Hegarty, Carpenter, and Just (1991) classified scientific diagrams into three categories: iconic, schematic, and charts and graphs. Iconic diagrams, such as photographs and line drawings, provide a depiction of concrete objects in which the spatial relations in the diagram are isomorphic to those in the referent object. Accurate representation of spatial relations can be critical, for example to distinguish the venomous coral snake from the similarly-colored non-venomous Arizona mountain king snake. In the life sciences, iconic representations help students understand the structure of objects that are not easily open to visual inspection. For example, side-by-side drawings of the stomachs of people and cows, with the parts labeled, would provide insight into why digestion works differently in these two taxa.

Charts and graphs are diagrams that depict a set of related, typically quantitative, data. For example, a line graph can depict the relation between temperature and the rate of fermentation of yeast. A pie chart can show the percentage of extant vertebrate species that belong to various taxonomic groups (e.g., fish, birds and reptiles, amphibians, mammals). A bar graph can show the relation between age (e.g., teenagers, 20s, elderly, etc.) and the percentage of fatal crashes as a function of blood alcohol concentration.

Finally, schematic diagrams are abstract diagrams that simplify complex situations by providing a concise depiction of their abstract structure (Lynch, 1990; Winn, 1989). Examples include circuit diagrams, Euler circles, matrices, networks, and hierarchical trees. There is some evidence that diagrams in general, and schematic diagrams in particular, have come to play an increasingly important role in science during modern history (Maienschein, 1991). In science education, schematic diagrams are important beginning at least in middle school. My examples here come from 6th-9th grade textbooks.

D. Barker-Plummer et al. (Eds.): Diagrams 2006, LNAI 4045, pp. 1–11, 2006.
© Springer-Verlag Berlin Heidelberg 2006

The periodic table represents relations among the chemical elements in a matrix format. In all areas of science, matrices are used to compare and contrast properties of important objects and concepts. Electric circuit diagrams are common in physics. Networks are used to represent biological systems (e.g., cellular respiration, food webs) and cycles of matter (e.g., the water cycle). Cladograms—hierarchical diagrams that depict the distribution of characters (i.e., physical, molecular, and behavioral characteristics) among taxa (see Figure 1)—are perhaps the most important tool modern biologists use to reason about evolutionary relationships. Cladograms are hypotheses about the degree of relationship among a set of taxa that are supported by shared, derived evolutionary novelties called synapomorphies (Hennig, 1966). For example, a synapomorphy that supports the evolutionary relationship between birds and crocodiles is having a gizzard. This synapomorphy tells us that birds and crocodiles are more closely related to each other than either is to lizards.

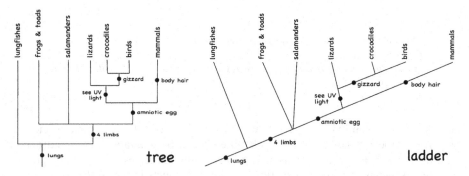

Fig. 1. Isomorphic tree and ladder versions of a cladogram representing evolutionary relationships among a set of vertebrate taxa

Diagrams are ubiquitous in scientific fields in part because they depict the way things are (or are hypothesized to be) and in part because they are important tools for learning and reasoning about (as well as communicating) structures, processes, and relationships (e.g., Day, 1988; Kindfield, 1993/1994; Whitley, Novick, & Fisher, 2006). For example, Kindfield studied how people who varied in their amount of formal training in genetics used chromosome diagrams to reason about the process of meiosis. She found that although all the subjects spontaneously generated diagrams while solving the problems, the more advanced subjects' diagrams focused more on meiosis-relevant features of chromosomes, whereas the less advanced subjects' diagrams focused more on perceptual appearance (i.e., how chromosomes look under a light microscope). Not only did the more advanced subjects draw diagrams that were more abstract, their diagrams were fine-tuned for the requirements of the particular problem they were trying to solve.

1   Diagram Literacy

Given that diagrams of various types play a critical role in science, it is important to understand what constitutes diagram literacy in science. In this section, I provide a brief consideration of diagram literacy for each of the three classes of scientific diagrams. For all three classes, it appears that developing diagrammatic competence goes hand-in-hand with acquiring conceptual knowledge in the domain of application. Put another way, general diagrammatic knowledge is not sufficient for literacy. Domain-specific knowledge in the area of application is required as well.

1.1 Iconic Diagrams

Iconic diagrams are very effective for (a) teaching students how different kinds of physical systems work (e.g., Hegarty et al., 1991) and (b) facilitating the task of object construction (e.g., Gauvain, de la Ossa, & Hurtado-Ortiz, 2001). They are useful for these purposes exactly because they look like what they represent (e.g., pulley systems, chromosomes), so there is little overhead in understanding the diagram prior to using it to support learning and performance. Nevertheless, it is not a one-way street from diagrammatic knowledge to scientific understanding. For example, Kindfield (1993/1994) argued that pictorial skill and conceptual knowledge of meiosis are mutually reinforcing: Increasing conceptual understanding enables students to efficiently represent important aspects of meiosis in their diagrams, and better pictorial representations support the construction of an accurate mental model of the biological situation under consideration.

1.2 Charts and Graphs

In contrast to iconic diagrams, the other types of scientific diagrams rely on convention for their interpretation. For example, putting the independent variable (IV) on the X-axis and the dependent variable on the Y-axis when graphing experimental results is a convention that must be learned. Students who fail to abide by this convention may find that their graphs are misinterpreted; and they may misinterpret the graphs of others. Creating a bar graph when the IV is categorical but a line graph when it is continuous is another convention. There is a sizeable literature on these conventions and on how students interpret different kinds of graphs (e.g., Gattis & Holyoak, 1996; Zacks & Tversky, 1999).

Some research by Mevarech and Stern (1997) suggests that knowledge of the conventions is necessary but not sufficient for comprehending graphs. They examined seventh graders' and college students' understanding of line graphs that were presented abstractly versus embedded in a real world context. In the abstract version, the axes were labeled "X" and "Y", the axis values arbitrarily ranged from 0-12, and the two lines were labeled "Line A" and "Line B." In the contextual versions, the abstract content was replaced with real-world labels and values—e.g., income in millions of dollars for the years 1980-1992 for Company A and Company B. Across four studies, Mevarech and Stern found that the abstract graph led to better understanding of abstract mathematical concepts (e.g., properties of slopes of lines) than did the real-world contexts.

1.3 Schematic Diagrams

Schematic diagrams also rely on learned conventions. Understanding these conventions is the difference between seeing the diagram as a representation of abstract structure—something that has meaning—versus seeing it simply as a design composed of various graphic elements. The conventions specify both the applicability conditions for each diagram (i.e., the types of situations, or problem structures, for which each diagram is best suited) and the rules for mapping components of a problem situation onto components of the diagrams. The necessity of learning the requisite conventions to understand and use schematic diagrams is often taken to be a defining feature of such diagrams (e.g., Dufour-Janvier, Bednarz, & Belanger, 1987; Hegarty et al., 1991).

Because many schematic diagrams are mathematical in nature, initial competence in understanding and using them ideally should come in math classes. In fact, the National Council of Teachers of Mathematics (NCTM), in their *Principles and standards for school mathematics* (NCTM, 2000), has argued strongly for the importance of diagrammatic representation: Children should learn how to create and use both domain-specific and domain-general diagrams to model phenomena in the world (including, and I would add perhaps especially, scientific phenomena) and to communicate the mathematical ideas underlying those phenomena. Such instruction should begin in late elementary school (grades 3-5) and continue through high school.

In biology, there has similarly been a recent call to incorporate instruction in diagrammatic reasoning into the curriculum, particularly for the study of macroevolution (Catley, Lehrer, & Reiser, 2005). Catley et al. argue for introducing students to Euler circles in the primary grades (K-2) and to cladograms in late elementary school. By middle school (grades 6-8), students should be able to use cladograms to reason about, for example, which characters resulted in widespread radiations of species (because they evolved only once a long time ago, in geologic time). This call is motivated by (a) the absence from current evolution curricula of species (the units of evolution) and their contributions to the formation of higher-order taxa and (b) the fact that cladograms are uniquely suited for making inferences about the evolutionary history of a species and about evolutionary relationships among groups of species and higher-order taxa.

Although learned conventions are clearly critical for understanding and using schematic diagrams, one does not attain expertise in using these diagrams for scientific representation solely by learning their conventions. Rather, domain-specific knowledge within the area of application is required as well. Kalyuga, Chandler, and Sweller's (1998) results support this claim with respect to comprehension of electric circuit diagrams.

In the next section, I discuss the structure of and conventions for a common schematic diagram—the hierarchy—that is very important in science, particularly biology. Then I present some data concerning both (a) students' knowledge of these conventions and (b) the additional importance of domain knowledge for diagrammatic competence.

2 The Structure of Hierarchies

Mathematically, a hierarchy is a special case of a network (Novick & Hurley, 2001). According to the NCTM (2000), students should learn to use these visual tools during middle school; and during high school they should develop facility in using them.

Novick and Hurley (2001) proposed that hierarchies can be distinguished from networks by their pattern of values on 10 structural properties—e.g., global structure, linking relations, traversal. These property values constitute the applicability conditions for each diagram. The applicability conditions discussed here are the ones that Novick (in press) determined to be most important for specifying each type of diagram and for distinguishing between the two types of diagrams.

One critical difference between hierarchies and networks concerns their basic structure—how they are put together. The global structure of a hierarchy is a multilevel organization in which a single root node branches out to subsequent levels such that the identities of the objects at one level depend on the identities of the objects at the preceding level. The building block for creating this structure is three items connected by two directional relations, arranged as a "V" in some orientation. In contrast, the network is unlike most other abstract diagrams in that it does not have any formal structure. In principle, any node may be linked to any other node. The fact that some nodes are connected and others are not in representing a particular problem is due to the semantics of the problem rather than the structure of the diagram.

Because of these differences in how the components of hierarchies and networks are arranged, the two types of diagrams differ in how information "moves" through them. The linking relations property specifies whether the links between nodes are one-to-many, many-to-one, or both. In a network, the links may be both one-to-many *and* many-to-one, as a given node may have multiple lines (i.e., relations) both entering it and leaving it. In a hierarchy, however, either one line enters each node and multiple lines leave it, or multiple lines enter and one line exits. These relations are viewed as one-to-many *or* as many-to-one depending on whether one progresses from the root to the leaves or vice versa. Because of this difference in local relations, there is only one path between any two non-adjacent nodes in a hierarchy, whereas there may be multiple paths connecting pairs of nodes in a network (traversal property).

3 Students' Knowledge of Hierarchies

3.1 Evidence for Understanding

College students have a fair amount of general knowledge about hierarchies. For example, Novick (2001) reported the results of a study in which students were given short scenarios set in a medical context and had to choose which of two diagrams (selected from hierarchy, matrix, and network) best captured the structure of the scenario. The scenarios were written to focus on one of the hypothesized properties from Novick and Hurley's (2001) structural analysis. For example, scenario #12 in Table 1 focuses on the global structure property for the hierarchy, and students had to choose between a hierarchy and a matrix. Scenario #16 focuses on the linking relations property for the hierarchy, and students had to choose between a hierarchy and a network. Novick compared the accuracy of diagram selection for two groups of students: a higher knowledge group consisting of advanced computer science and mathematics education students and a group of typical undergraduates. Although the higher knowledge students selected the correct representation for the six hierarchy scenarios more often than did the typical undergraduates—96% versus 87%—the latter group was correct a large majority of the time.

Table 1. Examples of medical scenarios used by Novick (2001) and Novick and Hurley (2001)

Scenario #12: Focuses on the hierarchy value of the global structure property	A certain skin disease is treated in several week-long phases, with a different medication being given in each phase. The particular medication used during the first week of treatment affects the kinds of medications that may be used in the second week of treatment, which in turn affects the kinds of medications that can be used during the third week, etc. The dermatology nurses would like a diagram showing the possible treatment regimens.
Scenario #16: Focuses on the hierarchy value of the linking relations property	The medical records staff has just started using a new software package for its filing system. No formal training program was implemented, but several employees were shown how to use the system. They were asked to instruct other employees, who in turn instructed still other employees. Each employee received instruction from only a single person, but may have tutored several other employees. The office manager would like a diagram showing who trained whom.

This high level of accuracy for the typical college students likely reflects the fact that hierarchies are fairly common in everyday situations. Unselected college students readily call to mind examples of situations for which a hierarchy can be used to organize information. In fact, they are able to generate more examples of this type of diagram in use than of either matrices or networks (Novick, Hurley, & Francis, 1999). Greater experience with hierarchies means more opportunities to induce aspects of this diagram's structure.

Evidence that typical college students have abstract rules concerning at least some hierarchy applicability conditions comes from a study by Hurley and Novick (in press). They compared the accuracy of diagram selection for medical scenarios like those shown in Table 1 to equivalent versions that were written completely abstractly. For example, the following abstract scenario is matched structurally to scenario #12: "Items from set R are selected in several phases, with a different item being selected in each phase. The particular item selected in the first phase affects which items can be selected in the second phase, which in turn affects which items can be selected in the third phase, etc. A researcher would like a diagram showing the possible selection patterns." Not only was selection accuracy for the abstract hierarchy scenarios significantly above chance (90% correct), it was equivalent to accuracy for the medical scenarios (89% correct).

Finally, let me describe one result from my research with Kefyn Catley on students' understanding of and ability to reason from cladograms. Consider the cladogram in Figure 2, which shows evolutionary relationships among seven placental mammal taxa. College students with weaker and stronger backgrounds in biology were asked several questions about this cladogram and several other cladograms. The weaker background students had taken an average of 0.8 biology classes in college, whereas the stronger background students had taken an average of 3.5 such classes. The two groups did not differ in year in school (overall $M = 2.8$). A manuscript reporting the full results of this study, as well as a second study, is in preparation (Catley & Novick, 2006). One question students were asked about the cladogram in Figure 2 was "what character was possessed by the most recent common ancestor of cows and whales?" The correct answer, multi-chambered stomach, was given by 93% of the weaker background students and 100% of the stronger background students. In this instance, the weaker background students clearly were able to apply their general knowledge of hierarchical structure to this diagram to extract the needed information.

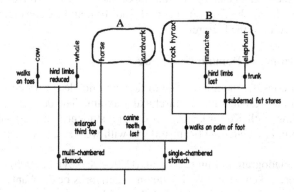

Fig. 2. A cladogram representing evolutionary relationships among 7 placental mammal taxa

3.2 Evidence for Deficiencies in Understanding

Although the results just described indicate that typical college students know quite a lot about hierarchies, other data reveal some deficiencies in their understanding. These deficiencies specifically concern the linking relations (one-to-many or many-to-one relations, but not both) and traversal (only one path between any two nodes) properties. For example, Novick (2001) found that the typical undergraduates were less likely than the more knowledgeable students to select the correct representation when hierarchy and network diagrams were pitted against each other and the scenario focused on the linking relations property (e.g., scenario #16 in Table 1): 67% versus 91% correct.

Additional evidence comes from students' verbal justifications for why the type of diagram they selected was best and why the other type of diagram was not as good (Novick, 2001; Novick & Hurley, 2001). Novick's analysis of these data revealed a difference in how often the subject groups referred to structural properties of a hierarchy in their justifications: The more knowledgeable students referred to a mean of 5.91 property values (applicability conditions), compared with a mean of 4.67 for the typical students. This difference between the two subject groups is primarily due to just two properties—linking relations and traversal. For linking relations, 78% of the more knowledgeable students mentioned the hierarchy value on this property, compared with only 17% of the typical undergraduates. For the traversal property, these percentages were 83% versus 42%.

Novick (in press, Exp. 2) gave subjects the hypothesized applicability conditions for a hierarchy diagram and asked them to rate each property value in terms of how diagnostic it is for a hierarchy. Two groups of subjects participated—advanced computer science and engineering majors and typical undergraduates. There was a reliable interaction between subject group and structural property, as the rank ordering of the property values in terms of diagnosticity differed for the two subject groups. Most relevant here, traversal and item/link constraints were rated as the fourth and sixth most diagnostic properties (out of 11), respectively, by the more knowledgeable students, but they were eighth and ninth in the ordering, respectively, for the typical undergraduates. Novick argued that these differences suggest that typical undergraduates are deficient in their understanding of the importance of (a) the traversal property for distinguishing the hierarchy from the network and (b) item/link constraints for specifying the rigid structure of the hierarchy.

3.3 The Importance of Domain Knowledge

In this section, I consider the importance of domain knowledge for correctly interpreting the information presented in a hierarchical diagram. The data come from research I have been doing with Kefyn Catley on college students' understanding of cladograms. As before, I present data from students with weaker and stronger backgrounds in biology.

Consider the cladogram in Figure 2 again (Catley & Novick, 2006). Another question students were asked was: "Which taxon—aardvarks or elephants—is the closest evolutionary relation to the rock hyrax? What evidence supports your answer?" Although the students in both groups were approximately equally likely to select the correct answer (elephants), the overall level of performance—67% correct—indicates some difficulty on the part of students in both groups in interpreting this hierarchy. Students' answers to the evidence question more strongly show the importance of domain knowledge. The correct answer is that rock hyrax and elephant share a more recent common ancestor than rock hyrax and aardvark. This justification for a correct response was cited by none of the weaker background students but 50% of the stronger background students.

A third question asked about the cladogram in Figure 2 was: "Do the circled taxa labeled 'A' constitute a clade?" Subjects indicated their answer by circling either "yes" or "no". A clade was defined for subjects as "a group that includes the most recent common ancestor of the group and all descendants of that ancestor." The stronger background students gave the correct answer (no) more often than the weaker background students: 85% versus 57%. Subjects who responded "no" were further asked which taxa need to be removed and/or added to the group to make it a clade. The difference between the knowledge groups was even greater for this question: 68% versus 19% correct, respectively. (The answer is to add rock hyrax, manatee, and elephant.)

Additional evidence for the necessity of domain knowledge comes from a comparison of reasoning from tree and ladder versions of cladograms. The ladder diagram is a hierarchical form that is specific to biological classification based on common ancestry and synapomorphies. Our data indicate multiple instances of ladders being more difficult to comprehend than trees, especially for weaker background students. Consider again the question about whether aardvarks or elephants are more closely related to the rock hyrax. As noted earlier, the two knowledge groups had comparable accuracy rates when given the tree diagram shown in Figure 2. Other students were asked a structurally-equivalent evolutionary relation question about a ladder version of a vertebrates cladogram similar to that shown in Figure 1: "Which taxon—frogs or mammals—is the closest evolutionary relation to lizards?" For this question, there was a large difference in accuracy favoring the stronger background students: 80% versus 23% correct. Looking at the results from another perspective, the stronger background students were similarly accurate for both the tree and ladder cladograms, whereas the weaker background students did much better with the tree than the ladder.

In another part of our project (Novick & Catley, 2006), students were asked to translate the evolutionary relationships depicted in one hierarchical form to another, structurally-equivalent, hierarchical form—either from a tree to a ladder or from a ladder to a tree. Translation from one diagrammatic form to another is an important component of diagram literacy (e.g., Kozma & Russell, 1997). The stronger background students did fairly well at this task, with a mean of 76% correct across four problems. The weaker background students, however, had considerable difficulty, whether they were given a ladder as the reference diagram or they had to draw a ladder—mean of 23% correct.

4 Discussion

I have focused on diagram literacy for schematic diagrams, particularly the hierarchy. I showed that college students know quite a lot about the structure of this type of diagram, which leads to generally high accuracy in distinguishing whether a hierarchy or a related diagram (either a network or a matrix) is most appropriate for a given situation. I also showed that when determining the correct type of diagrammatic representation is not an issue—because the correct hierarchy diagram is given—knowledge of the diagram's conventions is not necessarily sufficient for correctly interpreting the diagram. Rather, domain knowledge also is important for making sense of the scientific relations depicted in the diagram. The supporting evidence came from students' attempts to answer questions about, and translate between alternative versions of, cladograms, hierarchical diagrams used in biology to depict evolutionary relationships among taxa. As noted earlier, domain knowledge also plays an important role in expert interpretation and use of iconic diagrams and of charts and graphs, the other two types of scientific diagrams.

The important lesson here is that diagram literacy in science must be taught. Deficiencies in typical college students' knowledge of hierarchies exist because their knowledge was induced over normal school and life experiences rather than acquired through instruction (Novick, 2001). Computer science students, who are taught structural properties of hierarchies (and networks and matrices) in their classes on data structures, consistently did better in my studies. According to the NCTM (2000), students are ready to begin learning about schematic diagrams in late elementary or middle school. Meeting the goal of implementing this curriculum standard will require providing pre-service and in-service teachers with appropriate diagrammatic knowledge (see Novick, 2004). At the same time, science textbook authors and K-16 science teachers need to understand that schematic diagrams are tools, and like other educational tools (e.g., ruler, calculator, dictionary), students need to be taught how to use them. This may involve instruction in their conventions and certainly will involve instruction in how they are applied in the relevant area of science—e.g., evolutionary biology.

Acknowledgements

I would like to thank Kefyn Catley and Susan Gelman for their comments on an earlier version of this article.

References

Catley, K. M., Lehrer, R., & Reiser, B. (2005). *Tracing a prospective learning progression for developing understanding of evolution*. Paper Commissioned by the National Academies Committee on Test Design for K-12 Science Achievement. (a copy can be downloaded from http://www7.nationalacademies.org/bota/Evolution.pdf)

Catley, K. M., & Novick, L. R. (2006). *Reasoning From Cladograms: A Comparison Across Levels of Biological Knowledge*. Manuscript in preparation.

Day, R. S. (1988). Alternative representations. In G. H. Bower (Ed.), *The psychology of learning and motivation* (Vol. 22, pp. 261-305). San Diego, CA: Academic Press.

Dufour-Janvier, B., Bednarz, N., & Belanger, M. (1987). Pedagogical considerations concerning the problem of representation. In C. Janvier (Ed.), *Problems of representation in the teaching and learning of mathematics* (pp. 109-122). Hillsdale, NJ: Erlbaum.

Gattis, M., & Holyoak, K. J. (1996). Mapping conceptual to spatial relations in visual reasoning. *Journal of Experimental Psychology: Learning, Memory, and Cognition, 22,* 231-239.

Gauvain, M., de la Ossa, J. L., & Hurtado-Ortiz, M. T. (2001). Parental guidance as children learn to use cultural tools: The case of pictorial plans. *Cognitive Development, 16,* 551-575.

Hegarty, M., Carpenter, P. A., & Just, M. A. (1991). Diagrams in the comprehension of scientific texts. In R. Barr, M. L. Kamil, P. Mosenthal, & P. D. Pearson (Eds.), *Handbook of reading research* (Vol. 2, pp. 641-668). NY, NY: Longman.

Hennig, W. (1966). *Phylogenetic systematics*. Urbana, IL: University of Illinois Press

Hurley, S. M., & Novick, L. R. (in press). Context and structure: The nature of students' knowledge about three spatial diagram representations. *Thinking & Reasoning*.

Kalyuga, S., Chandler, P., & Sweller, J. (1998). Levels of expertise and instructional design. *Human Factors, 40,* 1-17.

Kindfield, A. C. H. (1993/1994). Biology diagrams: Tools to think with. *The Journal of the Learning Sciences, 3,* 1-36.

Kozma, R. B., & Russell, J. (1997). Multimedia and understanding: Expert and novice responses to different representations of chemical phenomena. *Journal of Research in Science Teaching, 34,* 949-968.

Lynch, M. (1990). The externalized retina: Selection and mathematization in the visual documentation of objects in the life sciences. In M. Lynch & S. Woolgar (Eds.), *Representation in scientific practice* (pp. 153-186). Cambridge, MA: MIT Press.

Maienschein, J. (1991). From presentation to representation in E. B. Wilson's *The Cell*. *Biology and Philosophy, 6,* 227-254.

Mevarech, Z. R., & Stern, E. (1997). Interaction between knowledge and contexts on understanding abstract mathematical concepts. *Journal of Experimental Child Psychology, 65,* 68-95.

National Council of Teachers of Mathematics. (2000). *Principles and standards for school mathematics*. Reston, VA: Author.

Novick, L. R. (2001). Spatial diagrams: Key instruments in the toolbox for thought. In D. L. Medin (Ed.), *The psychology of learning and motivation* (Vol. 40, pp. 279-325). San Diego, CA: Academic Press.

Novick, L. R. (2004). Diagram literacy in pre-service math teachers, computer science majors, and typical undergraduates: The case of matrices, networks, and hierarchies. *Mathematical Thinking and Learning, 6,* 307-342.

Novick, L. R. (in press). Understanding spatial diagram structure: An analysis of hierarchies, matrices, and networks. *The Quarterly Journal of Experimental Psychology.*

Novick, L. R., & Catley, K. M. (2006). *The role of perceptual and conceptual principles in students' ability to translate between alternative hierarchical forms: Evidence from clado-grams in biology.* Manuscript in preparation.

Novick, L. R., & Hurley, S. M. (2001). To matrix, network, or hierarchy: That is the question. *Cognitive Psychology, 42,* 158-216.

Novick, L. R., Hurley, S. M, & Francis, M. (1999). Evidence for abstract, schematic knowledge of three spatial diagram representations. *Memory & Cognition, 27,* 288-308.

Whitley, K. N., Novick, L. R., & Fisher, D. (2006). Evidence in favor of visual representation for the dataflow paradigm: An experiment testing LabVIEW's comprehensibility. *International Journal of Human-Computer Studies, 64,* 281-303.

Winn, W. (1989). The design and use of instructional graphics. In H. Mandl & J. R. Levin (Eds.), *Knowledge acquisition from text and pictures* (pp. 125-144). Amsterdam, The Netherlands: Elsevier.

Zacks, J., & Tversky, B. (1999). Bars and lines: A study of graphic communication. *Memory & Cognition, 27,* 1073-1079.

Learning by Producing Diagrams

Daniel L. Schwartz

Stanford University, School of Education,
485 Lasuen Mall, Stanford, California, USA 94305-3096
Daniel.Schwartz@Stanford.Edu

Abstract. When done well, diagrams can support comprehension, inference, and learning. How about the case when learners create their own diagrams instead of just view them? Though novices typically enjoy and have some natural facility at creating spatial representations, they can easily create flawed representations. They need feedback to help them make their diagrammatic understanding more like that of experts. In this talk, I present three models of feedback. A global model where students simply see a correct diagram after they create their own; a social model where students receive feedback from one another while creating their diagrams; and, an automated model where the feedback is tightly coupled to the learner's own diagram. I will describe the learning benefits of having students generate their own diagrams, and how different types of feedback help maximize those benefits.

Keywords: Diagrams, learning, feedback, interactivity.

D. Barker-Plummer et al. (Eds.): Diagrams 2006, LNAI 4045, p. 12, 2006.
© Springer-Verlag Berlin Heidelberg 2006

Eye Fixations and Diagrammatic Reasoning

Mary Hegarty

University of California, Santa Barbara Santa Barbara, CA 93106
hegarty@psych.ucsb.edu

Abstract. In this tutorial, the attendees will gain an understanding of the main methods of measuring eye fixations on diagrams and how these data are coded and analyzed to make inferences about internal cognitive processes. This will enable attendees to better interpret and critically evaluate the results of studies that use this measure. It will also give them an introduction to the advantages of using eye fixation data and the effort involved in setting up an eye-tracking laboratory and analyzing eye fixation data. The tutorial will also identify problems in the analysis and interpretation of eye fixations that might lead to the development of new software tools for analyzing and interpreting this type of data.

1 Introduction

Although there is a long history of eye-tracking research on such topics as reading, scene perception and visual search, this method has only recently been applied to the study of problem solving and reasoning with diagrammatic representations (Rayner, 1998). There has been considerable interest in the Diagrams community in the use of eye fixations to study diagrammatic reasoning, and there have been several presentations in previous conferences that have used this and related methodologies. With the increase in availability of eye tracking technology and the interest in this methodology, it seems timely to devote a tutorial to the use of eye tracking research on diagrams. The overall goal of the tutorial will be to give attendees an understanding of what is involved in measuring eye fixations during cognitive processing of diagrams and what types of inferences can and cannot be made from these data.

The tutorial will begin with a brief introduction to the history of eye-movement studies and a description of the main methods of collecting eye fixation data that are currently in use. This will include a videotape of a typical eye tracking experiment.

Then we will discuss basic research on how the eyes move during comprehension tasks (fixations, saccades etc.) and the relationship between eye fixations and attention. This will also include a discussion of the assumptions underlying interpretation of eye fixations, such as the eye-mind hypothesis (Just & Carpenter, 1976).

Then we will review some of the main ways in which eye fixations are coded and analyzed. It will cover two main stages of eye-fixation analysis (1) visualization of the data to gain a general understanding of the patterns of fixations and (2) derivation of quantitative methods from the data, which can be entered into statistical analyses. We will discuss some of the main measures that have been derived from eye-fixation studies

D. Barker-Plummer et al. (Eds.): Diagrams 2006, LNAI 4045, pp. 13–15, 2006.

with diagrams in the past. For example, we will discuss the identification of regions of interest (ROIs) in a diagram that represent different meaningful entities (e.g., the legend in a graph, a component of a machine, a particular region of space in a map). Then we will examine how these are used to derive quantitative measures, for example:

- Gaze duration on regions of interest in the of the diagram at different stages of task performance.
- Number of fixations on different regions of interest at different stages of task performance.
- Number of transitions between different regions of interest.
- Sequence of fixating different regions of interest.

This review will focus on measures used in studies of comprehension, reasoning and problem solving with diagrammatic displays, broadly defined (including diagrams, maps and graphs). For example, representative tasks to be discussed might include chess problem solving, (e.g., Charness, Reingold, Pomplun & Stampe, 2001) graph comprehension (Shah & Carpenter, 1998; Peebles & Cheng, 2003), mechanical reasoning (Hegarty, 1992), mental rotation (Just & Carpenter 1985) and insight problem solving with diagrams (Grant & Spivey, 2003; Knoblich, Ohlsson, & Raney, 2001). In reviewing and evaluating these studies, the focus will be on identifying the types of inferences that we can make from eye fixation data and also the types of inferences that one cannot make from eye fixations. For example in a given study, it might be possible to identify which region of a map a person is thinking about at a given time, but not exactly what they are thinking about that region.

The presentation will end with a discussion of how eye fixations are jointly influenced by top-down factors (such as knowledge, expectations and task demands) and bottom up factors such as relative salience of elements in a display.

The goal would be to use an interactive format in the tutorial, in which I do not just lecture, but bring examples of eye-fixation data and invite participants to interpret these data either alone in small groups, and then share their interpretations. I would also like to invite a discussion on the state of the art in eye-tracking analysis and perhaps identify ways in which our analysis tools could be improved.

Audience: The main goal of the tutorial is to enable researchers to interpret and critically evaluate the results of eye tracking studies. As such, the tutorial is of interest to any attendees who read the empirical literature on cognitive processing of diagrammatic representations. The tutorial should be of particular interest to cognitive psychologists, education researchers, and researchers in human-computer interaction, because eye tracking is used in all of these disciplines. I would also hope that it would be attended by researchers interested in information visualization, software development, and analysis of spatially distributed data, because I believe that there is great potential for improving our methods of visualizing and analyzing eye tracking data, and it would be outstanding if the tutorial could inspire some work on these topics. Finally, although I have been conducting eye fixation studies on diagrammatic representations for some time, I certainly do not consider myself to be an expert on all of the potential uses of these data or all of the techniques used to analyze these data, so I would welcome participation by other individuals who have used this technique in their research.

Acknowledgements. Dr Hegarty's research on eye fixations and diagrammatic reasoning is supported in part by grant N00014-03-1-0119 from the Office of Naval Research.

References

Carpenter, P. A. & Shah, P. (1998). A model of the perceptual and conceptual processes in graph comprehension. *Journal of Experimental Psychology: Applied, 4,* 75-100.

Charness, N. Reingold, E. M. Pomplun, M. & Stampe, D (2001). The perceptual aspect of skilled performance in chess: Evidence from eye movements. *Memory & Cognition, 29,* 1146-1152.

Grant, E. R. & Spivey, M. J. (2003) Eye movements and problem solving: Guiding attention guides thought. *Psychological Science, 14, 462-466.*

Hegarty, M. (1992). Mental animation: Inferring motion from static diagrams of mechanical systems. *Journal of Experimental Psychology: Learning, Memory and Cognition, 18(5)* 1084-1102.

Just, M. A. & Carpenter, P. A. (1976). Eye fixations and cognitive processes. *Cognitive Psychology, 8, 441-480.*

Just, M. A. & Carpenter, P. A. (1986). Cognitive coordinate systems: Accounts of mental rotation and individual differences in spatial ability. *Psychological Review, 92. 137-172.*

Knoblich, G., Ohlsson, S. & Raney, G. E. (2001). An eye movement study of insight problem solving. *Memory & Cognition, 29,* 1000-1009.

Peebles, D. & Cheng, P. C-H. (2003). Modeling the effect of task and graphical representation on response latency in a graph-reading task. *Human Factors, 45,* 28-46.

Rayner, K. (1998). Eye movements in reading and information processing: Twenty years of research. *Psychological Bulletin, 124,* 372-422

Cross-Cultural User-Experience Design

Aaron Marcus

Aaron Marcus and Associates, Inc. (AM+)
1196 Euclid Avenue, Suite 1F, Berkeley, California 94708-1640, USA
`Aaron.Maracus@AMandA.com`

Abstract. Designers of information visualization and user interfaces must take account of culture in the design of metaphors, mental models, navigation, interaction, and appearance. Culture models define dimensions of difference and similarity among groups of people regarding signs, rituals, heroes/heroines, and values. Examples on the Web reveal these dimensions. Developers will increasingly need to take into account culture and other factors in development to better ensure usability, usefulness, and appeal.

1 Introduction

Modern technology and commerce enable global distribution of products and services to increasingly diverse users. User-interface (UI) and information-design/usability disciplines have the objectives of improving performance and productivity, i.e., usability. UI and information-design include design of information-visualization. The International Standards Organization in Switzerland defines usability as effectiveness, efficiency, and satisfaction. Recently, designers and analysts have enlarged their objectives to include more complex user-experience (UX) issues, which are even more complex and challenging. Culture analysis offers *a* way to understand, even measure, differences and similarities of user experience. To help professionals understand the dimensions of culture as they apply to UI and information design, some definitions are helpful derived from the author's writings [8, 9, 10, ,11, 12, 13]:

UI (including information design) development includes the following tasks:

Plan: brainstorming information sonification
Research: technology, design issues, strategies
Analyze: user profiles, use scenarios, prototypes
Design: content, applications, brand, storyselling
Implement: scripting, coding, final production
Evaluate: focus groups, user tests, heuristic evals.
Document: guidelines, patterns, specifications
Train: courseware, tutorials, mentoring
Maintain: continuing client relations

UI (including information-visualization)components are the following:

Metaphors: Essential concepts in words, images, sounds, touch
Mental Models: Organization of data, functions, tasks, roles

D. Barker-Plummer et al. (Eds.): Diagrams 2006, LNAI 4045, pp. 16–24, 2006.

Navigation: Movement through mental models via windows, dialogue boxes, etc.
Interaction: Input/output techniques, feedback, overall behavior of systems
Appearance: Visual, verbal, acoustic, tactile

Information visualizations/sonifications are representations of structures and processes. They may be abstract or representational. Traditional techniques include tables, forms, charts, maps, and diagrams; many innovations are now available.

User-Experience (UX) Design includes an enlarged scope of objectives for products/services:

Usability: Efficient, effective, satisfying (ISO definition)
Usefulness: Fits the user's needs/desires well
Appeal: Delight, fun, engagement, emotions, branding

Experience covers all stakeholder "touch-points" such as the places where a buyer, customer, user, learner, expert, advocate, staff member, journalist, or investor comes into contact with the product or service, or its sponsoring company/organization. There is a greater focus on content, brand, and emotions. As an exemplar of good user experience is the Apple iPod is often cited; the product had enormous success worldwide within one year. When the user's experience is taken into account, evaluation techniques may shift. Ethnographic analysis, shadowing, and focus groups, as well as traditional usability tests become more important. Learning the underlying emotional motivations of "non-rational" customers may become more relevant. Users are involved within a socio-cultural context. The cultural context becomes more important.

2 Culture

Culture includes large-scale and small-scale group behaviors (rituals), leaders/followers, values, artifacts, signs. Many culture models exist as bases for analysis, design, evaluation. Culture analysis is related to semiotics/semiologie, the science of signs, which asks what do things "mean"?. "Meaning" derives both from the designer and what the user brings to the artifact. Culture affects every aspect of tool- and sign-making. Culture-centered design seems "inevitable". Designers/analysts are aware of culture, but may not be informed of specific dimensions by which cultures can be described and measured. In era of instant, global media, cultures are always being affected and evolving. For example, the Chinese culture, which is today considered "Confucian" was not always so, certainly not before Confucius. Cultures do absorb aspects of other cultures, but these are sometimes transformed: USA golf as a sport was imported to Japan but in some ways is more like a religion.

It is difficult for designers/analysts to escape being biased culturally. All designed artifacts are cultural objects. Websites are one set of examples; they are immediately accessible by people worldwide and offer design challenges of "localization" that go beyond translation. Is a traditional DaimlerChrysler Website in German and English, with a Prussian blue background and clean, spare layout suitable for all people? Perhaps not. Is a FedEx Website offering to send packages to Saudi Arabia, but showing an Asian woman with her face and arms bare appropriate for Saudi Arabia? Not likely. Is a Jordanian Website from Arabia.On.Line with a layout in English that reads right to left appropriate? Probably not. These are examples of potential culture clashes between the designed forms and the users.

Some nations have many recognized national languages. For example, South Africa has 11, and India has 23; English, even if official, is not the language spoken by the most people. Localization [7] goes beyond languages and translation . Many nations have multiple races, ethnic groups, religions, and languages. Designing for such groups presents many challenges but also opportunities. Sets of well-known differences include multi-lingual requirements (e.g., Canadian English and French), currency, time, and physical measurements. In some Arab countries, Thursday and Friday constitute 'the weekend" instead of Friday and Saturday in Israel or Saturday and Sunday in Western countries. Many national flag color sets reflect these differences, for example, those that are predominantly green and represent countries that are primarily Moslem. In China, certain numbers are considered especially lucky; telephone numbers with these numbers are more expensive to obtain. In some countries and cultures, bright, high-chroma colors are normal; others prefer more subdued combinations.

Small-scale communities with preferred jargon, signs, and rituals can constitute a "cultural group." This definition is different from traditional definitions of culture that more typically refer to longstanding historical differences established over many generations and centuries. The modern cultural group may be considered more a social group or "life-style group." They include the following examples:

Affinity group example: USA Saturn owners of VW "bug" owners
Social group example: Japanese house-wives or Swedish house-husbands
Web group example (geo-dispersed): MP3 music file downloaders.

The challenge for business it how to account for different user experiences that are culturally differentiated in a cost-effective manner. Developers may need to rely on market and user data to achieve short-term success and to avoid wasting money and time on too many variations. Paying attention to culture models and culture dimensions can assist.

3   Culture Models and Culture Dimensions

Professional researchers, analysts, designers, and educators have written about culture for many decades. Among others Geert Hofstede's [3] cultural dimensions are well known and established, although controversial for some anthropologists and ethnographers. Hofstede examined IBM employees in more than 50 countries during 1978-83 and was able to gather, then analyze enough data to be statistically valid. His definition of culture (in his model each country has one dominant culture) concerns patterns of thinking, feeling, and acting that are "programmed" by a particular group into their children. Culture differences manifest themselves in rituals, symbols, heroes/heroines, and values. Hofstede's five dimensions of culture are the following:

Power Distance (PD) is the extent to which less powerful members expect and accept unequal power distribution. High PD countries tend towards centralized power in few hands, tall hierarchies in organizations, rulers/bosses who are viewed as benevolent autocrats, and subordinates are expected to be told what to do. In low PD countries people consider themselves more as equals, may exchange roles, power is decentralized, and hierarchies are flatter.

Implications for global UX design for high PD cultures include the following:

More highly structured, guided access to information
Emphasis on larger social/ moral order (e.g., nationalism/ religion)
Focus on expertise (authoritative content) and leaders
Integrated security, unhidden "restrictions" to access
Importance of certifications, awards, logos
Social role used to organize information (e.g. special managers' sections)

Examples of such PD attributes were observed in university Websites with similar content. A Malaysia (PD = 104) Website (www.uum.edu.my) featured axially symmetric layout, a central official logo, and images of male leaders and monumental buildings. A Netherlands (PD = 38) Website (www.tue.nl) featured asymmetric layout, images of male and female students, and no monumental buildings.

Individualism vs. Collectivism refers to the looseness vs. closeness of ties between people. In individualistic cultures everyone is expected to ook after one's self or his/her immediate family (nuclear families). In collective societies, people from birth are integrated into strong, cohesive in-groups, which continue to protect them in exchange for unquestioning loyalty (extended families). For each culture dimension, Hofstede notes differences of attitudes toward work, family, and education. Some key differences between individualism vs. collectivism are the following:

Individualism:
Individual social/economic interests dominate
Right to privacy; private opinions expected
Strong political power of voters; greater press freedom
Ideology of freedom, self-motivation

Collectivism:
Collective soc/econ interests dominate
State dominates the economy and the press
Consensus: ultimate goal
Ideology of equality, harmony

Some implications for global UX design are consequently the following:

Individualism:
Focus on maximizing personal achievement
Materialism and consumerism demonstrate individual success
Controversial speech and extreme claims encourage "truth"
Images of youth/activity, rather than age/wisdom/"being")

Collectivism:
Individual roles downplayed (e.g., just product); group focus
Preference for socially supportive and constrained claims
Controversy discouraged: tends to divide people
Respect for tradition (historical focus)

Examples of differences of individualism vs. collectivism were observed on the Web for national park Websites. That for the USA (IC = 91) showed (www.nps.gov/glba/evc.htm) messages from an individual (Superintendent) to the individual park visitor, images of individuals, and references to "your" park. A Website from Costa Rica (IC = 15) showed images of nature only and a full-screen page linked to What's New button that announced a social-political message in keeping with the ideology of the government.

Gender Role (GR)s: The Femininity vs. Masculinity dimension refers not to physical differences, but to traditional gender roles, in which men are traditionally assertive, competitive, and tough; while women are more people-oriented, tender, and home-oriented. Hofstede defines masculine cultures as those that maintain typical separation of the roles, while feminine cultures tend to overlap them more. Some GR examples of differences for countries include the following:

95 Japan
79 Austria
62 USA (South Africa = 63)
53 Arab countries (great variations)
47 Israel
43 France
39 South Korea
05 Sweden

Implications for global UX design include the consequent following differences:

Masculine:
Traditional gender/ family/age distinctions emphasized; work tasks/roles preferred
Mastery most important; Websites designed for exploration, control
Games/competitions held grab attention
Artwork may be utilitarian/ instrumental

Feminine:
Gender/work roles blurred
Mutual exchange and support more important than master
Website task-oriented and provide quick results for limited task
More emotional/aesthetic appeal

Web examples of gender-role differences were observed for portal sites. In Japan (GR=95), a portal (www.excite.co.jp) showed different versions for men and women. The men's site featured investments and autos; the women's site cosmetics and cooking. Sweden (GR = 05), had only one site (se.excite.com) for both men and women.

Uncertainty Avoidance: This dimension refers to the tolerance of ambiguity or anxiety from things that are uncertaion or unknown, as opposed to a feeling of fear from clear threats. Countries vary in their formality, punctuality, and behavior of government burocracies, legal systems, and religions as a consequence. Examples of differences include the following:

High:
People seem busy, emotional, aggressive, active
What is different is dangerous, dangerous and dirty related
Students expect teachers to be experts with all the answers
Shun ambiguous situations; may engage in risky behavior to reduce ambiguities

Low:
Quiet, easy-going, indolent, controlled, lazy
What is different is curious (or ridiculous)
Students respect plain language, teacher who may not know all
Definitions of clean/dirty; safe/dangerous differ widely by country

Implications for global UX design include the following:

High:
Keep it simple
Reveal results/ implications of actions
Make attempt to prevent looping/ becoming "lost in cyberspace"
Use constraints/task animations/models to reduce "user error"
Carefully encode meaning through multiple redundant cues

Low:
Complexity and risk more valued: less protection from failure
Less controlled navigation: Links may open windows, change site
Help system focuses on information; task orientation secondary
Coding of color/ shape/ texture cues used to maximize information; less redundant

Examples of differences of uncertainty avoidance on the Web that seemed to confirm these differences of culture were observed for national airline companies. The Belgium (UA = 94(Website for Sabena Airlines (www.sabena.com) featured a limited number of links (about seven) on the home page, a very simple image, and a simple dark background color. In contrast, the United Kingdom (UA – 35) Website for British Airways (www.sabena.com) showed many more links, multiple panes, and a less clear image. To study whether these differences persist over time, the Websites were observed three years later (2003). The same differences were observed. Both sites showed central data entry panes in which there were 19 and 16 links respectively, but in the areas outside of this central area, the Sabena Website showed 23 links, while the Bristish Airways Wesbsite showed 43, nearly double. Again, lower uncertainty avoidance measurement seemed to indicate greater tolerance for ambiguity.

Long- vs. Short-Term Time Orientation (LTO) is also called Confucian dynamism and is exhibited especially by those cultures/countries influence by this philosophy. The tenets of the philosophy include the following:

Stable society requires unequal relations
Family is prototype of all social organizations
Virtuous behavior to others = not treating others as one would not like to be treated
Virtue regarding one's task in life = trying to acquire skills/education, working hard, being frugal, being patient, persevering
Practice oriented, not belief (truth) oriented

Hofstede studied only 23 countries for this dimension. Some rankings for countries/cultures include the following:

01   China
04   Japan
17   USA
22   Nigeria
23   Pakistan

Implications for global UX design for LTO cultures include the following:

Long-Term Orientation:
Practice more important than theory
Accomplishing the task sufficient; expertise not required
Personal network provides resources for achievement
(cf. Chinese Guanxi principle or life-long relationship building)

Examples of LTO attributes in Websites were observed by comparing the Germany (LTO = 31) Siemens corporate Website (www.siemens.com/de) with the China (LTO =118) Website (www.siemens.com.cn). Typical differences included the following:

Germany:
Design that is appropriate just for now (will be outdated in time)
Concentration on just showing task or product
Function, mastery, organization-oriented

China:
Soft focus, with warm, fuzzy images
Timeless, classic design
Emphasis on people images

4   Cautions and Considerations

Other data sources exist that are just as extensive as Hofstede's, for example Schwartz's [14]. Although Hofstede's model is well established, and many studies have been based on it, there are also criticisms of the model:

Old data
Little emphasis on media, sociology of culture, politics of culture
Corporate subjects only, not farmers, or other laborers
Assumes one culture per country
Assumes fixed, unchanging relationships
Gender roles, definitions debatable
Seems too general, stereotypical

There are an increasing number of studies of culture and how to use culture models. Honold's study of German and Chinese mobile phone users that used a mixture of culture models [4], Lee's study of Confucian virtual workspaces [6], and Choong and Salvendy's study of mental model performance differences between North American and Chinese users [2] are examples.

The author's firm published a study of 12 corporate international Websites in relation to Hofstede's model [11]. The study showed differences of Website design that seemed to map well to distinguishing culture attributes. For example, a Siemens Website whose corporate design standards required the use of an image for Home would use an anonymous female for the Netherlands (low PD) but leaders and official buildings in Malaysia (high PD). Patterns emerged in which all countries with no picture on the Home page have low PD values while the eight countries with the highest PD show a picture of a male. Similarly, the individualism of images at the PeopleSoft Website increases with the amount of IC value toward individualistic cultures. Among the 15 lowest-rated countries regarding individualism, there are no people shown in the Siemens localized Website imagery, but one can find images of people in those countries that have a higher IC values.

The author's firm also published a study of "best-of-breed" culture dimensions culled from 29 dimensions in nine models from 11 authors rated by 57 experts [12]. The top-rated dimensions are the following:

Context: high/low contexts in communication
Technological development: high/low development, favorable/unfavorable attitudes
Uncertainty avoidance: cultures have high/low tolerance for uncertainty
Time perception: long/short and future/present/past orientation
Authority conception: high or low power distance

5 Future Developments and Conclusions

Based on these alternative or best-of-breed dimensions, many previous studies of culture might be reconsidered; other new studies need to be done using these new dimensions. At the very least, awareness of culture models and culture dimensions enlarges the scope of issues concerning each development task cited earlier. For example, these models challenge the professions of UI development to think about appropriate metaphors for different cultures, user profiles that are culture-sensitive, differing mental models and their influence on performance not only preference, alternate navigation strategies, evaluation techniques, attitude toward emotions, etc. An additional challenge is introducing culture considerations into corporate and organization frameworks for product/service development and into centers of user-centered design.

There are also very different views about culture models and their relation to ethnographic approaches that study local groups, behaviors, "situated practice," and action-oriented descriptions. Post-modernism, media studies, and the sociology of culture are currently critiquing the production of culture, politics of culture, viewing sociological phenomena, power inequality, social construction of technology, and other patterns of social organization. The university courses of Lamont [5] and Brain [1] are examples of curricula devoted to these studies. There are other sources of insight into UX and culture, each of which has formulated models and 7±2 dimensions: persuasion, trust, intelligence, personality, emotion, and cognition. Each gives rise to further issues.

With the rise of India and China as sources of software and hardware production and innovation, it is inevitable that a technological and economic evolution or revolution emerges in this century, perhaps in this decade. Computer-mediated communication and interaction occurs in a context of culture. It is inevitable, also, that user-experience development must account for cultural differences and similarities. Models, methods, and tools exist, but many research issues lie ahead. Design/analysis professionals cannot afford to ignore the resulting issues. Future development of tools, templates, and treasure chests of patterns will provide a body of knowledge in the future for more humane, cultured design of computer-based artifacts.

References

1. Brain, David. New College, Florida, sociology of culture course: http://www.ncf.edu/brain/courses/culture/culture_syl05.htm
2. Choong Y., Salvendy, G. (1998). "Designs of icons for use by Chinese in mainland China. In Interacting with Computers." International Journal of Human-Computer Interaction, 9:4, February 1998, pp. 417-430. Amsterdam: Elsevier.
3. Hofstede, Geert. (1997). Cultures and Organizations: Software of the Mind. New York: McGraw-Hill.
4. Honold, Pia (1999). "Learning How to Use a Cellular Phone: Comparison Between German and Chinese Users," Jour. STC, Vol. 46, No. 2, May 1999, pp. 196-205. (now Pia Quaet-Faslem)
5. Lamont, Michèle , Princeton Univ., contemp. sociological theory: http://www.princeton.edu/~sociolog/grad/courses/fall1995/soc502.html.
6. Lee, Ook (2000). "The Role of Cultural Protocol in Media Choice in a Confucian Virtual Workplace," IEEE Trans. on Prof. Comm., Vol. 43, No. 2, June 2000, pp. 196-200.
7. Localization Industries Standards Association (LISA), Switzerland: www.lisa.org.
8. Marcus, Aaron (2005). "User-Interface Design and Culture." In Aykin, Nuray, ed., Usability and Internationalization of Information Technology, Chapter 3. New York: Lawrence Erlbaum Publishers, 51-78.
9. Marcus, Aaron. (2002) "Globalization, Localization, and Cross-Cultural Communication in User-Interface Design," in Jacko, J. and A. Spears, Chapter 23, Handbook of Human-Computrer Interaction, Lawrence Erlbaum Publishers, New York, 2002, pp. 441-463.
10. Marcus, Aaron, and Emilie W. Gould (2000). "Cultural Dimensions and Global Web User-Interface Design: What? So What? Now What?" Proc., 6th Conference onHuman Factors and the Web, 19 June 2000, University of Texas, Austin, TX.
11. Marcus, Aaron, and Baumgartner,Valentina-Johanna (2004). "Mapping User-Interface Design Components vs. Culture Dimensions in Corporate Websites," Visible Language Journal, MIT Press, 38:1, 1-65, 2004.
12. Marcus, Aaron, and Baumgartner, Valentina-Johanna (2004). "A Practical Set of Culture Dimension for Evaluating User-Interface Designs" in Proceedings, Sixth Asia-Pacific Conference on Computer-Human Interaction (APCHI 2004), Royal Lakeside Novotel Hotel, Rotorua, New Zealand, 30 June-2 July 2004, 252-261.
13. Marcus, Aaron, and Emilie W. Gould (2000). "Crosscurrents: Cultural Dimensions and Global Web User-Interface Design," Interactions, ACM Publisher, www.acm.org, Vol. 7, No. 4, July/August 2000, pp. 32-46.
14. Schwartz, S. H. (2004). Mapping and interpreting cultural differences around the world. In H. Vinken, J. Soeters, & P Ester (Eds.), Comparing cultures, Dimensions of culture in a comparative perspective (pp.43-73). Leiden, The Netherlands: Brill.

Communicative Signals as the Key to Automated Understanding of Simple Bar Charts*

Stephanie Elzer[1], Sandra Carberry[2], and Seniz Demir[2]

[1] Dept of Computer Science, Millersville Univ., Millersville, PA 17551 USA
elzer@cs.millersville.edu
[2] Dept of Computer Science, Univ. of Delaware, Newark, DE 19716 USA
{carberry, demir}@cis.udel.edu

Abstract. This paper discusses the types of communicative signals that frequently appear in simple bar charts and how we exploit them as evidence in our system for inferring the intended message of an information graphic. Through a series of examples, we demonstrate the impact that various types of communicative signals, namely salience, captions and estimated perceptual task effort, have on the intended message inferred by our implemented system.

1 Introduction

Information graphics such as bar charts, line graphs and pie charts are an important component of many documents. As noted by [1], [7], and [24], among many others, a set of data can be presented in many different ways, and graphs are often used as a communication medium or rhetorical device for presenting a particular analysis of the data and enabling the viewer to better understand this analysis. Although some information graphics are only intended to display data, the majority of information graphics that appear in formal reports, newspapers, and magazines are intended to convey a message, thus leading us to consider information graphics as a form of language. As Clark noted, language is more than just words. It is any "signal" (or the lack of a signal when one is expected), where a signal is a deliberate action that is intended to convey a message [6]. Clark's expanded definition of language includes very diverse forms of communication (such as gesture and facial expression) and the common factors among these varied forms of expression is the communicative intention underlying them and the presence of deliberate signals to aid in recognizing these intentions.

The design choices made by the designer when constructing an information graphic provide the communicative signals necessary for understanding the graphic. The design choices include selection of graphic type (bar chart, line graph, pie chart, etc.), organization of information in the graphic (for example, aggregation of bars in a bar chart), and attention-getting devices that highlight

---

* This material is based upon work supported by the National Science Foundation under Grant No IIS-0534948.

D. Barker-Plummer et al. (Eds.): Diagrams 2006, LNAI 4045, pp. 25–39, 2006.
© Springer-Verlag Berlin Heidelberg 2006

certain aspects of a graphic (such as coloring one bar of a bar chart different from the others, mentioning data elements in the caption, etc.). This paper focuses on the communicative signals that result from these design choices. It begins by presenting a very brief overview of our system for inferring the intended message of a bar chart. It then describes the types of communicative signals that our system extracts from simple bar charts, and provides several examples that illustrate how different communicative signals impact the message that is recognized by our implemented and evaluated system. It concludes by discussing the applications that we envision for our system.

2   A Bayesian Approach to Graphic Understanding

Figure 1 shows the architecture of our system for inferring the intended message of an information graphic. The visual extraction module (VEM) analyzes the graphic and produces an XML representation containing information about the components of the information graphic including the graphic type (bar chart, pie chart, etc.) and the caption of the graphic. For a bar chart, the representation includes the number of bars in the graph, the labels of the axes, and information for each bar such as the label, the height of the bar, the color of the bar, and so forth [5]. The XML representation is then passed to the caption tagging module (CTM) which extracts information from the caption (see Section 3.2) and passes the augmented XML representation to the intention recognition module (IRM). The IRM is responsible for recognizing the intended message of the information graphic, which we hypothesize can serve as the basis for an effective summary of the graphic. The scope of the work currently implemented is limited to the processing of simple bar charts. By simple bar charts, we mean bar charts that display the values of a single independent attribute and the corresponding values for a single dependent attribute. Although the type of information graphics is limited, we believe that the concepts, mechanisms and framework of our methodology is broadly applicable and extensible to other types of graphics.

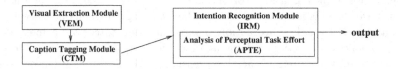

Fig. 1. System Architecture

2.1   Extending Plan Inference to Information Graphics

In our research, we have extended plan inference techniques that have been used successfully in identifying the intended meaning of an utterance to inferring the message conveyed by an information graphic [10]. Our goal is to identify the message that the graphic designer intended to convey, by recognizing his plan for the viewer — i.e., by recognizing the perceptual and cognitive tasks that the viewer is intended to perform in deciphering the graphic's intended message. By *perceptual tasks* we mean tasks that can be performed by simply viewing the graphic, such as finding the top of a bar in a bar chart; by *cognitive tasks* we mean tasks that are done via mental computations, such as computing the difference between two numbers [16]. Of course, we realize that not all graphics are well designed, and in the case of a poorly-designed graphic, we may not be able to infer the message that the graphic designer intended; in this case, our goal is to infer the same message that a human would get by viewing the graphic.

Following the work of Charniak [4] and others, we capture plan inference in a Bayesian network. In all plan inference systems, there is some explicit plan structure which defines the relationships among goals, subgoals and primitive actions. In Bayesian networks, the plan structure is captured by the network itself; each goal, subgoal, and primitive action is represented as a node in the network. If a goal can be decomposed into a particular set of subgoals or primitive actions, an arc from the goal to each subgoal (or primitive action) is used to represent this causal relationship. Our Bayesian network captures knowledge about how the graphic designer's goal of conveying a message can be achieved via the viewer performing certain perceptual and cognitive tasks, as well as knowledge about how perceptual and cognitive tasks decompose into sets of simpler tasks. For example, the graphic designer might intend for the viewer to find the relative difference between two data elements represented by bars in a bar chart — i.e., whether the value represented by the top of one bar is greater than, less than, or equal to the value represented by the top of the second bar. The viewer might achieve this goal by locating the bar that represents the first data element (by looking at the labels of the bars), locating the bar that represents the second data element, and then perceptually comparing the heights of the bars. This task decomposition is captured in our network, as shown in Figure 2.

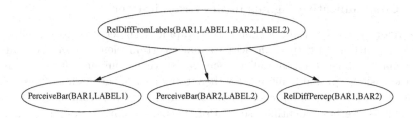

Fig. 2. A Task Decomposed Into Its Subgoals

In addition to nodes capturing the task structure, our Bayesian network for recognizing the message conveyed by an information graphic contains evidence nodes that reflect communicative signals in the graphic. Each evidence node captures the causal relationship between its parent node(s) being (or not being) part of the plan for the viewer to decipher the graphic's message and the presence of a particular kind of communicative signal in the graphic.

The arcs between the nodes in a Bayesian network capture causal dependencies using conditional probability distributions that represent the likelihood of a proposition given the various values of its parent node(s). Bayes' rule is then used to compute the probability of each proposition given causal evidence (from its parents) and diagnostic evidence (from its children) [9, 21]. In a Bayesian network used for plan inference, the root nodes represent various high-level hypotheses about an agent's plan, and the probabilities computed for those nodes represent the likelihood of the hypotheses given the available evidence [2]. In many domains, it is difficult to empirically determine the probabilities that should be used in the conditional probability tables; however, the probabilities used in our network have been obtained through an analysis of a corpus of 110 bar charts that were previously annotated with their intended messages. Since the focus of this paper is on identifying and exploiting the communicative signals present in simple bar charts, space precludes further discussion of the Bayesian network itself; however, greater detail can be found in [13].

3   Communicative Signals in Bar Charts

A key component of a plan inference system is the evidence that is used to guide the inference process, so that one hypothesis might eventually be preferred over another. Therefore, in extending plan inference techniques to the recognition of intentions from information graphics, we need to identify the types of evidence present in information graphics and exploit them in our Bayesian network. The evidence that we utilize in our system consists of communicative signals resulting from the design choices made by the graphic designer. Section 3.1 discusses the types of communicative signals that we extract from the graphic itself (such as highlighting, annotations, and perceptual task effort), while Section 3.2 discusses the communicative signals we extract from captions.

3.1   Communicative Signals in the Graphic Itself

Salience: Our contention is that if the graphic designer goes to the effort of employing attention-getting devices to make certain elements of the graphic particularly salient, then the salient elements serve as communicative signals — i.e., the designer probably intends for them to be part of the intended message of the graphic. By examining a corpus of bar charts gathered from magazines and newspapers, we have identified several of the most common design techniques that graphic designers employ to increase the salience of an element or elements in simple bar charts.

In order to draw attention to a particular element (or elements) of a bar chart, the graphic designer may choose to *highlight* it. From our examination of a corpus of bar charts, we have found that graphic designers typically highlight an element or elements of a bar chart by drawing the viewer's attention to the bar itself or to an attribute of the bar, such as its label or its annotated value. For example, the designer could highlight a bar in the graphic by making it a different color, shade or texture than the other bars. This is a communicative signal, conveying to the viewer of the graphic that the bar (and thus the data element that it represents) is of significant import to the message of the graphic. Consider the graphic in Figure 3 which appeared in a local newspaper. The graphic appeared in shades of gray, as it is depicted here, with the bar representing June of 2001 in a darker shade than the other bars. The design choice to highlight this bar by making it a darker shade of gray than the other bars seems to signal the importance of the unemployment rate in June of 2001 to the message that the designer is attempting to convey with the graphic — ostensibly, the contrast between the unemployment rate a year ago and the most recent unemployment data. In addition, instead of highlighting the bar itself, the designer could highlight attributes of the bar such as its label or its annotated value in order to increase the salience of the represented element. Thus the highlighting of either 1) the bar itself or 2) attributes of the bar (such as the label or annotated value) serves as a communicative signal regarding the salience of the element to the message of the graphic.

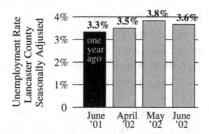

Fig. 3. Bar Chart Showing Unemployment Data[1]

A graphic designer can also convey the significance of an element in an information graphic by annotating the salient element in some way. The most common form of annotation in our corpus of information graphics is the annotation of an element with its exact value. Annotating individual elements with their exact values can signal salience if the annotations are *not* a general design feature of the graphic. If all of the elements are displayed with their exact values (as is the case in Figure 3), then we consider this to be a general design feature of the graphic since the annotations do not draw attention to a specific subset of elements. However, if only a subset of the elements are annotated with their values, the annotations signal the salience of those elements. This is the case in Figure 4 where only the first and last elements are annotated with their values. Annotations of bars in a bar chart are not limited to the exact value represented by the bar — they can also include content such as dates or other additional notes. In the graphic shown in Figure 3, the bar representing June 2001 has a note annotation of "One year ago". The fact that this bar is annotated with additional information further indicates its importance to the message the graphic designer is trying to convey with the graphic.

---

[1] This is based on a bar chart from a local newspaper (the Lancaster Intelligencer Journal on July 30, 2002).

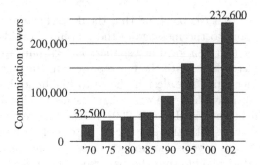

Fig. 4. Bar Chart Showing the Number of Communication Towers[2]

We have also identified several factors that increase the salience of an element in a graphic without the application of any particular design techniques. Although no specific action is required on the part of the graphic designer to make these elements salient, we posit that it is mutually believed by both designer and viewer that such elements will be salient to the viewer. These elements include any element that is significantly taller than all of the other elements in the graphic and the most recent date on a time-line, since the viewer will certainly notice the height of a bar that is much taller than all of the others, and will naturally be interested in what has occurred most recently.

The Visual Extraction Module (VEM) of our system produces an XML representation of a bar chart that includes information about each bar such as bar height, annotations, color, and so forth [5]. By analyzing this XML representation, our system is able to identify any particularly salient elements of the graphic according to the aforementioned criteria for salience.

Perceptual Task Effort: Given a set of data, the graphic designer has many alternative ways of designing a graphic. As Larkin and Simon note, information graphics that are *informationally* equivalent (all of the information in one graphic can also be inferred from the other) are not necessarily *computationally* equivalent (enabling the same inferences to be drawn quickly and easily) [18]. Peebles and Cheng [22] further observe that even in graphics that are informationally equivalent, seemingly small changes in the design of the graphic can affect viewers' performance of graph reading tasks. Much of this can be attributed to the fact that design choices made while constructing an information graphic will facilitate some perceptual tasks more than others. Following the AutoBrief work on generating graphics to achieve communicative goals, we hypothesize that the designer chooses a design that best facilitates the tasks that are most important to conveying his intended message, subject to the constraints imposed by competing tasks [16, 14].

In order to identify the perceptual tasks that the graphic designer has best enabled in the graphic, our methodology was to construct a set of rules that estimate the effort required for different perceptual tasks within a given information

---

[2] This is based on a bar chart from the newspaper USA Today.

graphic. To develop these rules, we applied the results of research from cognitive psychology. In doing this, we constructed a model representing the relative ease or difficulty with which the viewer of a graphic could complete various perceptual tasks. The component of our system that is responsible for estimating effort is called APTE (Analysis of Perceptual Task Effort). The goal of APTE is to determine whether a task is easy or hard to perform with respect to other perceptual tasks that could be performed on an information graphic.

In order to estimate the relative effort involved in performing a task, we adopted a GOMS-like approach [3], decomposing each task into a set of component tasks. Following other cognitive psychology research, we take the principal measure of the effort involved in performing a task to be the amount of time that it takes to perform the task, and our effort estimates are based on time estimates for the component tasks.[3] Wherever possible, we utilize existing time estimates (primarily those applied in Lohse's UCIE system) for the component tasks. For example, the rule shown in Figure 5 estimates the effort required to determine the exact value represented by the top of a bar in a bar chart, given that the viewer is already focused on the top of the bar.[4] In the case of condition-computation pair B1-1 (finding the exact value for a bar where the bar is annotated with the value), the effort is estimated as 150 units for discriminating the label (based on work by Lohse [19]) and 300 units for recognizing a 6-letter word [15]. In the case of B1-2 (finding the exact value for a bar where the top of the bar is aligned with a tick mark on the axis), the effort estimate includes scanning over to the dependent axis (measured in terms of distance in order to estimate the degrees of visual arc scanned [17]) in addition to the effort of discriminating and recognizing the label. Our eye tracking experiments showed that when the top of the bar is aligned with a tick mark, participants frequently repeat the task of scanning to the axis and reading the label (presumably to ensure accuracy), so our effort estimate also includes 230 units [23] to perform a saccade back to the top of the bar before repeating the task. Our set of APTE rules for estimating the effort of tasks in bar charts and our eye tracking experiments that validated those rules are described in [12, 10].

The evidence provided by perceptual task effort can sometimes be subsumed (probabilistically) by other communicative signals, such as a bar being highlighted or the presence of helpful words in a caption. However, this communicative signal plays an extremely important role in our system; in the absence of all other evidence, we always have communicative signals provided by perceptual task effort. For example, an information graphic might not have any salient elements and may be lacking a helpful caption, but our system can still reason about the relative ease or difficulty with which tasks can be performed on the given graphic and may, therefore, be able to draw useful inferences about the intended message of the graphic designer.

---

[3] The units of effort estimated by our rules roughly equate to milliseconds.

[4] *Rule-B1* does not estimate the effort required to get the value represented by the top of a bar in the case where the viewer must scan to the axis and interpolate an estimated value. This task is represented by a separate rule in our system.

Rule-B1:Estimate effort for task
       PerceiveValue(<viewer>, <g>, <att>, <e>, <v>)

Graphic-type: bar-chart

Gloss: Compute effort for finding the exact value <v> for attribute <att>
       represented by top <e> of a bar <b> in graph <g>

Conditions:
    B1-1: IF the top <e> of bar <b> is annotated with a value,
          THEN effort=150 + 300
    B1-2: IF the top <e> of bar <b> aligns with a labelled tick mark on
          the dependent axis, THEN effort=230 + (scan + 150 + 300) x 2

Fig. 5. APTE Rule for Estimating Effort for the Perceptual Task *Perceive Value*

3.2 Communicative Signals in Captions

When considering the communicative signals available in an information graphic, one might suggest relying on a graphic's caption to capture its primary message. However, the work of Corio and LaPalme [8], who studied captions with the objective of categorizing the kinds of information contained in captions in order to form rules for generating captions to accompany graphics, as well as our own corpus study of captions associated with bar charts, show that captions are often missing, ill-formed, or too general to be solely relied on for summarizing an information graphic. In our corpus study, almost half the captions (44%) failed to contribute at all to understanding the graphic's intended message, and only 34% of the captions were judged to convey most of the intended message. Details of our corpus study can be found in [11].

Although captions cannot be relied upon as the sole mechanism for recognizing a graphic's message, we do want to exploit any evidence that is contained in a caption. However, our corpus analysis also showed that full understanding of a caption through a general natural language understanding system would be problematic. This is due to the fact that many captions are ill-formed (often involving ellipsis or sentence fragments), or would require extensive domain knowledge or analogical reasoning to understand. Moreover, once the caption was understood, we would still need to relate it to the information extracted from the graphic itself, which appears to be a difficult problem. Thus we began investigating whether shallow processing of the caption might provide evidence that could be effectively combined with the other communicative signals gleaned from the graphic itself. Our analysis provided the following observations:

- Verbs in a caption often suggest the general category of message being conveyed by the graphic. An example from our corpus is *"American Express total billings still lag"*; the verb *lag* suggests that the graphic conveys that some entity (in this case *American Express*) falls behind some others.
- Adjectives in a caption also often suggest the general category of message being conveyed by the graphic. An example from our corpus is *"Soaring*

Demand for Servers" which is the caption on a graphic that conveys the rapid increase in demand for servers. Here the adjective *soaring* is derived from the verb *soar*, and suggests that the graphic is conveying a strong increase.

- Words that usually appear as verbs, but are used in the caption as a noun, may function similarly to verbs. An example is *"Cable On The Rise"*; in this caption, *rise* is used as a noun, but suggests that the graphic is conveying an increase.
- Nouns in a caption sometimes refer to an entity that is a label on the independent axis of the graphic. When this occurs, the caption brings the entity into focus and suggests that it is part of the intended message of the graphic. An example from our corpus is *"Germans miss their marks"* where the graphic displays a bar chart in which Germans correlates with a label in the graphic and the graphic is intended to convey that Germans are the least happy with the Euro.

Based on these observations, we designed and implemented a type of shallow processing of captions to extract communicative signals consisting of 1) *helpful* verbs and adjectives (identified through our corpus study, WordNet [25] and a thesaurus [20]) and 2) nouns which match the label of a data element in the bar chart (this is actually a way for a designer to emphasize the salience of an element or elements through the caption.)

4 Exploiting Communicative Signals in an Automated System

The communicative signals that we have discussed (such as highlighting, perceptual task effort and caption evidence) are utilized as evidence nodes in our network; for each perceptual task node in the network (such as PerceiveBar in Figure 2), evidence nodes representing the various communicative signals are added as children of the perceptual task node. Evaluation of our system using leave-one-out cross validation[5] on a corpus of 110 bar charts showed that our system had a success rate of 78.2% at inferring the message conveyed by the graphic. The system was judged to fail if either its top-rated hypothesis did not match the intended message that was assigned to the graphic by the human coders or the probability rating of the system's top-rated hypothesis did not exceed 50%. The output of our system consists of a logical representation of the top-rated hypothesis and its relevant parameters from which a natural language gloss can be derived.

In order to demonstrate how our Bayesian system utilizes the communicative signals, this section presents several examples which illustrate how different kinds of evidence impact our network's hypothesis as to the intended message of a bar

---

[5] Leave-one-out cross validation in the context of a Bayesian network means that for each test graphic, all conditional probability tables are computed anew with the data from the test graphic itself excluded.

chart. In order to clearly show the impact of the choices made by a graphic designer, we will consider multiple variations of bar charts displaying the same underlying data.[6]

4.1   Bar Chart with Minimal Communicative Signals

As our first example, consider the bar chart shown in Figure 6, which shows data regarding the average purchases by customers of various credit card companies. The input to the Intention Recognition Module (IRM) is a file containing an XML representation of the information graphic. Before being processed by

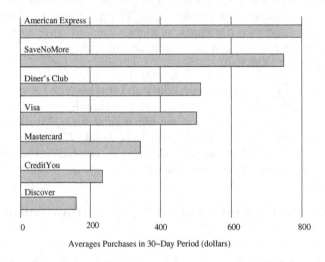

Averages Purchases in 30–Day Period (dollars)

Fig. 6. Information Graphic Example

the Intention Recognition Module, the XML representation is augmented by the Caption Tagging Module (CTM) to include information about any relevant verbs, adjectives, and nouns. Since the bar chart shown in Figure 6 does not have a caption, no additional caption information will be recorded in the XML; that is, there are no communicative signals provided by a caption. Also, according to our criteria for salience of an element, there are no particularly salient elements in the bar chart shown in Figure 6. Thus the only communicative signals in this graphic are the relative effort of the different perceptual and cognitive tasks. Given the XML representation of the bar chart, our system hypothesizes that the graphic is intended to convey the relative rank in amount of purchases for the various credit card companies displayed in the bar chart and assigns this intention a probability of 88.6%. Other possibilities also have some probability

---

[6] The data displayed in the example graphics is based on a bar chart that appeared in the September 13, 2004 issue of Business Week. However, the original graphic had only five bars; we have added two bars to enrich the complexity of the graphics. To our knowledge, CreditYou and SaveNoMore are fictitious entities.

assigned to them. For example, the intention of conveying that American Express has the highest average customer purchases is assigned a probability of 9.9% because the bars are in sorted order according to height, thus making it relatively easy for the viewer to recognize the maximum, and because finding the entity in the graphic with the maximum value is a fairly common intention (occurring approximately 22.7% of the time in our corpus). However, there is no other evidence suggesting that the bar representing the maximum value is salient (such as that bar being highlighted), so the system hypothesizes that the primary intention of the graphic is to convey the relative rank of all of the companies listed. Several other hypotheses have nearly negligible probabilities (less than 1%), including the hypotheses that the graphic is intended to convey that Discover has the lowest average customer purchases, the relative difference between two elements, or the rank of a particular element.[7]

4.2   A Bar Chart with a Single Salient Element

Now suppose that the bar representing Diner's Club was darker than the other bars in the bar chart (as shown in Figure 7), thus making this element salient. The fact that the bar is highlighted provides strong evidence that it plays a role in the intended message of the graphic. Our system takes this into account, along with the relative effort of the different perceptual tasks, and the Bayesian network that is constructed for this information graphic hypothesizes that the intended message of the graphic shown in Figure 7 is to convey that Diner's Club ranks third among the companies shown in the bar chart with respect to average customer purchases over a 30-day period. This hypothesis is believed to be extremely likely, as reflected by its calculated probability of 97.26%.

4.3   A Bar Chart with Two Salient Elements

Elements of the graphic could also be made salient in other ways, such as through annotations. Suppose that the bar representing Diner's Club was still darker than the other bars, but that the bars representing Diner's Club and Mastercard (and only those bars) were annotated with their exact values, as shown in Figure 8. Here the evidence still suggests the salience of Diner's Club, as in the previous example, but also suggests that Mastercard is salient. The fact that two bars are now salient will provide evidence against intentions involving only Diner's Club and will favor hypotheses involving both bars. Thus it is not surprising that the system hypothesizes the intention of the graphic to be finding the relative difference (and the degree of that difference) between the average customer purchases of Diner's Club and Mastercard and assigns it a likelihood of 88.8%.

4.4   A Bar Chart with a Helpful Caption

So far, none of our examples have included a caption. Suppose, however, that the caption of our previous example was "Diner's Club Beats Mastercard" In

---

[7] For the latter two messages, there were several instantiated hypotheses with nearly equal probabilities.

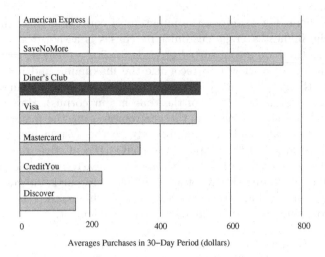

Fig. 7. Information Graphic Example with a Single Salient Element

this case, the initial XML representation would be augmented by the Caption Tagging Module to include additional information about the caption. The verb "beat" (the root form of "beats") is considered to be a helpful verb, and the nouns "Diner's Club" and "Mastercard" match the labels of bars in the bar chart, so this information is included in the augmented XML representation that is input to the Intention Recognition Module (IRM). The inclusion of "Diner's Club" and "Mastercard" in the caption provides additional evidence of the importance of these elements to the intended message of the graphic, and the presence of the verb "beat" provides evidence regarding the general category of message of the graphic. These additional communicative signals strengthen the system's belief in its hypothesis of the relative difference and degree message, and it now assigns it a probability of 99.8% (as opposed to 88.8% without the caption).

4.5   Conflicting Communicative Signals

Now consider a significant variation of the graphic design. Suppose, again, that the bar representing Diner's Club was darker than the other bars. But now suppose that the bars were sorted by the alphabetical order of their labels, rather than by descending order of their height. This variation is shown in Figure 9. The perceptual task of determining the rank of Diner's Club is now estimated as being *hard* (or difficult) to perform. This new evidence results in the system assigning a probability of only 5% to the GetRank message. In fact, the system hypothesizes the most likely message (with a probability of 63.1%) to be that American Express has the highest customer purchases. This is a dubious conclusion, but it illustrates the impact of the conflicting communicative signals on the inference process. Diner's Club is salient, but any tasks that involve this element (such as getting the rank) are difficult to perform or do not match the salience evidence (for example, we could compare Diner's Club with another element, but

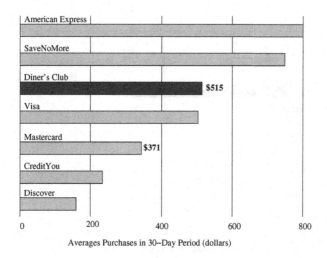

Fig. 8. Information Graphic Example with Two Salient Elements

we have no evidence suggesting the salience of another element). In contrast, finding the maximum is a much easier task to perform on this graphic. However, this is clearly a poorly designed graphic, and the conflicting communicative signals regarding salience and perceptual task effort present in this graphic result both in our system having relatively little confidence in its conclusion, and also make it difficult even for humans to hypothesize an intended message of the graphic (as reflected by the probability of 63.1%).

5   Conclusion

This paper has demonstrated how we exploit communicative signals from bar charts in our automated system. We have presented illustrative examples of simple bar charts which display the same underlying data, but contain very different communicative signals regarding the message that the viewer is intended to recognize. The messages inferred by our system for these different bar charts show the impact that various forms of evidence, namely salience, captions and estimated perceptual task effort, can have on the inference process.

In the future, our intention recognition system will become part of several larger projects. We hypothesize that the core message of an information graphic (the primary overall message that the graphic conveys) can serve as the basis for an effective summary of the graphic. This summary could then be used in a variety of ways. For digital libraries, the summary of the information graphic could be used to appropriately index the graphic and to enable intelligent retrieval. If there is accompanying text, the summary of the graphic can be used in conjunction with a summary of the document's text to provide a more complete representation of the document's content. For individuals who are sight-impaired or who are using low bandwidth devices, using the core message of the information graphic as the basis

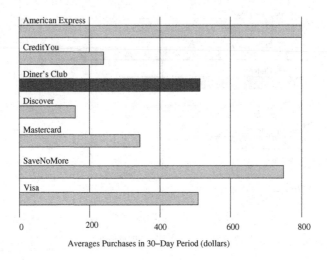

Averages Purchases in 30–Day Period (dollars)

Fig. 9. Information Graphic Example with Conflicting Communicative Signals

for a summary would provide access to the informational content of the graphic in an alternative modality. For graphic designers, the core message inferred by the system might indicate problems with the design of a graphic and thereby provide a tool for improving the graphic.

References

1. Jacques Bertin. *Semiology of Graphics.* The University of Wisconsin Press, Madison, Wisconsin, 1983.
2. Sandra Carberry. Techniques for plan recognition. *User Modeling and User-Adapted Interaction,* 11(1–2):31–48, 2001.
3. S. K. Card, T. P. Moran, and A. Newell. *The Psychology of Human-Computer Interaction.* Lawrence Erlbaum Associates, Inc., Hillsdale, NJ, 1983.
4. Eugene Charniak and Robert P. Goldman. A bayesian model of plan recognition. *Artificial Intelligence,* 64(1):53–79, November 1993.
5. Daniel Chester and Stephanie Elzer. Getting computers to see information graphics so users do not have to. In *Proceedings of the 15th International Symposium on Methodologies for Intelligent Systems (ISMIS), Lecture Notes in Artificial Intelligence 3488,* pages 660–668. Springer-Verlag, 2005.
6. Herbert Clark. *Using Language.* Cambridge University Press, 1996.
7. William S. Cleveland. *The Elements of Graphing Data.* Chapman and Hall, New York, 1985.
8. Marc Corio and Guy Lapalme. Generation of texts for information graphics. In *Proceedings of the 7th European Workshop on Natural Language Generation EWNLG'99,* pages 49–58, 1999.
9. Marek J. Druzdzel and Linda C. van der Gaag. Building probabilistic networks: 'where do the numbers come from?'. *IEEE Transactions on Knowledge and Data Engineering,* 12:481–486, 2000.

10. Stephanie Elzer. *A Probabilistic Framework for the Recognition of Intention in Information Graphics*. PhD thesis, University of Delaware, December 2005.
11. Stephanie Elzer, Sandra Carberry, Daniel Chester, Seniz Demir, Nancy Green, Ingrid Zukerman, and Keith Trnka. Exploring and exploiting the limited utility of captions in recognizing intention in information graphics. In *Proceedings of the 43rd Annual Meeting of the Association for Computational Linguistics (ACL)*, pages 223–230, June 2005.
12. Stephanie Elzer, Sandra Carberry, Nancy Green, and James Hoffman. Incorporating perceptual task effort into the recognition of intention in information graphics. In *Diagrammatic Representation and Inference: Third International Conference, Diagrams 2004, LNAI 2980*, pages 255–270, March 2004.
13. Stephanie Elzer, Sandra Carberry, Ingrid Zukerman, Daniel Chester, Nancy Green, and Seniz Demir. A probabilistic framework for recognizing intention in information graphics. In *Proceedings of the Nineteenth International Joint Conference on Artificial Intelligence (IJCAI)*, pages 1042–1047, July 2005.
14. Nancy Green, Giuseppe Carenini, Stephan Kerpedjiev, Joe Mattis, Johanna Moore, and Steven Roth. Autobrief: an experimental system for the automatic generation of briefings in integrated text and information graphics. *International Journal of Human-Computer Studies*, 61(1):32–70, 2004.
15. B. E. John and A. Newell. Toward an engineering model of stimulus response compatibility. In R. W. Gilmore and T. G. Reeve, editors, *Stimulus-response compatibility: An integrated approach*, pages 107–115. North-Holland, New York, 1990.
16. Stephan Kerpedjiev and Steven F. Roth. Mapping communicative goals into conceptual tasks to generate graphics in discourse. In *Proceedings of Intelligent User Interfaces*, pages 157–164, 2000.
17. Stephen M. Kosslyn. Understanding charts and graphs. *Applied Cognitive Psychology*, 3:185–226, 1989.
18. Jill H. Larkin and Herbert A. Simon. Why a diagram is (sometimes) worth a thousand words. *Cognitive Science*, 11:65–99, 1987.
19. Gerald Lee Lohse. A cognitive model for understanding graphical perception. *Human-Computer Interaction*, 8:353–388, 1993.
20. Merriam-Webster On-Line Thesaurus, http://www.webster.com.
21. J. Pearl. *Probabilistic Reasoning in Intelligent Systems: Networks of Plausible Inference*. Morgan Kaufmann Publishers, 1988.
22. David Peebles and P. C-H. Cheng. Modeling the effect of task and graphical representation on response latency in a graph reading task. *Human Factors*, 45:28–46, 2003.
23. J. E. Russo. Adaptation of cognitive processes to eye movement systems. In J. W. Senders, D. F. Fisher, and R. A. Monty, editors, *Eye movements and higher psychological functions*. Lawrence Erlbaum Associates, Inc., Hillsdale, NJ, 1978.
24. Edward R. Tufte. *The Visual Display of Quantitative Information*. Graphics Press, Cheshire, CT, 1983.
25. Wordnet, http://wordnet.princeton.edu/main/.

On Line Elaboration of a Mental Model During the Understanding of an Animation

Emmanuel Schneider and Jean-Michel Boucheix

LEAD/UMR CNRS 5022, Université de Bourgogne, Pôle AAFE, Esplanade Erasme,
BP 26513, 21065 Dijon Cedex, France
emmanuel.schneider@leadserv.u-bourgnogne.fr,
jean-michel.boucheix@u-bourgogne.fr

Abstract. This experiment examines how, high and low mechanical and spatial abilities, learners understand an animation. Two variables were manipulated: the controllability of the animations and the task type of the learners to study the device. The comprehension test results indicated a positive effect of a fully controllable animation and also a positive effect of task type, when the attention of the learner is focused on the functional model and on local kinematics. The eye tracking data indicated that the learners attend more to the areas of the animations where a great amount of motion is involved along the causal chain of events. We show an effect of the controllability of the system and of the task type of the learner on the amount of eye fixations and on the number of transitions between areas that included the causal chain.

1 Introduction

Today, the rapid growth of technology produces a large variety of animations in learning and educational areas. In spite of this amazing development, the cognitive benefits of animated illustrations for the better comprehension of technical or scientific documents still remain hypothetical (Bétrancourt & Tversky, 2000; Tversky, Bauer-Morrison & Bétrancourt, 2002). Some of the research shows a positive effect of animation in complex system comprehension (Mayer; 2001; Hidrio & Jamet, 2002; Rieber, Tzeng & Tribble, 2004; Boucheix & Schneider, submitted), but much other research shows no effect (Palmiter & Elkerton, 1993, Rouet, Merlet, Ros, Richard & Michaud, 2002) or a negative effect (Lowe, 1999; Schnotz & Grzondziel, 1999 and recently Mayer, Hegarty, Mayer & Campbell, 2005). Methodological and cognitive explanations have been proposed to explain the contradictions which exist in this large body of results (Tversky, Bauer-Morrison & Bétrancourt, 2002; Boucheix & Guignard, 2005). An interesting explanation for the lack of cognitive benefits of animations appear to be the frequent violation of the apprehension principle defined by Tversky, Bauer-Morrison & Bétrancourt (2002): "the structure and content of the external representation should be readily and accurately perceived and comprehended" (p. 256). From this perspective, attentional regulation appears to be a central notion in the comprehension processes of a complex system. Taking this point of view, an overly realistic behavioural animation does not seem to be the best way to assist the subject during their building of a model of the

D. Barker-Plummer et al. (Eds.): Diagrams 2006, LNAI 4045, pp. 40–54, 2006.

described system. Realism could be the wrong condition of delivery for spatial and temporal changes of an animation.

The experiment presented in this paper tests two possible ways to increase the apprehendability of an animated mechanical system: The first attends to the level controllability of the animation; The second focuses on the orientation of attention by the specific's task.

1.1 Interactivity and Learner Control of the Animation

Recently, a number of researchers explored the "principle of interactivity" (Bétrancourt, 2005) by studying learner-control on the animation (Mayer & Chandler, 2001; Tassini & Bétrancourt, 2003; Schwan & Riempp, 2004; Lowe, 2004, Bétrancourt & Réalini, 2005; Boucheix & Guignard, 2005; Boucheix & Schneider, submitted). Among this research, some results show a benefit of interactive presentation in the case of a basic control of the narration and rhythm of the animated lesson (Mayer & Chandler, 2001; Boucheix & Guignard, 2005) and in the domain of procedural tasks (for example tying nautical knots, Schwan & Riempp, 2004). In some cases the benefits of the interactive presentation is limited to the group of participants with low spatial abilities and mechanical reasoning, or with weak previous knowledge (Boucheix & Schneider, submitted, Boucheix, submitted). However, a high level of interactivity is not always a beneficial method of presentation. Another body of research shows that the use of animation with full interactivity is not effective for the learner with little prior knowledge of complex systems (Lowe, 2004; Bogacz & Trafton, 2005), or may shows no (or negative) benefits of control on the animation (Tassini & Bétrancourt, 2004; Bétrancourt & Réalini, 2005). A fully interactive animation can be source of cognitive problems: the participants may not correctly use all the possibilities of a control during the animation, and the activity of control itself can also produce an added cognitive demand.

1.2 Attentional Orientation: Organisational or Functional

Another way to enhance the apprehension of animations could be to focus the orientation of learner's attention on specific information. For example, during the learning of a mechanical system, the subject has to focus his attention on the configuration of the main pieces of the system in order to realize a mental decomposition of the elements (Narayanan & Hegarty, 1998), in this case the learner lends great importance to the organisation of the diagram. But during the study of the mechanical diagram the subject should also favour the functional dimension of the system: The kinematics and causal chain function of the device. The effect of these two kinds of strategy on the elaboration of an accurate mental model of the system may not be equivalent.

We can hypothesize that as a function of the goal of the task specified to the learner (for example, focused on the organisation of the elements or focused on the functional causal chain of the system), their need for information is not equal. Interactive controllable animations could be more suitable than no-controllable and less apprehendable ones for their to precisely process causal chain and kinematics information. The learner focus of attention on the functional model with in a specific task could facilitate the information processing, and the user's control could improve the quality of this processing

In the experimental investigation presented below, we study three different levels of user's control during the learning of a diagram about three pulleys system by adults participants (figure 1). This three pulley system is similar to the famous and well-known system used by Hegarty (1992), Hegarty & Just (1993), Hegarty & Sims (1994), Ferguson & Hegarty (1995), and Sims & Hegarty (1997), and also, Hegarty (2004, 2005). The different user's control modalities for the course of dynamic simulation in the functioning of the three pulley system are: First, a no controllable animation; Second, a partial controllable animation, and; Third, a full manipulable one. For each level of control, we specified four different levels of task specification (plus a control one). The goal of this task specification is to influence the orientation of the learner's attention during their study time using the animated system. The three levels of task specification proposed to the participants are related to the different levels of processing of a mechanical device observed by Narayanan & Hegarty (1998, 2002): The first level concerns the configuration of the main elements of the system; The second level is related to local kinematics (rotation and direction of each pulley) of the system; the last level includes the entire causal chain of the system (Boucheix & Schneider, submitted). Two groups of subjects are contrasted in the experiment: a group of participants with high spatial abilities with a second group with low spatial abilities.

1.3 Off Line and One-Line Measures of Cognitive Processing During Comprehension

Except few pioneers works (for example, Hegarty & Just, 1989, 1993; Hegarty, 1992) the great majority of research in multimedia comprehension was focused on the animated illustrations used in off line written questionnaires to measure comprehension performance (this situation has begun to change with the growing use of eye tracking apparatus and software). The comprehension questionnaire can measure the comprehension level of the learner, but it is not adequate to understand the dynamic acquisition and integration of information. At this time, we ignore the nature of the on-line cognitive processing of animations. In the experiment presented below we measure the on-line processing of animations of the three pulley system, using an eye tracking technique, followed by off line comprehension measures. The previous research, that make use of eye tracking, has studied mainly text and static diagrams. Eye data about text and static diagrams depicting a simple pulley system show that a diagram contributes to the elaboration of a referential representation of information delivered by text, like a memory aid for the deeper processing of the information that has already been read (Hegarty & Just, 1989). Another experiment with eye data (Hegarty, 1992) suggests that, with a static diagram, the learner builds, step by step, a piecemeal mental animation of the system uses the causal chain of events to simulate more precisely the motion of the system. The learner begins to inspect local motions of the first, second, and third pulleys to construct meaningful relations between the components. Then, follows a more global inspection, the goal of which is the integration of the relations between all the elements to enable the elaboration of an animated mental model of the causal chain (Hegarty & Just, 1993,

Hegarty & Sims, 1994). We therefore expect that user's control features and specific orientation of the learner attention will lead to different eye tracking patterns.

1.4 Experimental Overview: Eye Tracking and Animated Diagrams

In this study, eye movements are captured during the presentation of an animated three pulleys system without text (see figure 1). How an internal mental model is built with an external animated diagram? Do controllability of the animation and task specifications facilitate? Several eye movement indicators are used to answer this question. The first eye data indicator was the number of fixations in different areas of interest (AOI) in the diagram (each pulley, ropes, load) fixed in the function of the casual chain. The second eye data indicator was the number (and orientation) of the transitions between the different AOIs. This indicator could bring data about the information research strategy of the learner and suggest the nature of the comparative and inferential activity during the study time of the system. The last eye data indicator monitored the "scan path" of each subject, which could give precise information about the dynamic cognitive activity of the learner, related to the causal chain functioning of the mechanical system (Baccino, 2004; Hegarty & Just, 1993). One interest of this study concerns the possibility of comparisons between the off line comprehension performance and the one-line eye tracking data. Are there best strategies of information processing to elaborate an efficient dynamic mental model of the system?

2 Method

2.1 Subjects

One hundred and twenty one undergraduate students from the University of Burgundy participated.

2.2 Stimuli and Apparatus

2.2.1 Experimental Task Materials

The material used in this study was adapted from Hegarty's (1992) experiment and consisted of an animated three pulley system (figure 1). This system was chosen because its development is continuous with different states. We supposed that animations could have an effect for this specific type of system. We used three versions related to the three levels of user control. The subjects were presented with a dynamic animation, or a sequential dynamic animation, or finally, an interactive animation that can be totally controlled[1].

The non-controllable version consisted of an animated diagram of the pulley system. Animation started when the subjects clicked with the mouse in the diagram area. In the sequential dynamic version, the course of this animation was controllable

---

[1] In all versions, the name of all elements of the system was first displayed for sixty seconds (figure 1). During this time, the subject did not control the animation.

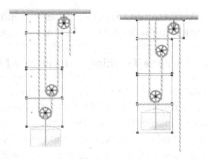

Fig. 1. The three pulleys system no pulled up and pulled up and Areas of interest used to analyze the eye fixations

sequence by sequence (five short animated states) when the subjects clicked with the mouse in the diagram area: one click, one sequence. Subjects could repeat the animation when all the sequences were finished. In the fully controllable version, the learner could manipulate the system, as they wanted by pulling on the end of the free rope, after they have clicked on the end of the high rope. Subjects had total control on the progress of the animation, like an entire simulation of the functioning of the pulleys that lifts the load.

2.2.2 Memorisation and Comprehension: Off Line Measures
After the study of the diagram, subjects were asked fourteen memorisation and comprehension questions. Three types of questions were elaborated from the basis of the model of Hegarty (1992, 2004) and Narayanan & Hegarty (1998, 2002). The first series of questions concerned the configuration of the elements of the pulleys system (for example: what elements touch the middle pulley?). The second series of questions concerned the local kinematics of the system (rotation and direction of the pulley). The third series of questions concerned the entire functional mental model of the system that involves the integration of all the components and local kinematics. We determined three comprehension sub-scores for each type of questions.

2.2.3 Individual Differences Measures
Before the experiment mechanical and spatial abilities of each participant were measured with a French abbreviated form of DAT, Differential Aptitude Tests (Bennett, Seashore & Wesman, 1973, 2002).

2.2.4 Apparatus
The stimulus was presented on a computer screen situated approximately sixty centimetres from the subject. The eye fixations of the subjects were monitored by an ASL (5000) corneal reflectance and pupil-centre eye tracker and were recorded with gaze tracker software, which allowed the processing of dynamic information (like videos, animations, or the internet). Moreover the system was combined with a magnetic head tracker, and therefore this system allowed the subjects to maintain (relatively) free head. ASL 5000 is a 50-hertz system recording the position of the subject's gaze every 20 millisecond.

2.3 Procedure

The overall procedure involved three stages.

During the first stage, the mechanical and spatial abilities were measured collectively with the differential aptitude test. This, then, allowed two groups of participants to be identified: subjects with high and low mechanical and spatial abilities. A mean was calculated for spatial and mechanical scores across all subjects, since we found a significant correlation between the measures (p <.01). On the base of this single score, we calculated the median score. The median score was 57 percent correct: 60 high (score superior to the median score) and 61 low (score inferior to the median score) mechanical and spatial abilities formed the two contrasted groups. The subjects were then equally distributed among the three type of control and the four level of task specification. Then, five subjects were assigned to each experimental condition. The validity of the distribution was tested with Neuman-Keuls tests.

During the second stage of the experiment, using a computer screen, subjects were presented with one of the three versions (no-controllable, partially controllable and fully controllable) of the animated pulley system. During the study period for the device, eye movements were captured. According to his or her specific task (attention focused on the configuration of the system only, on the local kinematics, on both configuration and overall kinematics of the system, or no specific orientation of attention –control group), the goal assigned to each participant was to understand the operating system. The subjects were informed that they had to use the animation at least three times.

After the consultation of the animated material, subjects answered the compre-hension questions test in a booklet.

3   Results

In order to stay within the space limits of this paper, the main result about the comprehension performances will be briefly outlined exposed, while the eye tracking data will be discussed in a little bit more detailed.

3.1 Comprehension Performance

The mean rates of correct answers picked up at the comprehension test for each indicator (configuration, kinematics, functional model) are presented table 1.

An ANOVA including the three control levels, the two spatial ability levels, the four-task types, and the three dependant comprehension indicators shows firstly a global effect of the spatial abilities. The learner with high spatial and mechanical abilities performed better than the learner with low abilities, $F(1, 97) = 21.91$, $p < .0001$. The learner with a fully controllable animated diagram performed slightly better than the learner with no controllable or partial controllable animated diagrams, $F(2, 97) = 2.97$, $p = .056$. The learners with a specific orientation of attention upon the functional model and, to a lesser extent, with local kinematics performed better than the learners with a specific orientation on the configuration elements or with no orientation of attention, F(3, 97) = 4.47, p = .006. These effects show that a specific task focuses attention on the functional model and local kinematics allow a better mental model of the system to be constructed. Also, the fully controllable animation allows, to a lesser extent, the building of a better mental model. These results suggest that the elaboration of a mental model can be improved, especially with the orientation specific of attention on a task and also with the user's control of animation.

Moreover, the effect of the comprehension indicators is significant: the scores of configuration are higher than the scores of kinematics which are higher than the scores obtained for the functional model, F (2, 194) = 555.54, p <.0001.The interaction between the spatial ability levels and the three indicators of comprehension is significant, F (2, 194) = 5.81, p = .004. This interaction indicates that the learners with high spatial abilities performed better than the learners with low spatial abilities only for the kinematics and functional model questions. The interaction between the task type and the three indicators of comprehension is also significant, F (6, 194) = 5.04, p = .0008. This interaction shows a higher benefit when the attention of subjects is focused on the functional model of the system (figure 2). These results have been examined in further detail by three other ANOVAs, one for each comprehension indicator.

For the configuration level, we observed a significant effect for the different orientation of attention, F (3, 97) = 3.47, p = .019. The learners with a specific orientation of attention about functional model performed better than the three others specific orientations.

For the kinematics level, we found a significant effect for the spatial ability levels, F(1, 97) = 25.33, p < .0001. The learners with high spatial and mechanical abilities performed better than the learners with low abilities for the kinematics questions. Also, we found a significant effect for the level of control, F (2, 97) = 4.13, p = .019. The learners with a fully controllable animated diagram performed better than the learners with a partially controllable or non controllable animated diagrams. Moreover, we observed an effect of interaction, that approaches significance, between the spatial ability levels and the user control level, F (2,97) = 2.51, p = .086. This interaction shows a weaker performance for the learners with low abilities in particular when the control of the animated diagrams was partial.

Finally, for the functional model level, we observed a significant effect for the spatial ability levels, F (1, 97) = 12.75, p = .0006. The learners with high spatial abilities performed better than the learners with low spatial abilities. Moreover, we found a significant effect for the task type, F (3, 97) = 7.90, p < .0001. The learners with a specific orientation of attention upon the functional model or kinematics performed better than the two others specific orientation (configuration and no-orientation).

In conclusion, the controllability of the animation allow a better comprehension of the local kinematics and the task type allows better comprehension of the functional model.

Table 1. Mean rates of correct answers at the three indicators of the comprehension test with each level of control of the pulley animation for each spatial and mechanical aptitude group and for each level of orientation of attention

	High Spatial			Low spatial		
	No-control	Partial control	Fully control	No-control	Partial control	Fully control
Configuration score						
No specific task	79.17	81.67	90.00	80.00	73.33	75.00
Config. Task	80.00	84.17	84.17	84.17	78.33	89.17
Kinematics task	81.67	89.17	73.33	66.67	84.03	83.33
Func. model task	91.67	91.67	89.17	83.33	85.83	89.17
Local kinematics score						
No specific task	66.90	84.14	91.72	77.93	68.67	71.03
Config. Task	74.48	71.72	88.28	69.66	47.58	68.97
Kinematics task	82.07	91.03	71.72	68.28	72.41	82.07
Func. model task	84.83	86.21	94.48	60.69	55.17	79.31
Functional model score						
No specific task	28.33	39.17	35.00	21.67	17.50	27.50
Config. Task	33.33	25.00	32.50	25.00	*19.17*	32.50
Kinematics task	45.83	45.83	40.83	39.17	37.50	43.33
Func. model task	51.67	46.67	65.00	37.50	25.00	34.17

3.2 Eye Tracking Measures

In order to assess the on-line processes in the comprehension of animation, we analyzed the eye fixations of one hundred and thirteen subjects. Areas of interest (AOI) are used to record the amount of time, and activity, the subject's gaze spends in a particular region of the image. We analysed the number of fixations in every AOI, the transition between the different AOI, and the scan path. So, we created four specific AOI, marked in figure 1. The first AOI included the upper pulley (the only pulley that doesn't move). This AOI correspond to the start time of the causal chain of the system. The second AOI included the middle pulley when this pulley was pulled up during the animated simulation phase. The third AOI included the middle pulley before the animation was pulled up and the lower during and after the simulation. In this way, the third AOI was an important area for the integration of the kinematics of the system: the direction and the rotation of two pulleys. In this area we expect a great number of eye movements (comparison transitions). These three AOI characterize the causal chain areas. Finally, the fourth AOI included the lower pulley before the starting of the animation and the area of the loading object during and after the simulation. This AOI seems less relevant for the integration of the kinematics than AOI two and three.

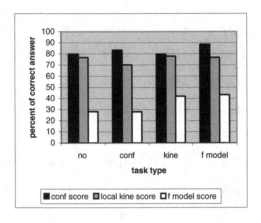

Fig. 2. Percent of correct response for the three comprehension indicators according to the task type

3.2.1 Number of Fixations

The number of fixations in the different AOI was the first eye movements measure to be analyzed. The results are displayed in the figures 3, 4. An ANOVA including the three control levels, the two spatial ability levels, the four orientation of attention, and the four AOI as dependant measures shows a significant effect of the control level, $F (2, 89) = 7.08$, $p = .001$. The number of fixations was more important in the partial control animation than for the full control animation, that was in turn more important than for the no controllable animation. Moreover, the effect of the type of AOI was significant, the number of fixations for the third AOI (area important motions) was higher than the first and second AOI (these two last AOI had a equivalent number of fixations) and higher than the fourth AOI, $F (3, 267) = 49$, $p<.0001$.

In order to go deeper into these results, we conducted four further ANOVAs, one for each AOI. For AOI 1, the partial controllable and the full controllable animation lead to a higher number of fixation than the no controllable animation, $F (2, 89) = 4.34$, $p = .016$. With the same tendency, we observe an effect of the control level for AOI 2, $F (2, 89) = 8.30$, $p < .001$ and for AOI 3, $F (2, 89) = 3.78$, $p = .027$. When the animation could be controlled, the learners watched the third AOI the most, which is more relevant in the understanding the causal chain of the system. During the course of the animation, these AOI deliver a great amount of crucial information about the kinematics of the pulleys. These effects indicates that the number of fixations in the partial control level and in the full control level were higher than the no control level for the AOI 1, 2, and 3. But, the number of fixations in the AOI 4 was equal for all of the three levels of control (figure 3).

Finally, an interesting result is the fact that, for the AOI 4, which could potentially represent the less pertinent area for the mental integration of the causal chain of the system, we observed a significant effect of the specification of the task, F (3, 89) = 2.70, p = 05. The number of fixation on this AOI is higher when there is no specification of the task. This result reveals an interesting fact: the AOI 4 appears less crucial in the processing of the functioning of the three pulleys system, and when the task is specified, the participants spend less time on irrelevant information (cf. figure 4). Moreover, we observe a significant interaction effect between the two spatial ability levels and the four orientations of the attention for this AOI, F(3, 89) = 2.83, p = .043. This interaction suggest that the number of fixations in AOI 4 were higher in the condition which no orientation of attention, but only for the subjects with high spatial abilities.

In summary, we observed a number of fixations more important for the AOI that contain a lot of movements and we also observed a number of fixations more important in these AOI for the full and partially controllable animations. Then, we can suppose that attending more to these AOI allows a best integration of the causal chain motions of the system because we have observed better performances for the comprehension test in the two controllable animations conditions. Nevertheless, these results could be also partially due to the activity of control of the device itself. Therefore, in order to allow deeper analysis, this is the reason why we centred the next analysis on the study of the eye transitions between the different AOI for a deeper analyse.

3.2.2 Transition Between the AOI

The results about the transitions are presented figures 5, 6. An ANOVA including the three control levels, the two spatial ability levels, the four orientations of attention, and the six possible transitions between the AOI (as dependant measures) shows a

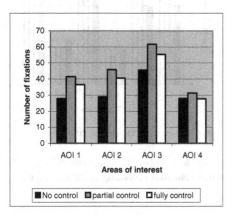

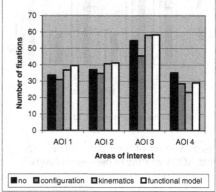

Fig. 3. Number of fixations for every AOI according to the control levels

Fig. 4. Number of fixations for every AOI according to the task type

significant effect of the control level, F (2, 89) = 4.01, p = .02. This effect points to the fact that the number of transitions for the partial control level is higher than for the full control level, and the number of transitions for this last level is higher than for the no control level. Moreover, the number of transitions observed between the AOI respectively 1&2, 1&3, 2&3, and 3&4 is higher than the number of transitions observed in the AOI 1&4; 2&4, F (5, 445) = 47.12, p<.0001. The transitions between AOI 1&2 and 2&3 imply the local piecemeal elaboration of the causal chain. The transitions between the look zones 1&3 suggest a more global integration of the entire functional model of the causal chain of the system. The interaction between the levels of control and the six transitions is significant, F (10, 445) = 2.06, p = .026. This interaction indicates a higher number of transitions between the AOI 1&2, 1&3, and 2&3 for the controllable levels than for the non controllable one. It can also be observed a few numbers of transitions between the AOI 1&4, and 2&4. At last, the number of transitions between the AOI 3&4 is medium (figure 5).

In order to more precisely examine these broad effects results, we conducted four another ANOVAs, one for each kind of transition area. For AOI 1&2, the number of transitions for the partial control level is higher than the full control level and the no control level, F(2, 89) = 4.2, p = .02. Also we observe a significant effect of the control level for the transition between the AOI 2&3, F(2, 89) = 4.61, p = .012. The number of transitions for the controllable conditions was higher than for the non controllable level. Moreover for the areas 2&3, the results show a significant effect of the attentionnal orientation conditions, F(3, 89) = 2.76, p = .046. The number of transitions between the AOI 2&3, is lower when attention was focused on the system configuration rather than the three others orientation of attention (figure 6).

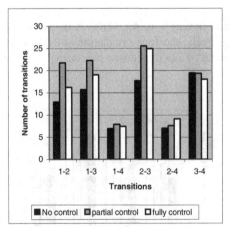

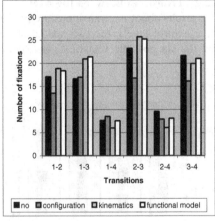

Fig. 5. Number of transitions between the different AOI according to the control levels

Fig. 6. Number of transitions between the different AOI according to the task type

3.2.3 Scan Path

The results previously described can be related to the on line time of processing the three pulleys system during the course of the animation, and for each level of controllability and each level of attentionnal orientation. To achieve this goal, the scan path that presents the on line dynamic trajectory of the participant can be used as an essential indicator to visualize the directions of the eye's trajectory and the different strategies used by the learner for the selection of the relevant information. In these proceedings, we will present only the first preliminary results because further analysis is still in progress.

In each learner scan path, we determined three types of transitions between the different look zones. First, we distinguished the global transitions. The global transitions concerned the transitions between two no neighbourhood AOI, for example the transitions between AOI 2 and 4. Second, we distinguished the local transitions. These local transitions concerned the transitions between two neighbourhoods AOI, for example the transitions between AOI 1 and 2. Finally, we distinguished the sequence of transitions (for example, AOI 1,2,3). These "causal chain" transitions concerned the transitions between three neighbourhoods AOI at less, for example the transitions between the AOI 1, 2, 3 or 1, 2, 3, and 4.

The results of these different kinds of transitions (local, global, sequences) are presented in figure 7. An ANOVA including the three control levels, the two spatial ability levels, the four orientations of attention, and the three types of the transitions between the look zones as dependant measures, showing a significant effect of the type of transitions, $F(2, 178) = 302.32$, $p < .00001$. The number of local transitions is more important than the number of global transitions and this last one is more important than the causal chain transitions. Moreover, results show a significant interaction effect between the different types of AOI and the task type, $F(6, 178) = 2.18$, $p = .047$. This effect shows that the number of local and causal chain fixations is less important for the configuration task (figure 7).

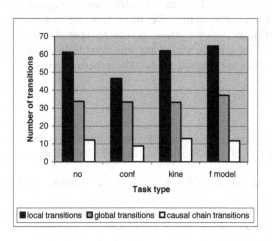

Fig. 7. Number of transitions between the different types of AOI according to the task type

4   Discussion

Our goal was to study the effect of a user-controllable animation and of specific orientations of attention in the elaboration of an efficient mental model of a mechanical diagram without text. The performances on the comprehension test showed a benefit of the controllable modalities, particularly when the task is oriented toward a mental model level. The full control animation seems to be the relatively most efficient type of control for an effective integration of the mental model of complex mechanical system presented without textual explanation. The orientation of attention by a specific task level showed that the building of the functional mental model was enhanced when the focus of this orientation was focused on the functional model and on the kinematics of the system. The orientation of attention could influence the organization of the integration activity of the different elements and of the relations between the elements according to the causal chain of the system. Although the results showed an effect of the control level and of the specification of the task, they did not show an effect of interaction between these two variables.

The eye tracking measures showed that the number of fixations was more important in the AOI showing the kinematics of the system. The results obtained from the study of the transitions between the different AOI suggest that the learners use the animated features and the controllability of the motions to process the relevant information, which in turn allows them to construct a model of the causal chain of the system. When diagrams are alone, without explanatory texts, interactive animations could enhance significantly the building of an better accurate mental model, and should support best internal simulations. Two sorts of transition were distinguished; transition to integrate local relations, and another to integrate entire functional system. The next analysis, which in progress, of the dynamic processing during the study time of the three pulleys by the scan path, should bring more detailed results about the iterative cycles of the integration: The local, global and sequences of local transitions.

Finally, can our results from the pulley system can be generalized to other complex types of devices or contents? The answer to this question appears to be crucial for the designing of educational systems. We think that the reasoning demands for the comprehension of such pulley system, involving spatial features (rotations, causal chain, speed, direction apprehension) are transferable to other kinds of mechanical systems. Moreover this speculation needs to be experimentally investigated.

References

Baccino, T. (2004). *La lecture électronique*: Presses Universitaires de Grenoble.

Bennet, G. K., Seashore, H. G.& Wesman, A. G. (1973,2002). Differential aptitude test. 5ème edition. *ECPA*.

Bétrancourt, M. & Réalini, N. (2005). Le contrôle sur le déroulement de l'animation. 11ème JETCSIC ;17 Juin, Nice.

Bétrancourt, M. (2005). The animation and interactivity principles in multimedia learning. In R. Mayer (Eds.), *The Cambridge handbook of multimedia learning* (pp. 287-296). Cambridge: Cambridge University Press.

Bétrancourt, M., & Tversky, B. (2000). Effect of computer animation on user's performance: a review. *Le travail humain, 4, 63*, 311-329.

Bogacz, S., & Trafton, J. G. (2005). Understanding dynamic and static displays: using images to reason dynamically. *Cognitive Systems Research, In Press, Corrected Proof*.

Boucheix, J.M. (submitted). Children interaction with technical animations: user control and attentionnal cues

Boucheix, J.M. ; & Guignard, H. (2005). Which animation condition can improve text comprehension in children? European Journal of Psychology of Education, 20(4), 369-388..

Boucheix, J.M. ; & Schneider, E.. (submitted). Designing "apprehendable" animation features to enhance mental animation in multimedia comprehension: micro steps format and user control.

Ferguson, E. L., & Mary, H. (1995). Learning with real machines or diagrams: application of knowledge to real-word problems. *Cognition and Instruction, 13*(1), 126-160.

Hegarty, M. (1992). Mental Animation: Inferring Motion From Static Displays of Mechanical Systems. *Journal of Experimental Psychology: Learning, Memory, & Cognition September, 18*(5), 1084-1102.

Hegarty, M. (2004). Mechanical reasoning by mental simulation. *Trends in Cognitive Sciences, 8*(6), 280-285.

Hegarty, M. (2005). Multimedia learning about physical system. In R. Mayer (Eds.), *The Cambridge handbook of multimedia learning* (pp. 447-465). Cambridge: Cambridge University Press.

Hegarty, M., & Just, M. A. (1989). Understanding machines from text and diagrams. In H. Mandl & J. R. Levin (Eds.), *Knowledge acquisition from text and pictures* (pp. 171-194). North-Holland: Elsevier Science.

Hegarty, M. & Just, M.A. (1993). Constructing mental models from texts and diagrams. *Journal of Memory and Language, 32,* 717-742.

Hegarty, M., & Sims, V.K. (1994). Individual differences in mental animation during mechanical reasoning. *Memory & Cognition, 22,* 411-430.

Hidrio, C., & Jamet, E. (2002). Compréhension d'un dispositif technique: apports d'une illustration dynamique et des traitements multiples. *Psychologie Française, 47*-1, 61-67.

Lowe, R.K. (1999). Extracting information from an animation during complex visual learning. *European Journal of Psychology of Education.* Special Issues: visual Learning, W. Schnotz, (Ed). XIV, 2, 225-244.

Lowe, R.K. (2004). Interrogation of a dynamic visualization during learning. *Learning and Instruction*, 14, 257-274.

Mayer, R.E. (2001) *Multimedia learning.* Cambridge: Cambridge University Press.

Mayer, R.E., & Chandler, P. (2001). When learning is just a clik away: does simple user interaction foster deeper understanding of multimedia messages. *Journal of Educational Psychology, 93*, 390-397.

Narayanan, N.H., & Hegarty, M. (1998). On designing comprehensible interactive hypermedia manuals. *International Journal of Human-Computer Studies*, 48, 267-301.

Narayanan, H. N., & Hegarty, M. (2002). Multimedia design for communication of dynamic information. *International Journal of Human-Computer Studies, 57*(4), 279-315.

Palmiter, S., & Elkerton, J. (1993). Animated demonstrations for learning procedural computer-based tasks. *Human-Computer Interaction, 8*, 193-216.

Rieber, L. P., Tzeng, S.-C., & Tribble, K. (2004). Discovery learning, representation, and explanation within a computer-based simulation: finding the right mix. *Learning and Instruction, 14*(3), 307-323.

Rouet, J.F., Merlet, S., Ros, C., Richard, E., & Michaux, C. (2002). Effects of animated illustrations on the comprehension of an expository text. Poster Session of *EARLI-SIG, "Comprehension of Verbal and Pictorial Information"*. Poitiers: august 29.

Schwan, S., & Riempp, R. (2004). The cognitive benefits of interactive videos: learning to tie nautical knots. *Learning and Instruction*, 14, 293-305.

Schnotz, W., & Grzondiel, H. (1999). Individual and co-operative learning with interactive animated pictures. *European Journal of Psychology of Education*, XIV, 245-265.

Sims, V. K., & Hegarty, M. (1997). Mental animation in the visuospatial sketchpad: evidence from dual-task studies. *Memory and Cognition, 25*(3), 321-332.

Tassini, S., & Bétrancourt, M. (2003). Le contrôle sur l'animation influence-t-elle le niveau d'efficacité cognitive de l'animation. *Neuvièmes Journées JETCSIC*, 21 juin : Dijon.

Tversky, B., Bauer-Morrison, J., & Bétrancourt, M. (2002). Animation : can it facilitate ? *International Journal of Human-Computer Studies, 57*, 247-262.

From Diagrams to Models by Analogical Transfer

Patrick W. Yaner and Ashok K. Goel

Design Intelligence Group
Artificial Intelligence Laboratory
Division of Interactive and Intelligent Computing
College of Computing, Georgia Institute of Technology
Atlanta, GA 30332-0280
{yaner, goel}@cc.gatech.edu

Abstract. We present a method for constructing a teleological model of a drawing of a physical device through analogical transfer of the teleological model of the same device in an almost identical drawing. A source case, in this method, contains both a 2-D vector-graphics line drawing of a physical device and a teleological model of the device called a Drawing-Shape-Structure-Behavior-Function (DSSBF) model that relates shapes and spatial relations in the drawing to specifications of the structure, behavior and function of the device. Given an almost identical target 2-D vector-graphics line drawing as input, we describe how an agent may align the two drawings, and transfer the relevant structural, behavioral and functional elements over to the target drawing. We also describe how the DSSBF model of the source drawing guides the alignment of the two drawings. The Archytas system implements this method in domain of kinematic devices that convert translational motion into rotational motion, such as a piston and crankshaft device.

1 Background, Motivation and Goals

Diagram understanding is a persistent and central issue in research on diagrams. Novak [20] explicitly viewed the task of understanding a diagram as one of constructing a model for it: his Beatrix program understood a textually-annotated schematic diagram kinematics problem by constructing a structural model. In cognitive science, Narayanan, Suwa and Motoda [19] describe a psychological study in which human subjects built behavioral models of a physical device from its labeled schematic diagram. The computational advantage of model construction is that the model enables inferences for higher-level tasks, e.g., analysis of the functions of the physical device in case of Narayanan, Suwa, and Motoda's human subjects. The type of the output model in the task of diagram understanding depends in part on the input diagram and partly on the inferences desired of the model.

Current AI methods for constructing models from diagrams, e.g., in the Beatrix and the GeoRep [9] programs, often apply domain-specific rules to extract the model from the diagrammatic representation. For example, GeoRep first uses a domain-independent diagrammatic reasoner that extracts shapes and spatial relations from a given diagram, and then applies domain-specific rules for extracting a behavioral model from the intermediate representation of shapes and spatial relations. Given the state of the art in

D. Barker-Plummer et al. (Eds.): Diagrams 2006, LNAI 4045, pp. 55–69, 2006.

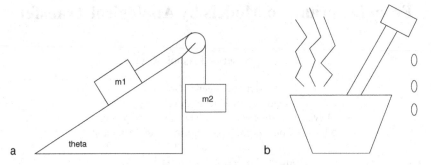

Fig. 1. (a) a sample input to Beatrix system (adapted from Novak [20]): a drawing of two block masses suspended over an inclined plane by a pulley; and (b) a sample input to the GeoRep system (adapted from Ferguson and Forbus [9]): a drawing of a cup of hot liquid with a metal bar and a melting ice cube at the other end of the metal bar

research on diagrams in general, current AI programs typically focus on very simple diagrams. Figure 1(a) illustrates a sample diagram input to Beatrix; figure 1(b) similarly illustrates a sample diagram input to GeoRep.

Now let us consider the 2-D vector-graphics line drawing illustrated in figure 2(b). It depicts a piston and crankshaft device that converts linear motion of the piston (shown on the right side of the figure) into rotational motion of the crankshaft (shown on the left side; the axis of rotation is perpendicular to the page). Note however that drawing itself does not contain any labels or annotations. Given this input drawing, let us consider the task of constructing a teleological model of the device depicted in the drawing, where the teleological model specifies both the output functions of the device (what does it do?) and the internal causal behaviors that accomplish the functions (how does it work?). Understanding this diagram by constructing a teleological model that can answer the above questions is a complex task because the functions and the behaviors of the device are not directly apparent from the drawing.

But let us also suppose that the agent is familiar with this domain, and has come across many drawings and their teleological models in the domain. In particular, let us suppose that the agent already knows of the teleological model for the drawing in figure 2(a). Note that this source drawing is almost identical to the target drawing in figure 2(b); in fact, the two drawings depict the same view of the same piston and crankshaft device in two different states. Our first hypothesis in this work is that, under these knowledge conditions, the agent may generate a teleological model for the target drawing by *analogical transfer* of the teleological model of the source drawing.

In general, analogical reasoning involves the steps of retrieval, mapping, transfer, evaluation, and storage, as illustrated in figure 3. In this paper, however, we focus only on the tasks of mapping and transfer. The design domain is that of kinematics devices, and, in particular, devices that convert linear motion into rotational motion (and vice versa). Archytas is a computer program that implements our theory of diagram understanding by analogical reasoning. The source and target drawings are created using XFig (see www.xfig.org/), a vector-graphics program. Archytas begins its processing of the target drawing with a representation of shapes and their locations generated by XFig.

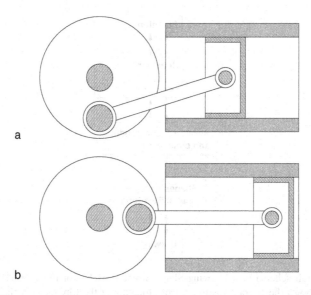

Fig. 2. (a) A sample source drawing of a piston and crankshaft assembly. The components depicted are the cylinder, the piston, the crankshaft, and the connecting rod. (b) A target drawing; the same device, with the piston now at the top of its range of motion ("top dead center").

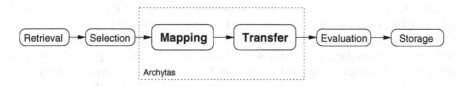

Fig. 3. The five major subtasks of analogy: retrieval, selection, mapping, transfer, and evaluation. This paper focuses on the mapping and transfer tasks implemented in the Archytas system.

Both mapping and transfer are complex tasks. The mapping task is complex because a given shape (or spatial relation) in the input target drawing may map into many similar shapes (or spatial relations) in the known source drawing. The transfer task is complex because based on a mapping between the shapes and spatial relations in the two drawings, the goal is to transfer the functions and behaviors of the device in the source drawing to the target drawing. This implies that we need some representation that relates shapes and spatial relations in a source drawing to the functions and behaviors of the device depicted in the drawing. In earlier work on adaptive design [10, 11, 12], we developed Structure-Behavior-Function (SBF) models of physical devices: an SBF model of a device uses the internal causal behaviors of the device as intermediate abstractions that relate the functions of device to its structure. Our second hypothesis in the present work is that the schema and ontology of SBF models can be productively expanded and extended into Drawing-Shape-Structure-Behavior-Function (DSSBF) models: a DSSBF model uses structure as an intermediate abstraction to relate shapes and spatial relations in a drawing to the behaviors and functions of the device depicted in the drawing.

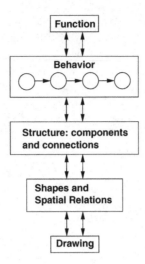

Fig. 4. The overall schema of a Drawing-Shape-Structure-Behavior-Function (DSSBF) model. This diagram shows the overall hierarchy, from function at the top, through the causal behavioral model, the structural components and connections, the shapes and spatial relations depicting them, and finally the original drawing.

Given the complexity of the tasks of analogical mapping and transfer, our work on the Archytas project so far has focused on mapping and transfer between two nearly identical drawings. Specifically, in Archytas, we assume that the target and the source drawings are so similar that any distinction between them at the shape level makes no difference at the structural level of abstraction. Note that while this assumption of near identicality of the source and the target drawings simplifies the transfer task, it makes little difference to the mapping task since the mapping occurs at the level of shapes and spatial relations. This implies that we need a method to control the complexity of mapping at the shape level. Our third hypothesis is that knowledge of the functions of the important shapes in the DSSBF model of the source drawing informs the mapping task at the shape level, and seeds the mapping with shapes that play a critical role in the functioning of the device.

2   DSSBF Teleological Models

As figure 4 illustrates, a DSSBF model of a physical device unifies the functional and spatial representations of the device. This unification results in a five-level model with function at the top and form (e.g., a drawing) at the bottom, with shape, structure and behavior as intermediate levels of abstraction. Note that in general there may be many drawings, and, hence, many shape representations of a single device (but figure 4 does not depict this for clarity).

The representations of any two consecutive levels in the five-level DSSBF model contain two-way pointers to each other. For example, as in SBF models, the specification of a function specifies the behavior that accomplishes it and the specification of

Table 1. Outline of the structural model of the piston/crankshaft and example

Component	Properties	Variable Quantities	Connected to
Piston	height, diameter	linear momentum	cylinder, connecting rod
Crankshaft	diameter	angular momentum	crankcase, connecting rod
Connecting Rod	length	angular, linear momentum	crankshaft, piston
Cylinder	diameter, length		piston, crankcase
Crankcase			cylinder, crankshaft

a behavior specifies the function (if any) that it accomplishes. Similarly, the specification of a behavior specifies the structural constraints (in the form of connections among components) that enable it, and the specification of a component specifies its functional abstractions and the role they play in a behavior. In addition, in a DSSBF model, the specification of a structural component or connection specifies the shape that depicts it in a drawing, and the specification of a shape specifies the component or connection it depicts. Thus, the organization of a DSSBF model of a physical device affords navigation of the entire model, and accessing of knowledge at one level of abstraction that is relevant to reasoning at another level.

2.1   An Illustrative Example

Figure 2 illustrates a piston and crankshaft device. In this device, there are five components, though only four are depicted in the figure: the (1) piston, (2) crankshaft, (3) connecting rod, (4) cylinder, and (5) crankcase (not shown). The function of this device is to turn the crankshaft. In the DSSBF language, this function is represented as a schema that specifies the state it takes as input, the state it gives as output, and the behavior that accomplishes the function.

In the DSSBF language, a behavior refers to an internal causal process, and is represented as a sequence of discrete states and state-transitions. The states in a behavior specify values of variables relevant to the behavior. The specification of the behavior of the crankshaft, for example, tracks the angular momentum of the crankshaft, which it gains from a downward force coming from the connecting rod through the joint with the connecting rod, and loses through friction. The annotations on a state transition specify the causes and the conditions of the transition.

In a DSSBF model, structure refers to the components and the connections among the components. Table 1 shows an outline of the specification of the components in the piston and crankshaft device. Briefly, each component schema specifies its properties, which take values, and quantities, which have a type of either scalar or vector, and which are variables whose values are changed by the causal processes in the behaviors of the device.

Figure 5 illustrates the connections in the piston and crankshaft example. In the DSSBF language, connections are represented as schemas. Connections also have types indicating their degrees of freedom, e.g., revolute has one degree of freedom (rotation), prismatic has one degree of freedom (translation), and so on.

Shapes and Spatial Relations: Let us consider the shape-level representation of the drawing of the source case illustrated in figure 2. Since this is XFig file, the properties

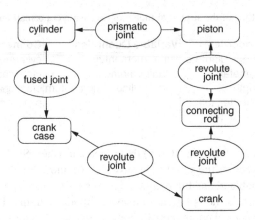

Fig. 5. A diagram of the connections in the structural model shown in table 1, where rounded boxes represent components and ovals represent connections between components. Properties and variable quantities of components are detailed in slot values for each component schema.

and locations of lines and their interconnections already are known. Thus, Archytas begins with whole shapes, such as rectangles and circles, and their geometric properties. In the DSSBF language, each shape schema has slots associated with it detailing specific dimensions and aspects of the shape, such as the height and width of rectangles and the diameter of circles.

But Archytas needs to infer the relevant interrelationships among the shapes. For XFig drawings, the DSSBF language uses a taxonomy of spatial relationships among the shapes in a drawing: parallel-ness and perpendicularity, end-to-end and overlapping connections between lines, collinearity, horizontal and vertical alignment and relative length, and containment. Figure 6 illustrates a partial specification of the spatial relations in the piston and crankshaft drawing. This specification is an abbreviation of the actual representation: for instance, the component lines of the rectangles and the part-whole relations with the rectangles, as well as the interrelationships between them, are not shown, and only some of the spatial relationships are shown. The complete description, including mereological (part-whole) relations of component lines to polygons, involves hundreds of terms: the actual relational description of this drawing has 644 relations, including all relations involving individual lines that are parts of polygons. Excluding these component lines and their interrelations, the description had 72 relations in it.

Relating Shape to Structure: In order to be useful, the representation of the shapes and the spatial structure of the drawing must enter into a relationship with the structural elements in the model. These relations take the form of links between the shape schemas and the component schemas in the model. At the most basic level, we need relations of the form A DEPICTS B from the shape schemas to the components schemas, A being a shape schema, B being a component or connection schema. It is important to note that only the shapes themselves enter into these relationships; the relations between them do not.

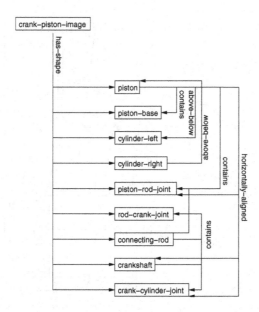

Fig. 6. Representation of most of the the shapes and only some of the principal spatial relations from figure 2(a). For clarity the labels on the shapes reflect the names of components depicted. This diagram shows only the properties and relations of the aggregate shapes, not the aggregation of lines into rectangles, or the interrelationships of these lines.

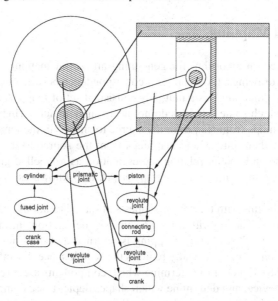

Fig. 7. The relation of shape to structure, showing the mapping between the shapes and the structure. The relations between the shapes and the components and connections are DEPICTS relations: *shape* DEPICTS *component*. In this figure, the drawing from figure 2(a) is shown for clarity, but in the actual representation the relations are between the shape nodes shown in figure 6 and the components and connections shown in 5.

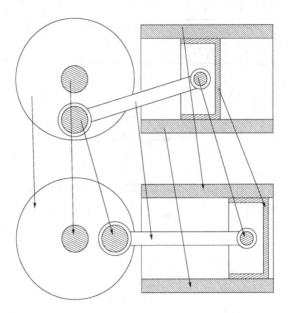

Fig. 8. The mapping between the images in figures 2(a) and 2(b). The actual mapping is between the representations of the sort shown in figure 6. Not all relations match: some relations in the source (on the top) will not be found in the target (on the bottom) and vice versa. This is only a single (best) mapping; the several smaller partial mappings are not shown.

3   Analogical Mapping

We have developed an algorithm for generating all partial mappings between given source and target drawings. (A "map" is a correspondence between individual entities or relations, and a "mapping" as a mutually consistent set of maps. A partial mapping is a mapping in which some relations do not necessarily match.) This algorithm gathers up individual maps from shapes in the source to shapes in the target drawing, and attempts to merge them into whole mappings from the source to the target drawing. Treating the shape and spatial relation representation as a labelled graph, this corresponds to the problem of maximal overlap set [4], also known as maximal common edge subgraph [21].

The algorithm begins with the source shapes and spatial relations and a target shapes and spatial relations. The algorithm marks some of the relations in the source representation as "important" so that no mapping is returned that does not involve at least one of these relations. This drastically reduces the search space. Specifically, the algorithm uses the DSSBF model to determine which components are directly involved in behaviors of the device, and determine which shapes depict those components. Certain individual relations involving these particular shapes are marked as "important".

In addition to this, there is a parameter to the algorithm that is a lower bound for the size of any mapping generated, so that Archytas will not return any mapping that is smaller than this bound (in terms of the number of relations mapped). There can be dozens or even hundreds of these small degenerate mappings of just one or two shapes, even when there's a single complete mapping.

The procedure runs as follows:

1. Gather up all maps between source and target relations. Each map between a pair of relations will entail two maps between the entities related, so that, if A contains B maps to X contains Y, then this means A maps to X and B maps to Y
2. Pull out those maps involving marked ("important") relations as "important" maps.
3. Choose one of the "important" term maps m_1. Now, recursively gather every other map m_i, for $i > 1$, such that m_1 and m_i are consistent. They are consistent when:
   - the source terms in m_1 and m_i are different
   - the target terms in m_1 and m_i are different
   - the associated entity maps are consistent (same source entity maps to the same target entity, and conversely the same target entity has the same source entity being mapped to it)

   These rules enforce a one-to-one mapping between both relations and entities.
4. When all of the mutually consistent individual maps have been gathered, save this as a single (partial) mapping if it exceeds the minimum size for an acceptable mapping (minimum bound).
5. Choose the next marked "important" term map m_2, and repeat steps 3 and 4, iterating through all of the marked "important" term maps.
6. Return the resulting list of mappings.

The third step—expanding the mapping from the "important" map—involves a recursive operator that expands a given mapping with a set of candidates and excluding those in another set, initially null. This procedure (not shown) is based on an algorithm of Bron and Kerbosh [1].

These mappings allow term-by-term comparison of the source and target, so that Archytas can reason about the similarities and differences between the two drawings with respect to a potential alignment of the drawings. Note that each individual mapping is a term mapping (associated relations), as described above, and must be converted into mappings between entities. The method returns all *maximal* complete mappings, and as such the largest of these will be the *maximum*; a maxim*al* mapping is one that cannot be expanded to include other terms without contradiction, and a maxim*um* mapping is the largest of all of the maximal mappings. Thus, if there is a subgraph isomorphism between source and target either way (source entirely aligns with target, or vice versa), the algorithm will return no partial mappings that are consistent with that particular mapping, but it may return others that are inconsistent with it.

In the context of figures 2(a) and 2(b), all the shapes from the source map onto shapes from the target, but the algorithm does discover that the target (figure 2(b)) has the connecting rod rectangle parallel with the cylinder and the piston, but the source (figure 2(a)) does not. It also returns several partial mappings: for instance, the top rectangle for the cylinder in figure 2(a) may map to the bottom rectangle for the cylinder in figure 2(b) or the top one. These are inconsistent with each other, but both would be maximal, and so the algorithm returns them both.

3.1 Transfer

Once we have a mapping between the source and target shapes, we need to transfer the teleological model. Since we have assumed that the source and target drawings are so similar that any distinction between them makes no difference at the structural level, the transfer task is straightforward. The basic problem is that Archytas has mappings between shapes in each drawing, and from those mappings Archytas can hypothesize that they should therefore depict the same component or connection. Thus, the procedure is to begin with the mapped shapes, and transfer the components and connections depicted by those mapped shapes, reconstructing the model iteratively. This process is shown in figure 9. Here is an outline of the basic algorithm:

1. Transfer DEPICTS relations, setting up relations to new components for each shape that depicts some component.
2. Transfer motion constraints (e.g. piston moves horizontally within the rectangles representing the cylinder).
3. From these, transfer the structural frames themselves:
   (a) Transfer depicted components, their properties and variable quantities
   (b) Transfer depicted connections between components (e.g. the piston/rod joint and the rod/cylinder joint, both represented explicitly by circles).
   (c) Transfer non-depicted connections amongst depicted components (e.g. piston/cylinder joint, which is merely a spatial relation).
   (d) Hypothesize new non-depicted components for each non-depicted component in the source (e.g. crankcase).

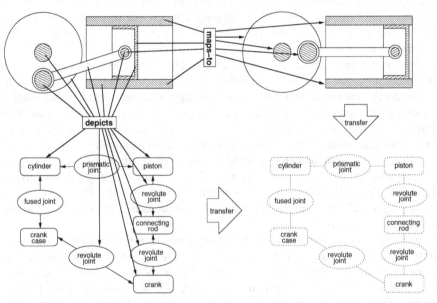

Fig. 9. The process of analogical transfer. For clarity, the mappings shown are partial. Given a mapping from the source (on the left) to the target (on the right), and a structural model for the source (bottom left), Archytas constructs new hypothetical structural model for the target (bottom right) by analogical transfer.

(e) For each component that maps onto a new hypothesized target component (depicted or undepicted), transfer all of the parameters, quantities, and primitive functions from the source to the target for that component.

(f) For each component that is involved in a behavior, transfer that behavior, iterating through each state and transition one by one.

(g) Some states are linked to the functional specification of the design, as are the behaviors of which they are part; for each behavior that realizes a function ("function F by behavior B"), transfer that functional specification from source to target.

Using this method we can hypothesize that each of the components depicted in figure 2(a) is also depicted in figure 2(b), and that the essential spatial constraints are not violated, and that therefore the undepicted component (the crankcase) and connections (those involving the crankcase) must be part of the target model as well.

4 Related Work

This research lies at the intersection of several lines of investigation in the cognitive sciences, including diagrammatic reasoning, analogical reasoning, qualitative models, and model construction. Here we relate this work only with the most similar AI work on diagram understanding by model construction, and analogical mapping with spatial representations.

Much earlier work on constructing models from drawings used forward-chaining rule-based reasoning. The early Beatrix system [20] for example, used domain-specific rules to construct a structural model of simple kinematics devices (e.g., a block on an inclined plane—see figure 1(a)) from a textually-annotated 2D diagram.

The more recent GeoRep system [9] takes as input simple a 2D vector-graphics line drawing depicting a physical process, e.g., a cup with steam coming out of it (figure 1(b)). It gives as output a symbolic description of the physical process depicted in the drawing, e.g., steam coming out of hot liquid contained in the cup. GeoRep is organized as a two-stage forward-chaining reasoner. First, a low-level, domain-independent relational describer recognizes shapes and spatial relations in the input line drawing, and then a high-level domain-specific relational describer applies domain-specific rules to produce an final description of the physical process in the diagram.

Joskowicz has done work on deriving behavioral descriptions from geometrical and topological descriptions of the parts of a device [16]. His work focussed on deriving constraints on possible motion in so-called "kinematic pairs" in a "Local Interactions Analysis", and then using constraint propagation to derive a qualitative "functional" description of the device in a "Global Interactions Analysis" (despite the use of the term "functional description", Joskowicz is speaking of what we are calling behavior). In more recent work, Dar, Joskowicz, and Rivlin [5] have derived similar descriptions from digital image sequences.

The SketchIT system [23] takes as input a 2D sketch of a physical device, and gives as output multiple designs of the physical device in the kinematics domain, where each design is augmented with a simple state-transition diagram to describe the device behavior. The system first produces "a behavior-ensuring parametric model" (or BEP model) of the components of the design, and from this determines geometric constraints on the motion of the parts, generating all qualitative configuration spaces consistent with the behavioral constraints. Next, it selects motion types for each component, and, finally, from the motion types and the geometric interpretations provided by a library of interaction types, it generates a BEP model for the design as a whole. In contrast to all these earlier AI projects on diagram understanding by model construction, our work uses analogical reasoning for constructing models from diagrams.

Evans' early ANALOGY program [6] solved four-term geometric analogies of the A : B :: C : ?? sort one might see on intelligence tests, where the task is to choose from several given alternatives the geometric figure that best completes the analogy. Given a multiple choice question of this kind, ANALOGY attempted to find a procedure for turning A into B, and then turned C into each of the (say) 5 answers, and whichever of those transformation procedures, represented as a labelled graph, was closest to the original was chosen. It used a simple language of primitives (dots, lines, closed polygons, etc.), relationships (above, left-of, and so on), and transformations (rotate, reflect, expand, contract, add, delete) for this task.

The more recent LetterSpirit program [17, 15] takes a stylized seed letter as input and outputs an entire font that has the same style (or "spirit"). The system understands "roles", such as the crossbar of an f or a t. It makes new fonts by determining some attributes such as "role traits" (e.g. crossbar suppressed), "motifs" (geometric shapes, such as a parallelogram, used over and over again), and "abstract rules" (e.g. no diagonal lines, only horizontal and vertical), and using these attributes builds an entire alphabet in some new style.

The Structure-Mapping-Engine (SME) [7] is a powerful, but content-free, analogical mapping system. SME first generates local maps between the target and the source graphs, then uses heuristics based purely on the structure of the graphs to select among the local maps, and finally to builds a consistent mapping. JUXTA [8] uses SME to compare two nearly identical drawings of a physical process, such as two drawings of a coffee cup with a metal spoon, but with a thicker metal bar in one drawing than in the other. JUXTA first uses GeoRep for deriving structure from shape, and then uses SME to compare the two structures, looking for alignable differences (that is, differences in associated attribute values in two schemas), and based on these differences drawing candidate inferences about the overall relationship between what is depicted in the two diagrams. In contrast, our work constructs a structural model from a diagram by analogical transfer.

The origin of our SBF models lies in Chandrasekaran's Functional Representation (FR) scheme for representing the functioning of devices [22, 3]. The Torque program [13, 14] completed a partial SBF model of the stretching of a spring system by analogy to the SBF model of a flexible beam. However, Torque did not contain any spatial representation of either the spring system or the flexible beam. The Archytas program's DSSBF models explicitly relate SBF models of physical devices with their spatial representations.

Narayanan and Chandrasekaran [18] have investigated the role of spatial representations in behavioral simulation of physical devices, and proposed the use of images and symbolic structures as representations that may enable both spatial and propositional inferences. More recently, Chandrasekaran [2] has proposed multimodal internal representations as a central element of the cognitive architecture. Our DSSBF models might be viewed as an example of such multimodal representations.

5 Conclusions

A persistent and central question in diagrammatic reasoning is how an agent understands a diagram, where we take understanding of a diagram to be the construction of a model. This is a very complex, multifaceted question. We posit that for domains which are familiar to an agent, the agent might construct a model of a diagrammatic representation by analogical transfer of the model of a similar diagrammatic representation. For familiar situations, the agent already may have constructed models of diagrammatic representations it has encountered in the past, and encapsulated the diagram and the model as a case in its memory. For example, an expert engineer who specializes in the design of kinematic devices already might have constructed functional models of many known designs. Further, for familiar situations, the agent may find a source case in its memory that strongly resembles a target problem. For instance, given a target diagram of a kinematic device, a design specialist in kinematics design might find in its memory a source case that is very similar to the target diagram. When this happens, the agent may construct a teleological model of the target by analogical transfer of the model from the source case.

Translating the above high-level account of construction of a model from a diagram into an operational computer program raises a number of challenging process and content questions. We do not presently have a complete answer to all the hard process or content issues. Instead, in this paper, we have sketched outlines of (1) a content account of DSSBF models of physical devices that relate drawings to shape, shape to structure, structure to behavior, and behavior to function, and (2) a process account of analogical mapping between the shape-level representations of the target and the source drawings, in which knowledge about the function of shapes helps guide the mapping process. We have also shown that this content and process actually works for transferring a teleological model from a source drawing to a target drawing at least for a problem in which the distinctions between the source and the target diagrams make for no difference in the structural models of the source and the target.

Acknowledgments

This research has been supported in part by a NSF (IIS) grant (Award number 0534266) on Multimodal Case-Based Reasoning in Modeling and Design. This paper has significantly benefited from critiques by anonymous reviewers of earlier drafts.

References

[1] C. Bron and J. Kerbosch. Algorithm 457: Finding all cliques of an undirected graph. *Communications of the ACM*, 16(9):575–577, 1973.
[2] B. Chandrasekaran. Multimodal perceptual representations and design problem. In J. S. Gero, editor, *Visual and Spatial Reasoning in Design*. Key Centre of Design Computing and Cognition, University of Sydney, NSW, Australia, 1999.

[3] B. Chandrasekaran, A. K. Goel, and Y. Iwasaki. Functional representation as design ratio- nale. In *Proc. European Workshop on Case-Based Reasoning*, pages 58–75, 1993.

[4] C.-W. K. Chen and D. Y. Y. Yun. Unifying graph matching problems with a practical solution. In *Proc. Int'l Conf. on Systems, Signals, Control, and Computers*, 1998.

[5] T. Dar, L. Joskowicz, and E. Rivlin. Understanding mechanical motion: From images to behaviors. *Artificial Intelligence*, 112:147–179, 1999.

[6] T. G. Evans. A heuristic program to solve geometric analogy problems. In M. Minsky, editor, *Semantic Information Processing*. MIT Press, Cambridge, MA, 1968.

[7] B. Falkenhainer, K. D. Forbus, and D. Gentner. The structure-mapping engine: Algorithm and examples. *Artificial Intelligence*, 41:1–63, 1990.

[8] R. W. Ferguson and K. D. Forbus. Telling juxtapositions: Using repetition and alignable difference in diagram understanding. In K. Holyoak, D. Gentner, and B. Kokinov, editors, *Advances in Analogy Research*, pages 109–117. New Bulgarian University, Sofia, Bulgaria, 1998.

[9] R. W. Ferguson and K. D. Forbus. GeoRep: A flexible tool for spatial representation of line drawings. In *Proc. 17th National Conf. on Artificial Intelligence (AAAI-2000)*, Austin, Texas, 2000. AAAI Press.

[10] A. K. Goel. A model-based approach to case adaptation. In K. J. Hammond and D. Gen- tner, editors, *Proc. 13th Annual Conf. of the Cognitive Science Society*, pages 143–148. Lawrence Erlbaum Associates, August 1991.

[11] A. K. Goel. Adaptive modeling. In Y. Iwasaki and A. Farquhar, editors, *Qualitative Rea- soning: Papers from the 10th Annual Workshop*, Technical Report WS-96-01, pages 67–73. AAAI Press, Menlo Park, California, 1996.

[12] A. K. Goel and B. Chandrasekaran. Functional representation of designs and redesign problem solving. In *Proc. 11th International Joint Conf. on Artificial Intelligence (IJCAI- 89)*, pages 1388–1394. Morgan Kaufmann, 1989.

[13] T. Griffith, N. Nersessian, and A. K. Goel. The role of generic models in conceptual change. In *Proc. 18th Annual Conf. of the Cognitive Science Society*. Lawrence Erlbaum Associates, 1996.

[14] T. Griffith, N. Nersessian, and A. K. Goel. Function-follows-form transformations in sci- entific problem solving. In *22nd Annual Conf. of the Cognitive Science Society*, pages 196–201. Lawrence Erlbaum Associates, 2000.

[15] D. Hofstadter and G. McGraw. Letter spirit: Esthetic perception and creative play in the rich microcosm of the roman alphabet. In D. Hofstadter and the Fluid Analogies Research Group, editors, *Fluid Concepts and Creative Analogies: Computer Models of the Funa- mental Mechanisms of Thought*, chapter 10, pages 407–466. Basic Books, 1995.

[16] L. Joskowicz. Shape and function in mechanical devices. In *6th National Conf. on Artificial Intelligence (AAAI-87)*, pages 611–615. AAAI Press, 1987.

[17] G. McGraw and D. Hofstadter. Perception and creation of diverse alphabetic styles. In S. Harrison, editor, *Artificial Intelligence and Creativity: Papers from the 1993 Spring Sym- posium*, Technical Report SS-93-01, pages 11–18. AAAI Press, Menlo Park, California, 1993.

[18] N. H. Narayanan and B. Chandrasekaran. Reasoning visually about spatial interactions. In *Proc. 12th International Joint Conf. on Artificial Intelligence (IJCAI-91)*, pages 360–365. Morgan Kaufmann, 1991.

[19] N. H. Narayanan, M. Suwa, and H. Motoda. How things appear to work: Predicting behav- iors from device diagrams. In *Proc. 12th National Conf. on Artificial Intelligence (AAAI- 94)*, pages 1161–1167. AAAI Press, 1994.

[20] G. S. Novak. Diagrams for solving physical problems. In J. Glasgow, N. H. Narayanan, and B. Chandrasekaran, editors, *Diagrammatic Reasoning: Cognitive and Computational Perspectives*, pages 753–774. AAAI Press/The MIT Press, Menlo Park, CA, 1995.

[21] J. W. Raymond, E. J. Gardiner, and P. Willett. Heuristics for similarity searching of chemical graphs using a maximum common edge subgraph heuristic. *J. Chem. Inf. Comput. Sci.*, 42:305–316, 2002.

[22] V. Sembugamoorthy and B. Chandrasekaran. Functional representation of devices and compilation of diagnostic problem-solving systems. In J. Kolodner and C. Riesbeck, editors, *Experience, Memory, and Reasoning*, pages 47–73. Lawrence Erlbaum, Hillsdale, NJ, 1986.

[23] T. F. Stahovich, R. Davis, and H. Shrobe. Generating multiple new designs from a sketch. *Artificial Intelligence*, 104(1–2):211–264, 2001.

The Mathematics of Boundaries: A Beginning

William Bricken

Boundary Institute
18488 Prospect Road, Suite 14, Saratoga CA 95070 USA
bricken@halcyon.com

Abstract. The intuitive properties of configurations of planar non-overlapping closed curves (boundaries) are presented as a pure boundary mathematics. The mathematics, which is not incorporated in any existing formalism, is constructed from first principles, that is, from empty space. When formulated as pattern-equations, boundary algebras map to elementary logic and to integer arithmetic.

1 *De Novo* Tutorial

Boundary mathematics is a formal diagrammatic system of configurations of non-overlapping closed curves, or *boundary forms*. Transformations are specified by algebraic equations that define equivalence classes over forms. As spatial objects, forms can be considered to be *patterns*, and transformations to be *pattern-equations* that identify valid pattern substitutions. The algebra of boundaries is novel, and is not incorporated within existing mathematical systems. Boundary mathematics provides a unique opportunity to observe formal structure rising out of literally nothing, without recourse to, or preconceptions from, the existing logical, set theoretic, numeric, relational, geometric, topological, or categoric formal systems that define modern mathematics. The strategy is to build a *minimal* diagrammatic language and an algebra for that language, using substitution and replacement of equals as the only mechanism of computation.

1.1 Language

I. Set aside a space to support drawing. Notice that it is framed.

II. Notice that the frame indicates an identifiable (i.e. framed) empty space, S. Following a minimalist strategy, draw the only observable thing (the frame) inside the only available space (S).

III. Call the representation of the frame a *mark*. Notice that S has changed from empty to not-empty, and that there are now three identifiable diagrammatic proto-structures: the mark, the inside of the mark, and the outside of the mark.

IV. Notice that replicate marks can now be drawn in two places, on the inside and on the outside of the original mark. Construct each variety. Drawing on the outside is SHARING; drawing on the inside is BOUNDING.

V. We have constructed a language consisting of three structurally different forms, and one absence of form. This language has two operators for constructing further forms. SHARING and BOUNDING can be applied indefinitely, each application adding one mark. Four forms can be constructed from 3 marks, nine forms from 4 marks.

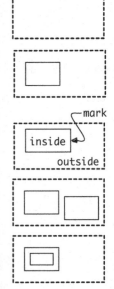

D. Barker-Plummer et al. (Eds.): Diagrams 2006, LNAI 4045, pp. 70–72, 2006.

Let's represent the mark more succinctly as $()$. The language of marks maps onto well-formed parenthesis structures without ordering. The dotted frame is part of the metalanguage that permitted description of mark drawings, and is no longer of use.

1.2 Algebra

VI. Let "=" mean *is-structurally-identical-to*: $()=()$, $()()=()()$, $(())=(())$.
Since marks are constructed in space, without ordering, note that $()(())=(())()$.

VII. Let "≠" mean *is-not-identical-to*: $()≠()()$, $()≠(())$, $()()≠(())$.

VIII. There are eight possible ways that the three forms can be collected into groups:
$\{\}$, $\{()\}$, $\{()()\}$, $\{(())\}$, $\{(),()()\}$, $\{(),(())\}$, $\{()(),(())\}$, $\{(),()(),(())\}$

IX. Let "=" also mean *in-the-same-collection*. There are four possible new equalities:
$()=()()$, $()=(())$, $()()=(())$, $()=()()=(())$

X. Discard the universal collection, since it does not distinguish between forms. There are then three possible arithmetics that can be constructed from the remaining three equalities.

Arithmetic I:	$()=()()$,	$()≠(())$,	$()()≠(())$	--	$\{(),()()\}/\{(())\}$
Arithmetic II:	$()≠()()$,	$()=(())$,	$()()≠(())$	--	$\{(),(())\}/\{()()\}$
Arithmetic III:	$()≠()()$,	$()≠(())$,	$()()=(())$	--	$\{()(),(())\}/\{()\}$

2 Interpretations

The minimalist mark-arithmetics map to elementary logic and to integer arithmetic, suggesting a diagrammatic foundation for conventional mathematics. Each mark-arithmetic can be generalized to an algebra by including variables that stand in place of arbitrary forms. Boundary logic is such a generalization [1][2]. Peirce first developed this logic in its implicative form as Entitative Graphs [3, 3.456-552 (1896)]. Spencer Brown [4] presents an algebraic version of *boundary logic arithmetic*, using Mark-Arithmetic I (below). Kauffman constructs *boundary integer arithmetic* from Mark-Arithmetic III [5].

2.1 The Map to Logic

The partitions of Mark-Arithmetic I, $\{(),()()\}/\{(())\}$, assert the *Call* rule $()=()()$; idempotency is the primary differentiator between logic and numerics. The mark is TRUE, while forms SHARING a space are interpreted as joined by disjunction. BOUNDING is negation.

f	*void*
t	$(\ )$
¬A	(A)
A→B	$(A)\ B$
A∨B	$A\ \ B$
A∧B	$((A)(B))$

CALL	$()() = ()$
CROSS	$(()) = void$
OCCLUSION	$(A\ ()) = void$
INVOLUTION	$((A)) = A$
PERVASION	$A\ \{A\ B\} = A\ \{B\}$

Spencer Brown's innovation was to equate $(())$ with nothing at all, that is, with the contents of the dotted frame prior to drawing the first mark. This created two Boolean equivalence classes while using only one symbol. Truth is confounded with existence, a capability unique to the spatial structure of the mark. The *Cross* rule generalizes to algebra as both *Occlusion*, which terminates proofs, and *Involution*, which enforces depth parity. The workhorse of boundary logic is *Pervasion*, which has no analog in conventional techniques. The curly brace is a *meta-boundary*, standing in place of any spatially intervening content, including none. Curly braces identify *semipermeable boundaries*, a defining characteristic of boundary logic. Boundary logic proof of

the *self-distributive law of the conditional*
is illustrated to the right. The three
reduction rules will reduce any TRUE
form to a mark. A logical interpretation
of *Pervasion* would delete deeply nested
connectives and their arguments that
match forms anywhere in their context.

```
(p→(q→r))→((p→q)→(p→r))   theorem
((p)(q) r) ((p) q) (p) r     transcribe
(  (q) ) (     q) (p) r       pervasion
(       ) (     q) (p) r      pervasion
(((      ) (   q) (p) r))     involution
(                       )     occlusion
```

2.2 The Map to Integers

The partitions of Mark-Arithmetic III, {()(),(())}/{()}, distinguish mark from
the other forms. Here, a mark is represented by a centered dot, • , for visual ease. It
is interpreted as the integer unit, 1. Forms SHARING the same space are interpreted as
added together, similar to stroke arithmetic used to tally units. Since by construction,
••=(•), BOUNDING is interpreted as doubling the unit, creating a boundary place notation.

0	*void*
1	•
2	•• = (•)
3	••• = (•)•
4	•••• = ((•))
...	

The *unit double* rule can be generalized to an algebraic
double rule. An elegant *merge* rule can be derived, as well,
by observing that forms can be partitioned for doubling in
two ways, the boundary analog of the rule of distribution.
This is illustrated by the structure of the number four:

$$•••• = (•) (•) = (••) = ((•))$$
$$1+1+1+1 = 2*1+2*1 = 2(1+1) = 2*2*1$$

UNIT DOUBLE   •• = (•)

DOUBLE   A A = (A)   Boundary multiplication is achieved by substituting forms

MERGE   (A)(B) = (A B)   for units. To multiply B by A, *substitute* A *for every* • *in* B.
Six is highlighted in this example, 5*6:

$$6=((•)•)   5=((•))•   5*6=((((\overline{(•)•)))}(\overline{(•)•)}))$$

The above result is immediately equal to 30, no further computation is required.
However, the reduction rules can be applied to reach a canonical shortest form of the
result. Both 5*6 and 6*5 are illustrated below:

```
30 = ((((•)•))) ((•)•)    5*6        ((((•))•)((•) )•)    6*5
     ((((•)•))   (•)•)    merge      ((((•))•  (•) )•)    merge
     ((((•)•)    •)•)     merge      ((((•)     •)•)•)    merge
```

Boundary integers differ from boundary logic in both type of boundary (impermeable
vs semipermeable) and in type of multiplication (substitution vs imposed structure).

References

1. Bricken, W., Gullichsen, E. Introduction to Boundary Logic. *Future Computing Systems* 2(4) (1989) 1-77.
2. Bricken, W. (2006) Syntactic Variety in Boundary Logic. Diagrams'06.
3. Peirce, C.S. (1931-58) *Collected Papers of Charles Sanders Peirce*. Hartshorne, C. Weiss, P., Burks, A. (eds.) Harvard Univ Press.
4. Spencer Brown, G. (1969) *Laws of Form*. George Allen and Unwin.
5. Kauffman, L.H. (1995) Arithmetic in the Form. *Cybernetics and Systems* 26: 1-57.

Syntactic Variety in Boundary Logic

William Bricken

Boundary Institute, 18488 Prospect Road, Suite 14, Saratoga CA 95070 USA
bricken@halcyon.com

Abstract. *Boundary logic* is a formal diagrammatic system that combines Peirce's Entitative Graphs with Spencer Brown's Laws of Form. Its conceptual basis includes boundary *forms* composed of non-intersecting closed curves, *void-substitution* (deletion of irrelevant structure) as the primary mechanism of reduction, and *spatial pattern-equations* that define valid transformations. Pure boundary algebra, free of interpretation, is first briefly described, followed by a description of boundary logic. Then several new diagrammatic notations for logic derived from geometrical and topological transformation of boundary forms are presented. The algebra and an example proof of *modus ponens* is provided for textual, enclosure, graph, map, path and block based forms. These new diagrammatic languages for logic convert connectives into configurations of containment, connectivity, contact, conveyance, and concreteness.

1 Introduction

Since antiquity, logical connectives have been presumed to be abstract; they are the *syncategoremata*, words that refer to nothing but themselves yet function to connect words that do have referents [1, p233]. Since logical connectives have no explicit form, they are represented by meaningless tokens. Peirce's Alpha Existential Graphs (AEG) [2 (1896)] introduces a radical re-conceptualization: the connectives of formal logic can take the form of diagrammatic structures consisting of closed non-intersecting curves, or *boundaries*. Composition of boundaries sharing the same space, and nested within each other, creates a spatial pattern language that is sufficient to express the sentences of propositional calculus. Traditionally, propositional rules of inference permit new sentences to be added to the collection of valid sentences, a strategy of accumulation. The diagrammatic reasoning in AEG follows a fundamentally different strategy: boundary patterns are transformed by rules that *add and delete* boundary structure. Peirce shows that inference can be achieved by creation and destruction, rather than by accumulation.

The *boundary logic* presented herein was introduced by G. Spencer Brown in *Laws of Form* (LoF) [3]. The mathematics is equational rather than inferential; like Boolean algebra, valid transformations are specified by equations. Boundary logic uses the boundary pattern language introduced by Peirce, but the add-and-delete rules of AEG are incorporated into *pattern-equations* that permit deduction to proceed solely though rewrite rules that delete structure. Pattern-equations define equivalence sets on boundary forms; the algebraic formulation replaces one-directional inference with the familiar bidirectional algebraic rules of substitution and replacement of equals.

The pure algebraic mathematics of boundaries [4] is based on the concept of *distinction*, or difference. It is constructed *de novo*, without reference to logical, set theoretic, relational, numeric, or categoric objects. Boundaries are strictly structural,

D. Barker-Plummer et al. (Eds.): Diagrams 2006, LNAI 4045, pp. 73–87, 2006.
© Springer-Verlag Berlin Heidelberg 2006

representing only the abstract concept of difference, without requiring identification of the type of object being differentiated. Thus, boundary mathematics differs substantively from the conventional mathematics of strings.

The abstract structure of boundary algebra is described first in Section 2, including the basic concepts of spatial operators with arbitrary numbers of arguments, spatial pattern-equations, and permeable boundaries. Boundary logic, the application of boundary algebra to logic, is then described in Section 3. Two new tools for deduction are introduced: void-substitution and boundary transparency. Section 4 shows a sequence of geometrical and topological transformations of boundary forms that generate over two dozen new diagrammatic notations for logic. Each of these notations provides potential new tools for Cognitive Science and for Computer Science. The structure of each notation suggests unexplored models of how we might read, analyze, manipulate, compute with, and think about deductive logic. The notations also suggest a wide diversity of data structures and algorithms for both hardware and software implementation of logic. Other than Peirce's original concept of a diagrammatic logic expressed as boundaries, and Spencer Brown's equational axiomatization of the same logic, Sections 2 and 4 are new. The axiomatization in Section 3 is the author's.[1]

2   Boundary Algebra

Composition of closed, non-intersecting planar curves, called *boundaries*, constructs a formal diagrammatic language, independent of an interpretation as logic. The alphabet is a singleton set of symbols consisting of the *empty boundary*, { O }, which is called a *mark*. A word consists of replicates of marks composed in a non-conventional manner. Since boundaries have both an inside and an outside, replicate symbols can be juxtaposed in two ways: on the inside of the original boundary and on the outside of the original boundary. Rather than one "concatenation" operator, there are two: SHARING is composition on the outside, while BOUNDING is composition on the inside. The formal language consists of the set of composable boundary forms.

2.1 Parens Notation

Delimiting tokens such as parentheses, brackets, and braces can be used as a shorthand notation for closed planar curves. Herein, parentheses that stand in place of boundaries are called *parens*. Some boundary structures and their parens abbreviations follow:

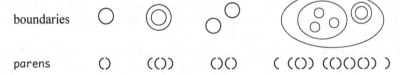

Above, the first form is the empty boundary; the second form is an elementary application of BOUNDING; the third, an elementary application of SHARING. The fourth

---

[1] This axiomitization of boundary logic has been extensively tested and applied to computer science problems over the last two decades, including theorem proving and satisfiability [5], minimization of software and knowledge-bases, parallel processing [6], visual languages [7], and logic synthesis of the over 200 small and large semiconductor designs included in the ISCAS'89 benchmarks. See the boundary mathematics section of www.wbricken.com.

parens form is compound. Well-formed parentheses are well understood; well-formed parens are identical but for the semantic constraint that parens do not have an addressable "left" and "right" portion, since this would violate the closure of the boundary. One accidental property [8, p3] of a parens is that it appears to be fragmented into two portions, another is that parens line up in an apparent sequential order, whereas boundaries can rest in any portion of the space they are drawn in.

2.2 Variary Operators

The functional basis set of the language of boundaries consists of two constructive operators, BOUNDING and SHARING. These diagrammatic operators are quite unconventional. An enclosing boundary can *bound* any number of forms, including none. Any number of forms can *share* a space, including none. In the example below, dots represent other single bounded forms. The explicit boundary encloses several forms, and while several others share the same external space.

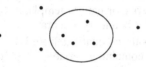

The operators BOUNDING and SHARING do not have a specific arity, both are *variary*. Absence of a specific arity strongly differentiates the boundary formalism from modern algebra. The closest conventional description for a variary operator is that of a "single argument set function". Conventionally, relations are defined by a set of ordered pairs. Relational properties such as commutativity and transitivity describe the symmetries that the relation imposes upon its ordered arguments. A boundary, however, can contain any number of forms. The collection of forms contained within a boundary is *a priori* neither countable nor orderable. Since forms are not taken two-at-a-time, there is no concept of associativity. By construction, boundary mathematics does not support the numerical concept of arity nor the proximal concepts of commutativity and associativity. Boundaries are neither functions nor relations.

The structure of a boundary form is defined explicitly by the boundaries themselves. The space upon which boundaries are imposed is void; *void space* is neither metric, nor geometric, nor topological. Thus, there is no concept of geometric or topological localization that applies to boundaries. Boundaries can reside anywhere so long as they do not intersect other boundaries. Another consequence is that the boundary language includes no "empty word" other than the entire space that supports all boundaries. Since there is no specifically defined relative position for forms SHARING a space, an empty placeholder is not necessary. The idea of a null boundary is wrapped up within the ground symbol; absence of form is simply the inside of the mark, ().

Seen as a relation, a boundary distinguishes what is inside from what is outside. The most natural conventional interpretation of boundary forms is as a partial ordering, with BOUNDING providing a strict ordering, and SHARING providing an equivalence class of forms in the same space. Even so, since both BOUNDING and SHARING are variary, a relational interpretation is difficult. The conceptualization represented by the boundary language does not support conventional relational properties such as reflexivity, symmetry and transitivity. Conventionally, variary relations are universal.

2.3 Pattern-Templates and Pattern-Equations

Let the set of capital letters, {A,B,...}, provide variables that can stand in place of any boundary form, including many forms and no forms, and the set of small letters, {a,b,...}, provide variables that can stand in place of either a mark, or the absence of a mark. When variables are included within boundary forms, for example (A ((B))), forms can be used as *pattern-templates* to identify specific structure within other forms. The pattern-template (A ((B))) matches ((()) (()())), with A=(()()) and B=*void*. Pattern matches are spatial rather than textual, while pattern variables can match many forms within a space, as well as matching no forms at all[2].

Similar to the idea of extensionality in set theory, two bounded forms are identical only if they enclose identical contents. A *pattern-equation* is an assertion that two structurally different patterns are equal; forms that match either pattern have the same value (modulo the asserted equation), partitioning the set of forms into equivalence classes. The *semantics* of the boundary language is defined by pattern-equations that assert specific patterns, or pattern-templates, to be equivalent. A set of pattern-equations create a particular boundary algebra.

Consider the following two pattern-equations from Laws of Form as providing an evaluation function for forms not containing variables. The two equations reduce any form either to a mark, (), or to the absence of a mark, thus establishing two equivalence classes, mark-equivalence and void-equivalence[3].

$$( ) ( ) = ( ) \qquad \text{SHARING EVALUATION}$$
$$(( )) = \qquad \text{BOUNDING EVALUATION}$$

2.4 Boundary Permeability

Since boundaries have two sides, they support two types of crossing over. *Impermeable* boundaries block crossing both from the inside outward and from the outside inward. *Semipermeable* boundaries block crossing from one side only, the other side is *transparent*. Boundaries are taken to enclose their contents, establishing the convention that semipermeable boundaries block crossing from inside outward. Fully permeable boundaries are imaginary, since a permeable boundary is transparent on both sides, and is thus indistinguishable from void space. Semipermeability is defined by the PERVASION pattern-equation, described below.

Thus far, features of a pure boundary algebra have been identified without placing an interpretation for conventional mathematics or logic on boundary forms. As is the case with any mathematics, an interpretation can be constrained to entry to and exit from the formalism. The mechanism of forms and pattern-equations can thus be used to compute over the structures of the interpretation.

---

[2] These same pattern matching options are provided in the mathematically comprehensive technical computing environment, *Mathematica*.

[3] The presentation in this paper is informal. Metatheory has been developed by Peirce [2], Spencer Brown [3], Kauffman [9], and Bricken [5]. Recent pioneers have legitimized the formal diagrammatic reasoning of AEG [10][11][12]; metatheory is invariant under provable equivalence. The equational structure of boundary logic supplies algebraic metatheory [13]; substitution and replacement are domain independent [14]. Void-equivalence is at the basis of boundary algebra, providing secure syntactic metatheory as a rewrite system.

3   Boundary Logic

Boundary logic is an interpretation as propositional logic of the abstract algebra of boundaries described above, using the following map:

LOGIC	BOUNDARY	ABSTRACTION
false		void
true	( )	ground
A	A	pattern-variable
not A	(A)	BOUNDING property
if A then B	(A) B	BOUNDING quasi-relation
A or B	A B	SHARING quasi-relation
A and B	((A)(B))	a compound form
A iff B	(((A) B)((B) A))	equivalence relation

The map from boundaries to logical connectives is *one-to-many*, implying that boundary logic is not isomorphic to propositional calculus or to Boolean algebra. This observation is also supported by comparing the basis constants of both systems; boundary logic has one while propositional logic has two. Transcribing the boundary void into a token (which is similar to representing the empty set as a set, {}) does establish isomorphism, however such a transcription undermines the fundamental mechanisms of boundary logic described below, those of void-equivalence and boundary transparency. Expressions that are structurally different in conventional logic may not be different in boundary logic. The following three transcriptions illustrate the condensation of logical structure into less boundary structure due to the one-to-many map:

Logic: $\neg f$ $((f \vee f) \to f) \vee f$ $A \wedge B = \neg(\neg A \vee \neg B)$

Parens: ( )   ( )   ((A)(B)) = ((A)(B))

Boundary logic is an amalgam of Peirce's Entitative Graphs [2, 3.456-552] and Spencer Brown's Laws of Form. Entitative Graphs are a dual variety of Existential Graphs, for which forms occupying the Sheet of Assertion (i.e. the blank page) are joined in disjunction rather than in conjunction. Forms in LoF map directly onto Entitative Graphs, however the transformation system in LoF is equational rather than implicative as in AEG. The axiomatization of boundary logic improves upon that of LoF by using only rules based on *void-substitution* (deletion of structure). The two pattern-equations that define valid transformations in boundary logic are:

(( ) A)   =   OCCLUSION

A {A B}   =   A {B}   PERVASION

OCCLUSION identifies void-equivalent structure. The *curly brace* in PERVASION is a *meta-boundary* indicating any number of intervening boundaries, including none. Curly braces represent semipermeable boundaries, permitting any form, A, on the outside to be arbitrarily anywhere on the inside, regardless of depth of nesting. Form B stands in place of the contents of the particular space containing the replicate of form A. PERVASION has no analogs in conventional mathematics. Two theorems that make the presentation of boundary proofs more succinct are:

$$((A)) = A \qquad\qquad \text{INVOLUTION}$$
$$(\ )\ A = (\ ) \qquad\qquad \text{DOMINION}$$

These pattern-equations map onto Peirce's rules of transformation for Alpha Graphs, with the exception that OCCLUSION combines the AEG rules "Erase at even depths" and "Insert at odd depths" into a single equation that eliminates depth specific transformation rules. The curly braces of PERVASION render boundaries transparent to outer forms, capturing the iteration/deiteration rules of AEG. INVOLUTION maps directly onto the double cut rule of AEG. DOMINION is the termination condition for algebraic proof, and is derived from OCCLUSION by calling upon Leibniz' inference rule for equational logic (functional substitution).

To use boundary logic, propositional sentences are first transcribed into their boundary form, then the algebraic pattern-matching mechanism of boundary logic is used to reduce the transcribed form. After reduction, what remains of the form is transcribed back into a propositional sentence. The boundary pattern-equations above are not inferential logic and they are not equational logic. Consider the analogy of matrix logic algebra [15]. The sixteen binary Boolean relations can be transcribed into 2x2 binary matrices. Deduction then occurs via the rules and structures of matrix algebra, with logical constants corresponding to scalars, variables corresponding to 2-vectors, and connectives corresponding to 2x2 matrix operators. Intermediate vectors that occur during matrix evaluation represent imaginary logical values, making the expressibility of matrix logic inherently greater than that of Boolean logic. Similarly, boundary logic uses extra-logical mechanisms to achieve deduction. Matrix logic provides *more* mechanism than inference to achieve deduction, boundary logic provides *less* mechanism than inference to achieve the same deductive results.

The pattern-equations of boundary logic have remarkable properties:

• An an axiomatization of elementary logic, the two equations are more succinct than any string-based axiomatization. Boundary logic itself maps *one-to-many* onto conventional string notation, making it a formally simpler system than conventional logic.

• Reduction takes place by *void-substitution*. The right-side of each pattern-equation is the same as the left-side but for some structure removed. The missing structure on the right-side is *void-equivalent*.

• Curly-braces render all intervening boundaries *transparent* to outside forms. PERVASION identifies a relativistic void-equivalence: relative to an outside form, both intervening boundaries and inner replicates are void-equivalent.

• Boundary forms can be *geometrically and topologically transformed* to generate different varieties of syntax. Reduction in each syntactic variety still consists of deletion of specific structures.

4 Syntactic Variety

The remainder of the paper presents over two dozen notations for boundary logic, most of them new[4]. The notations are purely syntactic varieties derived from geometric and topological transformation of boundary forms. Seven families of syntax (parens, circles, distinction networks, steps, centered maps, distinction paths, and blocks) differ topologically, the rest are simpler geometric reconfigurations. Each family could potentially shed light on the sub-structure of logic and of cognition.

Textual forms are one-dimensional token strings. Propositional calculus, Boolean algebra, and parens notation are examples. *Enclosures* are two-dimensional, AEG is the primary example. *Graph* and *map* forms are three-dimensional, requiring the use of depth cues such as crossing graph links. Distinction networks and rectangle maps are examples. *Steps* and *rooms* are anthropomorphized maps. *Paths* are one-dimensional forms spread over a two-dimensional space and may be seen as temporal or spatial traversal; distinction paths is the primary example. *Blocks* are three-dimensional spatial forms that can be physically manipulated. Importantly, each variety of syntax is accompanied by a mechanism for valid diagrammatic reasoning.

4.1 Display Conventions

Each syntactic variety is illustrated by a single example, the binary exclusive-or, XOR. Exclusive-or can be decomposed into simpler connectives, thus illustrating other logical connectives. The alignment between decomposed XOR and its parens form is:

$$p \text{ XOR } q = \neg(\neg(p \lor q) \lor (p \land q))$$ Conventional logic
$$((p \quad q) \quad ((p) (q)))$$ Parens notation

Syntactic varieties are loosely organized into topological families. The sequence of intermediate transformations that generate the new notations from prior notations is illustrated within each family. For one member of each family, the transcription of the boundary logic rules OCCLUSION, INVOLUTION and SHALLOW PERVASION into the new notation is shown, followed by a proof of *modus ponens* using the new notation. For all varieties, rearrangement after the application of a reduction step is not required, since each reduction step simply deletes structure. Some figures that follow have been rearranged slightly, solely to improve the aesthetics of the presentation. To reduce complexity and to improve readability, the deep boundary semipermeability indicated by curly braces has been simplified into a single semipermeable boundary[5]. The pattern-equation for SHALLOW PERVASION is:

$$A \text{ } (A \text{ } B) = A \text{ } (B)$$ SHALLOW PERVASION

Several notational guides have been incorporated:

- The meta-token ⟩⟨ indicates absence of form, and is used solely as a visual convenience. It has absolutely no interaction with forms.
- Square brackets, [ ], are highlighted parens. They are identical to rounded parens in all other respects.

---

[4] Many of the notations were inspired by the work of, and conversations with, Louis Kauffman.

[5] SHALLOW PERVASION can be composed in steps to generate PERVASION. LoF uses SHALLOW PERVASION; AEG uses PERVASION; neither fully develop the idea of permeable boundaries.

• A large dot, •, identifies the shallowest space (the outside) of a form. For some notations, this point-of-reference is embedded within the form.

• Multiple forms SHARING the shallowest space are always explicitly bounded, so that forms in the outermost space are singular. Multiple forms in the shallowest space are enclosed using INVOLUTION:

$$A\ B\ C\ =\ ((A\ B\ C))\qquad\text{Packaging via INVOLUTION}$$

4.2 String Varieties

For comparison, the pattern-equations of boundary logic are expressed below in parens notation and in conventional textual syntax:

Notation	OCCLUSION	INVOLUTION	SHALLOW PERVASION
Parens	(( ) A) = ⸓⸑	((A)) = A	A (A B) = A (B)
Implicative logic	(f→A)→f ⊨ f	(A→f)→f ⊨ A	((A→f)→B)→A ⊨ B→A
Boolean algebra	(t+A)' = f	A'' = A	A+(A+B)' = A+B'

In conventional logic, *modus ponens* is: $A \wedge (A \to B) \vDash B$

Modus ponens transcribed into parens notation is: [((A) ((A) B))] B = ( )

The double turnstile of logical implication becomes an assertion of equality in equational logic systems; both the Deductive Theorem and the steps of inference are absorbed into the match-and-substitute strategy of algebra. The directionality of implication is mitigated by collecting all forms on one side of the equation. The square bracket above highlights the transcribed logical implication. Conventional syntactic proof steps become a sequence of boundary logic void-substitution steps. Should the left-side reduce to mark, the asserted equality is logically valid.

In the following parens notation proof of *modus ponens*, a rule application on each line deletes void-equivalent structure. Since void-equivalent structure cannot impact values, the order in which structure is removed is irrelevant.

```
[ ((A) ( (A) B )) ] B = ( )     transcription
[ ((A) (          )) ] B = ( )     pervasion
[                    ] B = ( )     occlusion
[                    ]   = ( )     dominion, identity
```

4.3 Enclosure Varieties

AEG is almost always presented in the syntax of planar enclosures. The first two representations of XOR below show how parens delimiting tokens can be connected by *caps* to construct planar enclosures. The next two varieties geometrically modify the shape of the enclosures.

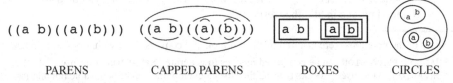

```
((a b)((a)(b)))    ((a b)((a)(b)))       a b   a b
```

PARENS   CAPPED PARENS   BOXES   CIRCLES

The three parens notation pattern-equations in the table above are textual abbreviations of boundary logic rules that are more naturally expressed in the diagrammatic syntax of enclosing circles below. A proof of *modus ponens* in circle notation then follows:

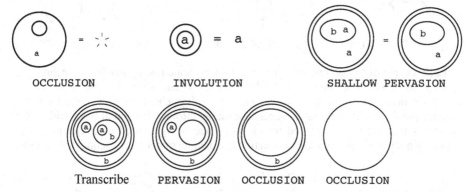

| OCCLUSION | INVOLUTION | SHALLOW PERVASION |

Transcribe   PERVASION   OCCLUSION   OCCLUSION

4.4 Graph Varieties

Extruding a parens form downward converts containment into graph connectivity. The first form below is an extruded parens form. In the second form, parens are capped to construct nodes. The third form shrinks capped parens to nodes of the same size, forming a boundary tree; nodes replace boundaries and directional links replace enclosure. The outside, or shallowest space, of the circle forms above becomes the top, or root, node of the tree. Replicates of variables are next joined together, constructing a directed acyclic graph that supports replication through multiple links rather than through multiple labels. The fifth form is a distinction network; the links have been aligned geometrically in circular patterns that emphasize graph sub-structures.

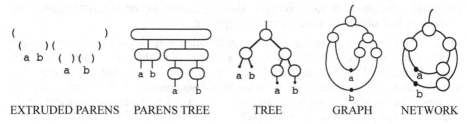

EXTRUDED PARENS   PARENS TREE   TREE   GRAPH   NETWORK

   Two conventions identify the shallowest space of a graph. North-south orientation places the root node towards the top of the page, making it the "top" node. This convention is quite contextual, since it requires meta-information about orientation in space. An alternative technique is simply to extend a link out of the graph, locating the context explicitly. Each variable node then identifies the bottom of the graph, its deepest content. When reading graphs as processing networks, variables are inputs and the root node is output. Although trees are two-dimensional, graphs are three-dimensional, since graphs violate planarity whenever combining multiple occurrences of variables into single nodes requires connective links to cross.

   The boundary logic transformation rules are expressed as distinction networks below. PERVASION is a path rule; when any two paths from a node join again at another node, the longer path can be detached from the lower node.

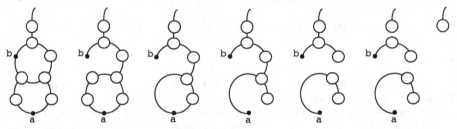

OCCLUSION   INVOLUTION   SHALLOW PERVASION

Each of the distinction network rules can be implemented by parallel asynchronous message-passing between nodes [6]. Global reduction occurs without global coordination. The distinction network proof of *modus ponens* displays characteristics of the localized message-passing regime. Reduction occurs primarily through deletion of links, with graph fragments that are not connected to the root node being treated as irrelevant. The implication for proof theory is that deduction is a parallel rather than a sequential process.

Transcribe PERVASION INVOLUTION PERVASION OCCLUSION OCCLUSION

4.5 Map Varieties

A distinction network can be converted into a map by enlarging each node until neighbors touch, constructing a common border in place of each link. Link connectivity becomes territorial contact. Unlike circle maps and rectangle maps, both steps maps and rooms maps are anthropomorphized.

Steps. The curvature of common borders in the steps map below incorporates visual cues in a third dimension that locate the top step, and show the relative depths of other steps. In the third form, background areas not part of the map but captured by the map are darkened to better contrast steps from background. A redundant point-of-reference dot is added to clearly identify the top step. Were the convex/concave depth cues to be lost by straightening the borders, the reference dot becomes the sole indicator of the top step.

DISTINCTION NETWORK   STEPS MAP   ORIENTED STEPS

For map varieties, reduction deletes borders. The boundary logic reduction rules are expressed below as steps forms. SHALLOW PERVASION is represented by slipping the pervaded step out from under the lower step. In steps maps, PERVASION deletes a common border, rather than disconnecting a link, or erasing a form.

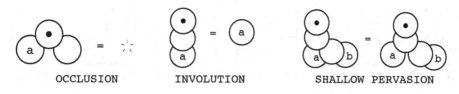

OCCLUSION   INVOLUTION   SHALLOW PERVASION

Although it is generally possible to construct a map representation of any form, it is not generally possible to maintain contiguity of territories in two dimensions. In particular, when a graph form requires crossing links, the associated map form requires a third, depth dimension. The proof of *modus ponens* for steps follows:

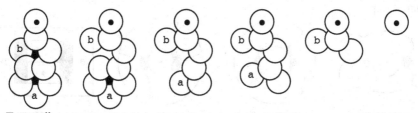

Transcribe PERVASION INVOLUTION PERVASION OCCLUSION OCC.

Below, the oriented steps map from above is geometrically rearranged in several ways. These map varieties offer visual rather than conceptual variation.

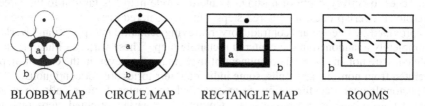

BLOBBY MAP   CIRCLE MAP   RECTANGLE MAP   ROOMS

Rooms. Rather than enforcing separation of non-adjacent territories with captured space, the walls of rooms enforce non-adjacency, while open doors represent shared borders. PERVASION simply closes a door, while OCCLUSION closes off several rooms, creating an "empty room" by denying access. INVOLUTION deletes walls. A room map is TRUE whenever there is direct access to an empty room from the outside.

OCCLUSION   INVOLUTION   SHALLOW PERVASION

In the proof of *modus ponens* below, rooms could of course be extruded into a third dimension, constructing a visceral proof technique based on walking around. Door closures can be determined subjectively, from within the representation, without a global perspective, analogous to the asynchronous reduction of distinction networks. Subjective PERVASION might be expressed as: "When you are in a room, R, with two inwardly open doors, and one door leads to a room that allows you to return to R through the second door, then close the second door."

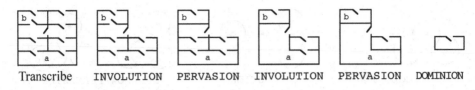

Transcribe   INVOLUTION   PERVASION   INVOLUTION   PERVASION   DOMINION

4.6 Centered Map Varieties

Centered maps put the reference point in an interior territory rather than a territory on the edge of the map. Centering often better displays symmetries. The first map below is a centered version of steps, the rest are geometric modifications of centered steps.

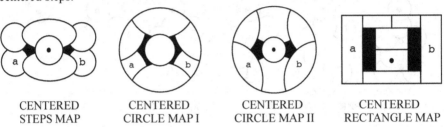

| CENTERED | CENTERED | CENTERED | CENTERED |
| STEPS MAP | CIRCLE MAP I | CIRCLE MAP II | RECTANGLE MAP |

The two versions of the centered circle map above illustrate the use of semantic visual cues to convey depth of nesting. Centered circle map I is faithful to the three-dimensional overlap cues of centered steps. In the centered circle map II, the depth cues provided by overlap are contradictory, removing semantic interpretation. Overlap is completely abandoned in the centered rectangle map. These varieties illustrate that depth cues are not an essential component of logic maps; however, in the case of maps generated from non-planar graphs, some individual territories are not contiguous.

The topological transformation from oriented to centered maps calls upon an unusual type of spatial rearrangement. Compare the earlier oriented steps map to centered steps maps above. In the steps map, the external space (adjacent both to the top step and to step b) becomes a captured interior space of the centered steps map. The surrounding space commutes from outside to inside. This is accomplished by drawing the arc-like step b of the steps map around the root step rather than around step a. An accidental property of these logic maps permits two planar forms, one read from the outside and one read from the inside. Although the logical semantics does not change, the diagrammatic form permits two different cognitive perspectives that echo familiar distinctions between objective/subjective, extrinsic/intrinsic, absolute/relative, and global/local. Point-of-view is a theme throughout the syntactic varieties.

4.7 Path Varieties

In the path varieties, structural containment, connectivity and contact have been replaced by conveyance along a path. The model is one of spatial and temporal transversal rather than reading an objective structure.

The bar tree below appears to be identical to a parens tree (earlier above) but for the shape of the nodes. Bar graphs and distinction networks share the same resemblance. However, this solely visual modification identifies significantly different implementations: the node/link model of distinction networks changes into the barrier/path model of bar graphs. Distinction network nodes actively enact transformation processes, while links are simple communication channels between nodes. In a bar graph however, the barrier is a simple communication interface, while the path represents the dynamics of an evaluation of the form. Nodes in a distinction network connect and functionally transform convergent links, while the convergence of paths in a bar graph is independent of the bars. A semiconductor design analogy is that distinction networks represent the circuit schematic model with nodes as logic functions and directional links as input and output, while bar graphs represent the transistor-level model with bars as inverting transistors and paths implementing disjunctive logic through physically wired signal convergence (wire-OR).

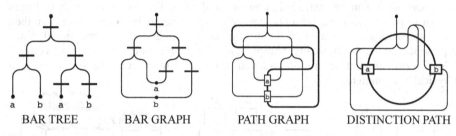

| BAR TREE | BAR GRAPH | PATH GRAPH | DISTINCTION PATH |

A bar graph converts into a path graph by connecting all bars by a single continuous loop[6]. The path graph is then modified geometrically so that the bar loop becomes the single circular boundary of a distinction path. Distinction paths convert connectivity into crossings and depth into number of crossings. All of the boundaries in the parens form collapse into the single distinction path boundary. BOUNDING by apparently distinct boundaries becomes crossing of a single boundary. Odd and even depths of nesting are replaced by the pure parity of locating a segment of the path on the inside or the outside of the distinction path boundary. SHARING of space by apparently distinct forms becomes divergence of traversal choices along the single path. Non-planar bar graphs require bridges (in a third dimension out of the plane of the paper) for paths to be able to cross over other paths.

Rules for distinction paths involve shifting and deleting path segments. Paths that terminate at the distinction path boundary are ground states: termination on the inside is FALSE while termination on the outside is TRUE.

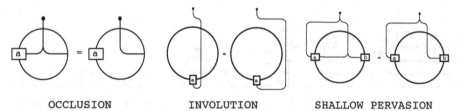

| OCCLUSION | INVOLUTION | SHALLOW PERVASION |

---

[6] Connecting bars into a single loop is an application of the Jordan curve theorem. The parens representation is first over-capped to convert each parens into a single bar. Then, rather than under-capping matching bar ends to construct the circle variety, under-caps are constructed iteratively from the deepest space to connect adjacent parens fragments: connecting)(in the case of ()(), and)) in the case of (()). This regime maps even depths to the outside of the curve and odd depths to the inside, with nesting depth counted by the number of transverse crossings of the Jordan curve [16, p 605ff].

A distinction path proof of *modus ponens* follows:

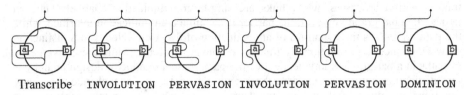

Transcribe   INVOLUTION   PERVASION   INVOLUTION   PERVASION   DOMINION

4.8 Block Varieties

A parens form can be extruded upward and capped to construct stacks of boxes similar to parens trees, but with contact replacing connectivity. These boxes can then be extruded again into a third dimension, constructing a three-dimensional block representation that is physically manipulable. Containment becomes concrete contact. Block forms are not anthropomorphized, they are simply concrete structures in a physical environment.

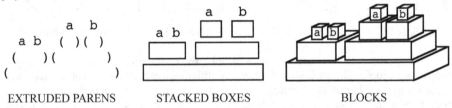

EXTRUDED PARENS   STACKED BOXES   BLOCKS

Reduction rules identify configurations of stacks from which blocks can be removed.

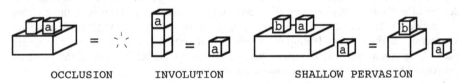

OCCLUSION   INVOLUTION   SHALLOW PERVASION

The proof of *modus ponens* that follows can take place through physical manipulation of concrete objects standing in place of abstract logical forms.

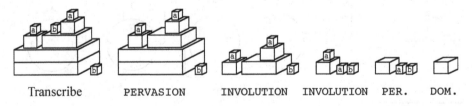

Transcribe   PERVASION   INVOLUTION   INVOLUTION   PER.   DOM.

Like Cuisenaire rods used to teach elementary numerical concepts to pre-schoolers through physical manipulation, logic blocks could be used to teach elementary logical concepts. The difficulty is that the concepts taught by logic blocks are those of diagrammatic boundary logic rather than the more familiar string-based logic embedded in language.

5   Conclusion

Algebraic formulation of Peirce's original Entitative Graphs provides a plethora of diagrammatic languages for logic. Diverse geometric and topological transformations of the spatial syntax result in several distinctly different two- and three-dimensional representations, all using the same three abstract pattern-equations to achieve form reduction. Underlying these syntactic varieties is a new set of mathematical concepts: void-equivalence, variary operators, boundary semipermeability, spatial pattern-equations.

That our familiar conversational logic and our formal typographic logic can both be rendered in a variety of structurally simpler diagrammatic and experiential representations raises interesting questions for cognitive science. How would a newly acquired ability to visualize or to physically manipulate logical form influence the quality of logical reasoning? More challenging, though, are the unfamiliar formal concepts that do not map onto conventional logic. The pattern of transformations that constitute reasoning in boundary logic is concise: a constructive proof can only proceed from mark to DOMINION to PERVASION, although the void-equivalent structure of OCCLUSION and the void-equivalent paired bounds of INVOLUTION can add syntactic complexity anywhere within a form, as illustrated below:

$$(\ ) \ \rightleftarrows \ A \ (\ ) \ \rightleftarrows \ A \ (A) \ \rightleftarrows \ ((A \ (A))) \ \rightleftarrows \ ((A \ (A))) \ (X \ (X))$$

For valid forms, PERVASION asserts that context creates content. The familiar idea that the antecedent (content) validates the consequent (context) is reversed. More fundamentally, deduction proceeds by the deletion of irrelevant structure rather than by the accumulation of facts. Like any mathematics, boundary logic is a way of looking at problems, a way of thinking and seeing. As a technique, it may seem incomprehensible at first. After some practice, syntactic manipulation becomes second nature. But, like Venn diagrams, Existential Graphs and other diagrammatic formalisms, can the formal techniques of boundary logic claim to represent cognitive processes?

References

1. Kneale, W., Kneale, M. (1962) *The Development of Logic*. Oxford Univ Press.
2. Peirce, C.S. (1931-58) *Collected Papers of Charles Sanders Peirce*. Hartshorne, C. Weiss, P., Burks, A. (eds.) Harvard Univ Press.
3. Spencer Brown, G. (1969) *Laws of Form*. George Allen and Unwin.
4. Bricken, W. (2006) The Mathematics of Boundaries: A Beginning. Diagrams'06.
5. Bricken, W., Gullichsen, E. (1989) Introduction to Boundary Logic. *Future Computing Systems* 2:4 1-77.
6. Bricken, W. (1995) Distinction Networks. in Wachsmuth, I., Rollinger, C.R., Brauer, W. (eds.) *KI-95: Advances in Artificial Intelligence*. Springer 35-48.
7. James, J. Bricken, W. (1992) A Boundary Notation for Visual Mathematics, *1992 IEEE Workshop on Visual Languages*, Seattle, IEEE Press, 267-269.
8. Shin, S. (1994) *The Logical Status of Diagrams*. Cambridge Univ Press.
9. Kauffman, L.H. and Varela, F.J. (1980) Form Dynamics. *J. Soc. Biol. Structures* 3:171-206.
10. Barwise, J., Etchemendy, J. (1996) Heterogeneous Logic. In: Allwein, G, Barwise, J.(eds.) *Logical Reasoning with Diagrams*. Oxford Univ Press.
11. Shin, S. (2002) *The Iconic Logic of Peirce's Graphs*. MIT Press.
12. Hammer, E. (1995) *Logic and Visual Information*. CSLI Publications, Stanford.
13. Halmos, P. and Givant, S. (1998) *Logic as Algebra*. Mathematical Assoc. of America.
14. Birkoff, G. (1935) On the Structure of Abstract Algebras. *Proc. Cambridge Phil. Soc*, 31 417-429.
15. Stern, A. (1988) *Matrix Logic*. North-Holland/Elsevier.
16. Kauffman, L.H. (1993) *Knots and Physics* (2nd edition). World Scientific.

Fixing Shin's Reading Algorithm for Peirce's Existential Graphs

Frithjof Dau

Technische Universität Dresden

Abstract. In her book "The Iconic Logic of Peirce's Graphs", S. J. Shin elaborates the diagrammatic logic of Peirce's Existential Graphs. Particularly, she provides translations from Existential Graphs to first order logic. Unfortunately, her translation is not in all cases correct. In this paper, the translation is fixed by means of so-called *single object ligatures*.

1 Introduction

The well-known system of Existential Graphs (EGs) by Peirce is divided into three parts which are called *Alpha*, *Beta* and *Gamma* which are built upon each other. The step from Alpha to Beta corresponds to the step from propositional logic to first order logic (FOL). In this step, a new syntactical element, the *line of identity (LoI)*, is added to EGs. LoIs are used to denote both the existence of objects and the identity between objects, and they are represented as heavily drawn lines. LoIs which are sometimes assembled to networks termed *ligatures*. Consider the four EGs of Fig. 1.

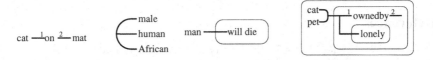

Fig. 1. Four Peirce graphs with so-called single-object-ligatures

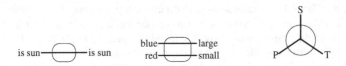

Fig. 2. Three Peirce graphs with non-single-object ligatures

The meaning of the graphs of Fig. 1 is 'a cat is on a mat', 'there exists a male, human african', 'there exists a man who will not die', and 'every pet cat is owned by someone and is not lonely'. In all these graphs, LoIs and ligatures, even if they cross cuts, are used to denote a single object. To put it more formally: We can provide translations of the EGs to formulas of FOL where we assign to each ligature one variable. In fact, we can translate the EGs of Fig. 1 to FOL as follows:

D. Barker-Plummer et al. (Eds.): Diagrams 2006, LNAI 4045, pp. 88–92, 2006.

1. $\exists x.\exists y.(cat(x) \wedge on(x,y) \wedge mat(y))$
2. $\exists x.(male(x) \wedge human(x) \wedge African(x))$
3. $\exists x.(man(x) \wedge \neg willdie(x))$
4. $\neg\exists x.(cat(x) \wedge pet(x) \wedge \neg(\exists y : ownedby(x,y) \wedge \neg lonely(x)))$

But the EGs of Fig. 2 show that this reading of ligatures is not always that simple: A ligature may stand for more than one object. For example, the first graph reads 'there are at least two suns'. Thus, in the translations of these graphs, we have to assign more than one variable to the ligatures. They are:

1. $\exists x.\exists y.(issun(x) \wedge issun(y) \wedge x \neq y)$
2. $\exists x.\exists y.\exists u.\exists v.(blue(x) \wedge large(y) \wedge red(u) \wedge small(v) \wedge \neg(x = y \wedge u = v))$
3. $\exists x.\exists y.\exists z.(S(x) \wedge P(y) \wedge T(z) \wedge \neg(x = y \wedge y = z))$

In [8], Shin thoroughly elaborates the diagrammatic logic of Peirce's EGs. But is has to be said that from a mathematical point of view, her treatise lacks preciseness (see [1] for a thorough discussion). Moreover, her elaboration unfortunately contains some flaws. Particularly, in her translation from EGs to FOL-formulas, she assigns in some cases too few variables to ligatures, which sometimes yields wrong formulas. In the next section, we clarify the terms "line of identity" and "ligature", and moreover, the notion of so-called "single object ligature" is introduced. The next section briefly discusses the flaw in Shin's translation, and then uses single object ligature to fix it. Further flaws in [8] can be found in her transformation rules. This will be briefly discussed in the outlook.

2 Lines of Identity and Ligatures

Peirce describes a LoI as follows: '*The line of identity is [...] a heavy line with two ends and without other topical singularity (such as a point of branching or a node), not in contact with any other sign except at its extremities.*' (4.116: We use the common notation to refer to [5]). It is important to note that LoIs do are neither allowed to branch nor they allowed to *cross* cuts. But it is allowed that they *touch* other elements at their extremities, i.e.:

1. Two or three LoIs may be connected at their endpoints. If three LoIs are connected, the point where they meet is a *branching point*.
2. LoIs may end on a cut. Particularly, it is allowed that LoIs are connected directly on a cut. Due to this possibility, we can have heavily drawn lines (composed of several LoIs) which cross a cut.

LoIs may be assembled to connected networks termed LIGATURES. Peirce writes in 4.407: '*A collection composed of any line of identity together with all others that are connected with it directly or through still others is termed a ligature. Thus, ligatures often cross cuts,[...]*', and in 4.416, he writes '*The totality of all the lines of identity that join one another is termed a ligature.*' So he explicit discriminates between one line of identity and a linked structure of lines of identity called ligature. Particularly, each LoI is a ligature, but not vice versa.

Consider the last graph of Fig. 1. This graph has two maximal ligatures. The left one is composed of (at least) seven LoIs. In the diagram below, these LoIs are numbered, and all endpoints of LoIs are indicated as bold spots. The right heavy line is a single LoI, but even single LoIs can be understood to be composed of smaller LoIs as well. This is indicated by breaking up this line into two LoIs.

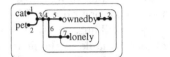

As Peirce terms the *totality* of all the LoIs that join one another a ligature, Peirce's understanding of a ligature is a *maximal* connected network of LoIs. To clarify matters, in this paper *each* connected network of LoIs is called ligature.

In all graphs of Fig. 2, a part of the ligature traverses a cut (i.e., there is a cut c and a heavily drawn line l which is part of the ligature such that both endpoints of l are placed on c and the remainder of l is enclosed by c). Such a device denotes non-identity of the endpoints of l, so a ligature containing such a device l usually denotes different objects. But if such a device does not occur, it has been shown in [2] that the ligature denotes a single object. For this reason, a ligature L such that no part of L traverses any cut will be called SINGLE-OBJECT-LIGATURE (SO-LIGATURE). Note that so-ligatures may contain cycles or may cross a cut more than once: An example for this is the right graph above.

3   Fixing Shin's Reading of Graphs

The clue to read arbitrary EGs is to break up non-so-ligatures into several so-ligatures by adding additional equality relations. Peirce writes in [6] that the second graph of Fig. 1 '*is a graph instance composed of instances of three indivisible graphs which assert 'there is a male', 'there is something human' and 'there is an African'. The syntactic junction or point of teridentity asserts the identity of something denoted by all three.*' That is, we can replace the branching point by a relation $\doteq_3$, termed TERIDENTITY, expressing that the objects denoted by the attached LoIs are all identical. The corresponding graph is the leftmost graph below (the index 3 on the relation-sign is omitted). Even simpler, when two LoIs meet in a point, we can replace this point by the usual binary identity relation $\doteq_2$. So, if an EG with a non-so-ligature is given, we can replace some branching points by $\doteq_3$ and some non-branching points by $\doteq_2$ until the non-so-ligature is split up into several so-ligatures. The graphs which correspond to the graphs of Fig. 2 are the second, third, and fourth graphs shown below.

Now we have two, four and three so-ligatures, respectively, that is why the above translations of these graphs to FOL need two, four and three variables.

Replacing branching points by a ternary identity relation has already been carried out by Zeman in [9] in his translation. But Zeman replaces *each* branching point by an identity relation, and moreover, he *always* splits heavily drawn lines if they cross a cut more than once (by adding the dyadic identity). Thus in nearly all cases, Zeman translation uses far more variables than necessary, and the resulting formulas are hard to read. This has been thoroughly discussed by Shin in [8]. She correctly points out that Zeman's reading algorithm for existential graphs is comprehensive and yields correct results (in contrast to Robert's reading in [7], as she argues), but it usually yields a '*translation that looks more complicated than the original graph*'. A main reason for her criticism is the '*mismatch between the number of lines in a graph and the number of the variables in the translation*'. Shin tries to overcome this problem in her reading algorithm, but unfortunately, sometimes she assigns *too few* variables to ligatures. To see this, consider the following graph $\mathfrak{G}_z$ and the model $\mathcal{M}$ with two objects a, b:

Now we have the following readings of $\mathfrak{G}_z$:

1. Zeman: $f_z := \exists x. \exists z. (Px \wedge Rz \wedge \neg \exists y. (Qy \wedge \neg(x = y \wedge y = z)))$
2. Shin: $f_s := \exists x. (Px \wedge \forall y. (x = y \vee \neg Qy) \wedge Rx)$ (see p. 128 of [8])

Note that f_z needs three variables, but f_s needs only two variables. We obviously have $f_s \models f_z$. But we have $\mathcal{M} \models f_z$ and $\mathcal{M} \not\models f_s$, thus $f_z \not\models f_s$. So, f_z is a strictly weaker formula than f_s. Particularly, Shin's reading is not correct.

With the observation that so-ligatures denote single objects, we can now fix Shin's reading of EGs in order to obtain a correct translation of EGs to formulas which, as Shin writes, '*respects Peirce's intuition about his graphical system*'. Shin's algorithm is provided in [8] on page 122ff. The first steps of Shin's algorithm have to be replaced by the following instructions:

1. Transform all ligs. into so-ligs. by appropriately adding relation signs '$\doteq$'
2. Erase all double cuts
3. Assign a variable to the outermost part of each so-ligature.
4. Continue with step 3. of Shin's reading algorithm.

Below, a sample graph for Zeman's and Shin's improved reading is provided. In the middle, the ligatures of the graph are split due to Zeman's algorithm, on the right, this is done according to Shin's improved algorithm. The corresponding translations f_Z of Zeman and f_S of Shin/Dau show the difference.

$$f_Z = \exists x_1, x_2, x_3, x_4 : [P(x_1) \wedge S(x_2) \wedge x_1 = x_2 \wedge x_2 = x_3 \wedge T(x_4) \wedge$$
$$\neg \exists x_5, x_6, x_7 : (x_3 = x_6 \wedge x_6 = x_5 \wedge x_5 = x_7 \wedge x_7 = x_4 \wedge \neg R(x_6, x_7))]$$
$$f_S = \exists x_1, x_2 : [P(x_1) \wedge S(x_1) \wedge T(x_2) \wedge (x_1 \neq x_2 \vee R(x_1, x_2))]$$

4   Future Research

Shin argues in [8] very clearly that in the system of EGs, the syntax, the reading, and the calculus of EGs have to respect the visual features of EGs.

Ligatures are a distinguishing element of EGs, thus we particularly need a precise understanding of them. As we have seen, even an expert as Shin sometimes fails in their handling. This unfortunately holds true for her transformation rules as well. For example, her rule NR3(iii) allows to extend a loose end of an LoI outwards from an even to an even area, if some restrictions are fulfilled. Below, a correct application of the rule is exemplified. Her rule NR5(b) allows to connect 'tokens of the same type'. This rule is exemplified as well. But in both examples, there are models in which the left graph is true, but the right is not (for example, the set $\mathbb{N}$ of natural numbers, where R is the relation $<$, and P is the predicate '$= 0$'). So these rules (and thus their counterparts NR4(iii) and NR6(a)) are *not sound*.

In [2], a mathematical elaboration of EGs, particularly of ligatures, is provided, and the results of [2] have been used in this paper to fix Shin's reading of EGs. I wholeheartly agree to Shin's approach that EGs should be redesigned in order to fully implement all their iconic aspects. So the next step in developing EGs as diagrammatic, efficient and formal logic system is to fix and formalize Shin's transformation rules in a way that Shin's intention is best reflected as possible.

References

[1] Frithjof Dau. Types and tokens for logic with diagrams: A mathematical approach. In Karl Erich Wolff, Heather D. Pfeiffer, and Harry S. Delugach, editors, *ICCS*, volume 3127 of *Lecture Notes in Computer Science*. Springer, 2004.

[2] Frithjof Dau. Mathematical logic with diagrams. based on the existential graphs of peirce. Habilitation thesis. To be published. Available at: http://www.dr-dau.net, 2006.

[3] H. Pape. *Charles S. Peirce: Phänomen und Logik der Zeichen*. Suhrkamp Verlag Wissenschaft, 1983. German translation of Peirce's Syllabus of Certain Topics of Logic.

[4] C. S. Peirce. *MS 478: Existential Graphs*. Harvard University Press, 1931–1935. Partly published in of [5] (4.394-417). Complete german translation in [3].

[5] Charles Sanders Peirce. *Collected Papers*. Harvard University Press, 1931–1935.

[6] Charles Sanders Peirce and John F. Sowa. Existential Graphs: MS 514 by Charles Sanders Peirce with commentary by John Sowa, 1908, 2000. Available at: http://www.jfsowa.com/peirce/ms514.htm.

[7] D. D. Roberts. *The Existential Graphs of Charles S. Peirce*. Mouton, The Hague, Paris, 1973.

[8] Sun-Joo Shin. *The Iconic Logic of Peirce's Graphs*. Bradford Book, Massachusetts, 2002.

[9] Jay J Zeman. *The Graphical Logic of C. S. Peirce*. PhD thesis, University of Chicago, 1964. Available at: http://www.clas.ufl.edu/users/jzeman/.

Canonical Correlation Analysis:
Use of Composite Heliographs for Representing
Multiple Patterns

Asaf Degani, Michael Shafto, and Leonard Olson

NASA Ames Research Center,
Mountain View, California,
U.S.A
{adegani, mshafto, lolson}@mail.arc.nasa.gov

Abstract. In a study of crew interaction with the automatic flight control system of the Boeing 757/767 aircraft, we observed 60 flights and recorded every change in the aircraft control modes, as well as every observable change in the operational environment. To quantify the relationships between the state of the operating environment and pilots' actions and responses, we used canonical correlation because of its unique suitability for finding multiple patterns in large datasets. Traditionally, the results of canonical correlation analysis are presented by means of numerical tables, which are not conducive to recognizing multidimensional patterns in the data. We created a sun-ray-like diagram (which we call a heliograph) to present the multiple patterns that exist in the data by employing Alexander's theory of centers. The theory describes 15 heuristic properties that help create wholeness in a design, and can be extended to the problem of information abstraction and integration as well as packing of large amounts of data for visualization.

1 Introduction

Canonical correlation analysis is a type of multivariate linear statistical analysis, first described by Hotelling [4]. It is currently being used in a wide range of disciplines (such as chemistry, meteorology, and artificial intelligence) to analyze the relationships between multiple independent and dependent variables. The information presented in Fig. 1 is derived from a canonical correlation analysis of a study of crew interaction with the automatic flight control system of the Boeing 757/767 aircraft. We observed 60 flights and recorded every change in the aircraft control modes, either manually initiated (e.g., the pilot selected a new mode) or automatically initiated (e.g., an automatic mode transition), along with all the settings relating to the flight control system status (e.g., waypoints and altitude values selected by the pilot). Likewise, every observable change in the operational environment (e.g., a new instruction from Air Traffic Control, or switching from one Air Traffic Control facility to another) was recorded, along with related variables such as the aircraft altitude, speed, and distance from the airport. In a way, it was like taking a snapshot of every

D. Barker-Plummer et al. (Eds.): Diagrams 2006, LNAI 4045, pp. 93–97, 2006.

change that took place both in and outside the cockpit. Overall, the dataset consisted of 1665 such snapshots, each characterized by 75 variables. Approximately half of the variables had to do with the operational environment and the other half had to do with pilot's responses [2].

In general, we were interested in identifying the relationships that exist between the state of the operating environment (independent variables) and pilots' actions and responses as represented thorough their interaction with the automatic flight control system and its mode and settings (dependent variables). The value of using canonical correlation in this case derived from its unique suitability for finding independent patterns in large datasets.

2 Representation of Statistical Patterns

Traditionally, the results of canonical correlation analysis are presented by means of numerical tables. However, a tabular format hinders the eye from recognizing and understanding the multidimensional patterns that exist in the data. Yet these patterns are extremely important, not only because they help the analyst characterize the most important environmental conditions and their corresponding effects on pilots' actions, but also because this method can reveal singular deviations from a well-established pattern (which is usually indicative of an operational error that can potentially lead to an incident or accident). Using structured correlations (the correlations of the X canonical variate with each of the original independent variables, and of the Y canonical variate with each of the original dependent variables), but seeking to avoid tabular representation of the data, we created a sun-ray-like diagram where all the independent variables (X1, X2, ...) are on the right side of the circle, and all the dependent variables (Y1, Y2, ...) are on the left. We chose a circle with rays to emphasize that "all variables are equal" (whereas employing a vertical and/or horizontal layout implicitly suggests some ordering). We call such a diagram a heliograph [5].

The canonical correlation analysis identified three sets of patterns that were operationally meaningful, statistically significant ($r = 0.95, 0.88, 0.72$; $p<0.001$), and independent (orthogonal) of each other. Each one of the three sets contains two patterns—one positive and one negative—depicted by dark and white bars respectively. For example, concerning the outer ring ($r = 0.95$) the positive pattern (dark bars) indicates that for all independent variables (X's) *when*

- altitude is high (above the average of 13,000 feet),
- the phase of flight is "descent,"
- the Air Traffic Control facility is "approach control,"
- and the vertical clearance is "descent to altitude,"

then the corresponding modes and settings selected by the pilots are most likely to be:

- autopilot "engaged,"
- pitch mode in "flight level change,"
- and thrust mode in "cruise."

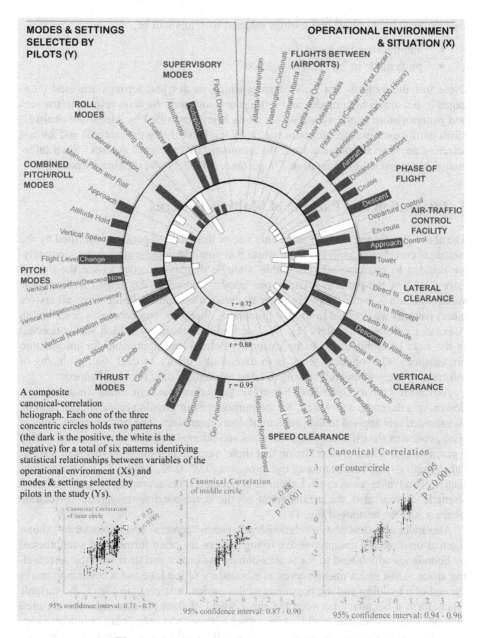

Fig. 1. A composite canonical-correlation heliograph

The reciprocal pattern (white bars) indicates that *when*

- the Air Traffic Control facility is "departure control," and
- the vertical clearance is "climb to altitude,"

then the most likely mode and settings selected by pilots will be:

- autothrottles "engaged,"
- pitch mode in "vertical navigation."

[Note that the patterns not only identify which modes and settings are used ("engaged"), but also which modes and settings are *not* used. So with respect to the second pattern (white bars), we know that while being controlled by "departure control," pilots hardly ever use the autopilot (i.e., they are hand-flying the aircraft) and are not selecting any lateral guidance from the automatic flight control system. Such information has considerable operational importance for safety and training purposes.]

3   Integration and Packing of Multiple Patterns

The above-mentioned r = 0.95 set is only one of three sets of patterns identified by the canonical correlation analysis. And while it is possible to present each set separately, we decided to combine all sets within a single display in order to see the overall "story" of how the patterns relate to one another and cover the range of all possible variables (both X's and Y's). In order to create a composite figure from all six (or more) patterns we decided to use the properties, operators, and processes described in Alexander's theory of centers [1]. We are finding this theory, which was conceived in the field of architecture, to be extremely helpful and applicable for information presentation. Our ongoing work is to extend this theory to deal with problems of information integration and packing of large amounts of data for visualization.

Alexander's theory describes 15 heuristic properties that help create wholeness in a design, or a diagram, and which, for the purpose of our ongoing research, can thereby be extended and applied to the problem of data integration. The first property, *level of scale*, concerns the different ranges of sizes and internal coherence of "centers" within a given design. Thus, after realizing that there were several different levels of statistical strength (significance) among the three sets (0.95, 0.88, and 0.72), it became geometrically advantageous to pack them as concentric rings according to their statistical strength. (Note also the arrangement of the corresponding bi-variate correlation plates across the bottom of Fig. 1).

Alexander also describes a relationship between "centers" and "boundaries" showing that inviting and comprehensive living centers are often formed and strengthened by boundaries which tend to focus attention on the center and tie it with the surrounding space. (Just like a marsh serves as a boundary of the lake and a colonnade marks the end of a building and the beginning of the garden or street). In Fig. 1, the variable labels form a boundary between the inner world of data (values, significance, etc.) and the outer operational world. *Alternating patterns* and *echoes* are two other properties present in the ray-like spokes that guide the reader's eye as the rings (and variables) become smaller and merge into the center.

The center of the figure forms a *void,* a profound property that is usually placed in the geometrical center of a design to draw the eye inward (e.g., the altar in a church or the empty space at the center of a mosque). We purposefully ordered the rings to (implicitly) suggest that as statistical significance decreases, the shrinking rings collapse into the void. Other utilized properties include *contrast* (between black and white bars), *interlock* (the overlap between black and white bars of the same variable) and *gradients* (in the magnitude of bar sizes, which, for the purpose of this display, was abstracted into three categories—strong, weak, and none).

The properties used to create the Figure act together to create a literal sense of wholeness. This allows the reader to inspect the sum total of the patterns in this dataset and identify regions where there is intensity of coverage (where bars of a certain cluster are juxtaposed and where interlocks exist along a certain variable axis), as well as regions on the circumference of the circle that are empty—indicating variables, mostly on the environmental (X) side, that are not important and do not contribute much to pilots' responses. For example, the fact that the "flights between airports" is not important provides a meaningful piece of the puzzle: It assures us, as the analysts, that there is nothing of major importance about the idiosyncrasies of particular flights. In other words, the patterns are consistent over different flight legs—an important fact about their generality.

To conclude, we believe that the concepts and methods of how to abstract, pack, and integrate data are key aspects for monitoring, analyzing, and interacting with data-rich environments such as networks, vehicle health monitoring systems, aircraft and spacecraft systems, and more [3]. Given the limited display "real estate," the (fixed) users' perceptual and cognitive resources, and the exponential growth in data availability, it is clear that visual displays and human computer interfaces represent a limitation which will only become more severe in the future. In response, our ongoing research is in developing a theory and formal methods for generating more capable data-rich interfaces.

References

1. Alexander, C.: *The Phenomenon of Life*. Berkeley, CA: The Center for Environmental Structure (2002).
2. Degani, A.: *Modeling human-machine systems: On modes, error, and patterns of interaction*. Unpublished doctoral dissertation. Atlanta, GA: Georgia Institute of Technology (1996).
3. Heymann, M., Degani, A.: Formal analysis and automatic generation of user interfaces: Approach, methodology, and an algorithm. *Human Factors* (paper accepted for publication).
4. Hotelling, H.: The most predictable criterion. *Journal of Educational Psychology 26* (1935) 139-142.
5. Shafto, M., Degani, A., Kirlik, A.: Canonical correlation analysis of data on human-automation interaction. *Proceedings of the 41st Annual Meeting of the Human Factors and Ergonomics Society*. Albuquerque, NM (1997).

Modularity and Composition in Propositional Statecharts

H.R. Dunn-Davies[1], R.J. Cunningham[1], and S. Paurobally[2]

[1] Department of Computing, Imperial College, London, UK, SW7 2BZ
[2] Department of Computer Science, University of Liverpool, Liverpool, UK, L69 7ZF

Abstract. Propositional Statecharts, described in [3], are a variation of David Harel's Statechart formalism [6] intended to enable both diagrammatic description of an agent interaction protocol, and interpretation as a theory in a dynamic logic. Here we provide an informal description of a diagrammatic extension to enable modular representation.

1 Introduction

Several diagrammatic methods have been proposed for the description of agent interaction protocols ([4], [2], [6], [1]). Although their visual basis allows a protocol to be described in a relatively easy-to-follow manner, it has been shown that these semi-formal methods of expressing protocols can lead to errors and ambiguities [7], with different interpretations among agent designers, so that different agents participating in the same interaction may in fact be using slightly different protocols. The Propositional Statechart formalism is intended to enable the visual intuition of a diagram to be underpinned by formal inferences from its interpretation as a theory in a dynamic logic such as ANML [7]. This is an on-going project, but it seems that Propositional Statecharts are capable of expressing many interaction protocols completely and unambiguously.

One noticeable drawback of the version of the Propositional Statechart formalism described in [3] is the lack of a mechanism for modularity, whereby Statecharts can be defined in terms of separate modular components. The Propositional Statechart formalism is intended to represent interaction protocols in their entirety, so can benefit greatly from this capability. In contrast, Harel's statecharts have chiefly been used for component description, for instance in an object oriented systems. Nevertheless, the modularity mechanism we illustrate appears to be of wider applicability.

The need for a modular approach to protocol definition has been highlighted by Vitteau and Huget [9], who point out that many real world interactions are composed of sub-interactions, which may be repeated during the main interaction. The ability to define a protocol in a modular fashion enables the designer to split large protocols into smaller, more manageable components, and provide a single definition of any part of the protocol that occurs more than once. This helps make definitions concise and makes it possible to create libraries of components, so that protocols do not always have to be built from scratch.

D. Barker-Plummer et al. (Eds.): Diagrams 2006, LNAI 4045, pp. 98–103, 2006.

A further advantage of modularity is that it becomes possible to represent recursively defined protocols which could not be described otherwise. This is an important step since many protocols, such as some versions of the multilateral protocol discussed in section 3, cannot be described adequately in any other way. Here we describe an extension to the Propositional Statechart formalism that allows the modular representation of interaction protocols[1].

2   Adding Modularity to Propositional Statecharts

The mechanism we propose to enable modular definitions takes advantage of Statecharts' use of topological enclosure of contours to represent state hierarchy. The principle behind it is as follows: since the superstate of any collection of atomic states represents the disjoint union of those states, any given group of contours describing states can be replaced by a contour representing their superstate, without loss of precision, as long as a complete description of all of the states contained within the superstate is provided in a separate Statechart.

This approach allows us to define recursive protocols, as superstates can contain skeleton representations of themselves. It is not in general possible to finitely enumerate all of the states of a recursive protocol, as the number of states may not be bounded. However, a modular representation may provide as complete a description as is required, provided more complete representations can be generated by replacing each skeleton representation with its definition.

The mechanism is shown in figures 1 and 2. Figure 1 is a simple Statechart, and figure 2 provides a modular definition with the same intended behaviour. Figure 2 can be generated from figure 1 by the following procedure. Firstly, add a superstate, named **H**, enclosing states **D**, **E**, **F**, and **G**, a process which is trivially behaviour preserving [8]. Secondly, copy state **H** and its contents (all of the labelled contours and arrows completely enclosed by the contour representing state **H**), to form a separate modular component[2]. Finally, replace the representation of state **H** in statechart **A** with a skeleton representation which includes only the states linked by a single transition to a state outside **H** (we refer to these as interface states)[3]. The recombination of the two components can easily be performed, essentially by reversing the above steps.

In order to ensure that a Propositional Statechart designer can refer to a component more than once in the same superstate while preserving uniqueness of the state labels we treat a component Statechart as a class of superstates,

---

[1] Thence it appears more expressive than the graphical representations cited above, which seem not to provide adequate descriptions of recursive protocols.

[2] In some cases it may be necessary to specify explicitly which of a number of components represents the top level of the combined statechart. In this example it is not, as there is only one possible combination of the two components.

[3] We have also shaded each interface state in **A** to mark **H** as a skeleton representation, but have neglected to shade any states in the modular component representing **H**. This is because we aim to create modular components which can be used in different contexts, and to provide definitions which are as general as possible.

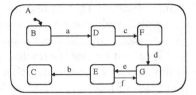

Fig. 1. A simple Statechart incorporating transitions between atomic states

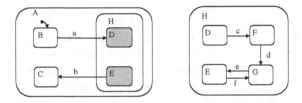

Fig. 2. A modular definition behaviourally equivalent to the Statechart in figure 1

in the same way that objects in Object Oriented programming languages are instantiations of classes of objects. So multiple references to a Statechart can be taken to refer to different instances of its class, and labelled accordingly.

Transitions which link atomic states that are more than one level apart can be specified if we allow skeleton representations of states in modular components to contain skeleton representations of other states. This is the case in figure 3, where a transition links states **C** and **I**, which are two levels apart.

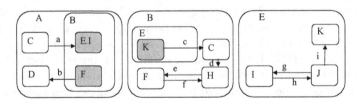

Fig. 3. A single transition can link states two levels apart

The mechanism described above allows a user to represent single transitions between single states in different modular components. It is also possible to represent regular sets of transitions from corresponding source states in different components to a single destination state, allowing an interaction to escape from a recursive protocol. This is done by allowing a single contour to specify a potentially unbounded set of source states, which are defined by means of a regular expression [5]. An example of this mechanism can be seen in the definition of **multilateral encounter 2** in figure 5, where one of the shaded contours is labelled using the regular expression **me3$^+$.me3.voting over**, which represents the set of strings consisting of a sequence of two or more instances of **me3.** followed by **voting over**.

3   A Modular Representation of a Multilateral Protocol

To illustrate how this extension to the Propositional Statechart formalism can be applied in practice we use it to describe a multilateral protocol, a simple recursive protocol which aims to reflect a group decision making process.

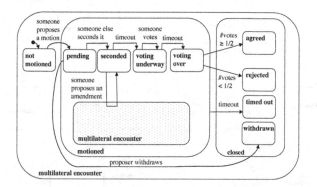

Fig. 4. A simple Statechart for the multilateral protocol

Figure 4 shows a recursive form of Statechart for a multilateral protocol. Despite giving the reader a rough idea of the protocol, it fails to provide a complete and unambiguous representation. The figure could be clarified by using precise state definitions and action semantics, although the representation would still be incomplete for two reasons. Firstly, it fails to specify exactly which states in the embedded protocol are linked to the state labelled **seconded** in the top level protocol. Secondly, it does not explicitly define the structure of the embedded multilateral protocol. These ambiguities are resolved in figure 5, which provides a more complete definition of one interpretation of the protocol, in which the action *amendment passed and motion amended* triggers a transition to the state labelled **seconded** in **multilateral encounter 2** from any state entitled **voting over** in a superstate of class **me3**.

The Propositional Statechart in figure 5 has informal action labels, although some of these represent complex actions. Complete protocol definitions would incorporate full definitions of each action, expressed using dynamic logic path labels as discussed in [3]. These would include details of the participants in the interaction, and of how the parameters specified in brackets after the name of each Propositional Statechart are manipulated.

4   Conclusion and Future Directions

The current paper provides an informal description of an extension to the Propositional Statechart formalism that allows it to represent interaction protocols in a modular fashion. We illustrate this mechanism by providing an unambiguous

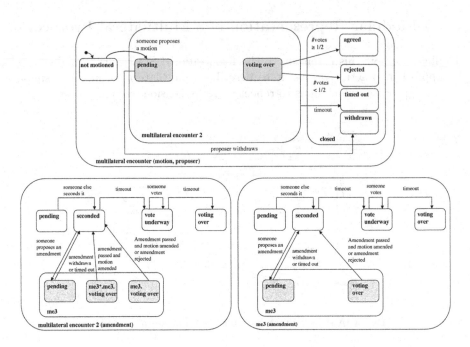

Fig. 5. Three Propositional Statecharts representing an interpretation of figure 4

definition of a recursive protocol. Although the extension described in this paper has been developed in the context of Propositional Statecharts, it could also be applied to conventional Statecharts, and the principle underlying it could be used to create modular extensions for other types of diagram which encode information via the topological enclosure of contours on a plane.

This work forms part of an attempt to create a clear and intuitive graphical formalism for the description of interaction protocols that can be automatically converted into ANML, so that properties such as soundness and completeness can be verified, and can be implemented automatically as executable program threads. We are in the process of developing a formal semantics for the extended Propositional Statechart formalism in order to achieve this.

References

1. Bauer, B., Muller, J.P. and Odell, J., *Agent UML: A Formalism for Specifying Multiagent Software Systems*, AOSE (2000), 91-104.
2. Cost, R., et al., *Modeling agent conversations with colored petri nets*, Workshop on Specifying and Implementing Conversation Policies (1999), 59-66.
3. Dunn-Davies, H.R., Cunningham, R.J. and Paurobally, S. *Propositional Statecharts for Agent Interaction Protocols*, ENTCS, 134, 55-75, 2005.
4. Fornara, N. and Colombetti, M., *Defining Interaction Protocols using a Commitment Based Agent Communication Language*, in Proceedings of AAMAS 2003.
5. Friedl, J.E.F. "Mastering Regular Expressions," O'Reilly, 2002.

6. Harel, D., *Statecharts: A visual formalism for complex systems*, Sci. Comput. Programming **8** (1987), 231-274.
7. Paurobally, S. and Cunningham, R.J., *Verification of Protocols for Automated Negotiation*, in Frank van Harmelen (Ed) Proceedings of ECAI2002, IOS Press, 43-47.
8. Sunyé, G. et al., *Refactoring UML Models*, in Proceedings of UML 2001, 134-148.
9. Vitteau B. and Huget, M.P., *Modularity in Interaction Protocols*, in Frank Dignum (Ed.) Advances in Agent Communication, ACL 2003. LNCS **2922**, 291-309, 2004.

Objects and Spaces: The Visual Language of Graphics

Yuri Engelhardt

Department of Media Studies, University of Amsterdam,
Turfdraagsterpad 9, 1012 XT Amsterdam, The Netherlands
engelhardt@uva.nl

Abstract. There is undoubtedly something like a 'grammar of graphics'. Various syntactic principles can be identified in graphics of different types, and the nature of visual representation allows for visual nesting and recursion. We propose a limited set of possible 'building blocks' for constructing graphic *spaces*, and a limited set of possible syntactic functions of graphic *objects*. Based on these ingredients, and the rules for their combination, the syntactic structure of any visual representation can be drawn as a *hierarchically nested tree*. We claim that the presented visual syntax applies to all types of visual representations.

1 Syntax in Visual Representations

In this project we are exploring the syntax of visual representations of information[1]. Why? Because understanding visual syntax can help us to understand the nature of – and the possibilities of – visual representation. Insights into the syntactic principles of visual representations can help us with the analysis, the comparison, and hopefully the automatic generation and parsing of graphics. In order to explore visual syntax, we are studying the nature of the 'ingredients' of visual representations – *objects*, *spaces*, and *properties* – and the nature of their relationships with each other. We find that *objects* may have different syntactic functions, and that *spaces* are constructed from basic 'building blocks'. We claim that the presented approach applies to all graphics.

2 Objects and Spaces

Any visual representation – and any meaningful visible component of a visual representation – may be referred to as a graphic **object**. This means that graphic objects can be distinguished at various levels of a visual representation. For example, a map or a chart in its entirety is a graphic object. In addition, the various symbols or components that are positioned within that map or chart are graphic objects as well.

We define a graphic **space** as the ('canvas'-)surface that is occupied by a graphic object. Any *object* may contain a set of *(sub-)objects* within the *space* that it occupies. When this principle is repeated recursively, the spatial arrangement of (sub-)objects is, at each level, determined by the specific nature of the containing space at *that* level.

---

[1] This paper extends our earlier work on visual syntax: Engelhardt et al. 1996, Engelhardt 2002.

D. Barker-Plummer et al. (Eds.): Diagrams 2006, LNAI 4045, pp. 104–108, 2006.

In other words, an *object* may contain an internal *space* and (sub-)*objects* that are arranged within that space (this is represented by the syntactic tree on the right). This principle can be repeated recursively: A (sub-)*object* may again contain an internal *space* and (sub-)*objects* that are arranged within that space (the syntactic tree 're-branches'.)

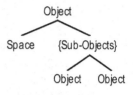

See figures 1-3 for examples of syntactic trees of such nestings of objects and spaces.

To make things complete, graphic objects may have meaningful graphic *properties*, such as color or size.[2] We propose a limited set of 'building blocks' for the construction of all spaces (section 3), and also a limited set of syntactic functions that an object may have in relationship to the space and the objects around it (section 4).

3   Construction of Spaces

Graphic spaces (Sp) may be constructed from the following basic set of possibilities. 'Integral' 2D or 3D projections of a physical space:

- **pictorial space** (the space inside a picture of a thing or a scene), or
- **cartographic space** (CartoSpace, the space inside a map – see fig.1 and fig.3).

Spatial dimensions (dim, 1D):

- axis (quantitative, allows for steps of visibly varying length): **time axis** (TimeAxis – fig.1) or **attribute axis** (AttribAxis, for all quantitative data other than time – fig.1,3).
- lineup (ordinal, all steps are shown at equal distance): **time-ordered lineup** (Time-Lineup – fig.3), **attribute-ordered lineup** (an ordering based on a criterion other than time), or **unordered lineup** (UnordLineup, order is not significant – fig.3).[3]

Spatial dimensions can run in different directions (horizontal, vertical, distal, circular, radial). The space of an analog clock face, for example, is defined by two or three superimposed *circular* time axes (for the hours, minutes, and maybe seconds). Spatial dimensions are often combined orthogonally to build more complex spaces (OrthoSpace – fig.3). A shared dimension (SharedDim – fig.1) can be used to create 'parallel graphics', for example several vertically stacked graphics that use the same horizontal time axis.

Other possibilities for constructing spaces are: **quantity space** (QuanSpace, in which surface area corresponds to some quantity, e.g. the space inside a pie chart or a tree-map – fig.2), **clustering space** (space that is used for the spatial grouping of objects), **Euler topology** (a space that is divided into overlapping regions, e.g. the space inside a Venn diagram), and – not to be forgotten – **uncoded space** (in which it would be possible to move objects around without changing the graphics' meaning).

This concludes our complete list of building blocks for constructing graphic spaces. These can be nested into each other as described (space inside object inside space...).

---

[2] *Objects*, *spaces* and *properties* as the ingredients of visual representations come from our earlier work (Engelhardt et al. 1996). Card, Mackinlay and Shneiderman (1999) build on Bertin and discuss spaces, "marks", and properties. However, they do not describe the nature of spaces and objects as fundamentally recursive and nested into each other, as we do here.

[3] *Axis*, *ordered lineup*, and *unordered lineup* can be and have been alternatively described as: "quantitative" dimension (= axis), "ordinal" dimension (= ordered lineup), and "nominal" dimension (= unordered lineup) in Engelhardt et al. (1996), and in Card et al. (1999).

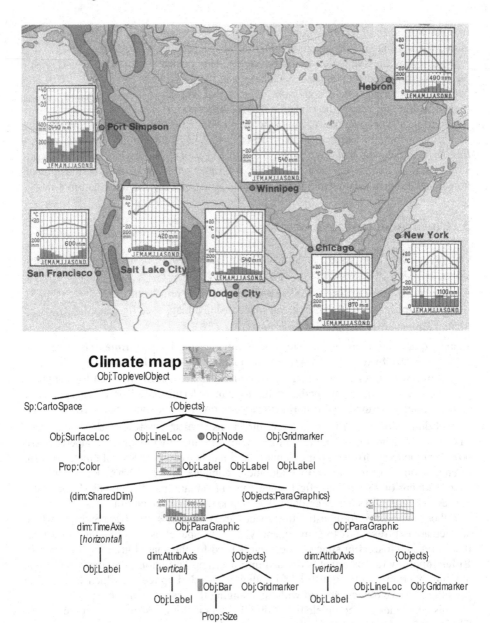

Fig. 1. This is an example of a visual representation with a deeply nested structure of objects and spaces. Compare the map with its syntactic tree: The map contains a *cartographic space* and *objects* of various sorts. The (red) dots function as *nodes* that locate cities. Each dot has two *labels*: a weather chart and the city's name. The weather chart consists of two stacked *parallel graphics* with a *horizontal time axis* as their *shared dimension*. One chart contains *bars* that vary in *size* along a *labeled vertical axis* representing rainfall, while the other contains a *line locator* that varies in height along a *labeled vertical axis* representing temperature. Both charts show a heavy pattern of *grid markers*. (Map fragment from Degn et al., Seydlitz, 1973)

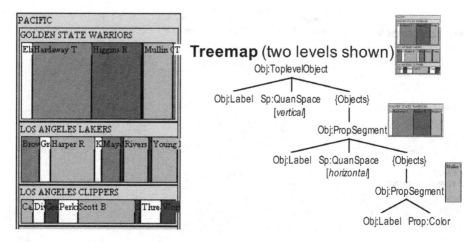

Fig. 2. The syntactic structure of a treemap: The treemap contains a *quantity space* that is filled with *proportional segments*. Each proportional segment is again filled with *proportional segments*. This may continue recursively. (Treemap image: www.otal.umd.edu/Olive/Class/Trees)

4   Syntactic Functions of Objects

The 'syntactic function'[4] of a graphic object concerns its function in relationship to the space and the objects around it. Frequently occurring functions of objects are **node**, **connector**, **frame**, and **label**. In some spaces nodes are 'point locators'. Other 'locator' functions are: **line locator** (LineLoc), **surface locator** (SurfaceLoc), and **volume locator** (VolumeLoc). A **grid marker** provides reference. 'Statistical' functions of objects include **bar**, **proportional segment** (PropSegment), and **iso-unit** (repeated pictogram in pictorial statistics). Inside pictures we find the **intrapictorial object** and maybe the **picture supplement**[5]. The **set-symbol** and **parallel graphics** complete the list.

Depending on its function (one of the above), an object is 'anchored' to the space or to the objects around it in a specific way. Having defined spaces and objects, we can make a list of specific *syntactic rules* that describe the possibilities and constraints for their nesting and 'anchoring'. A *connector* for example is loosely anchored to the two objects that it connects, while a *line locator* is tightly anchored to all points on a specific line in space. A *label* is loosely anchored to the object or the dimension that it labels, while a *bar* is tightly anchored both to a base line and to a height along an *axis*. An *intrapictorial object* is always anchored in a *pictorial space*, a *set-symbol* in an *Euler topology*. *Parallel graphics* are always lined up with each other along a *shared dimension*. *Size* (property) can be attached to *nodes* and *line locators*, but not to *surface locators* (surface locators are already fixed in their size). *Locators* cannot appear in *quantity space* or *uncoded space* (location is flexible), while *proportional segments* can *only* appear in *quantity space*. The complete set of these rules can be written out.

---

[4] Same as "grammatical function" in Engelhardt et al. 1996, "syntactic role" in Engelhardt 2002.
[5] Comics for example are *time-ordered lineups* of *nodes* (panels). Each of these has a *frame* and contains a *pictorial space*, *intrapictorial objects* and *picture supplements* (word balloons etc.).

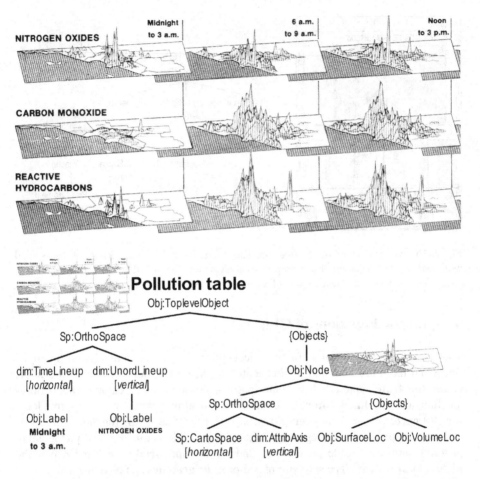

Fig. 3. Orthogonal constructions nested into a bigger orthogonal construction (L.A. Times, 1979)

Conclusion: The notions of a) the nested relationships between objects and spaces, b) the construction of spaces from basic 'building blocks', and c) the diversity of syntactic functions of objects, seem to offer a flexible approach to the nature of graphics.

References

1. Card, S.K., Mackinlay, J.D., Shneiderman, B.: Readings in information visualization: Using vision to think. Morgan Kaufman Publishers, San Francisco, CA (1999)
2. Engelhardt, Y., de Bruin, J., Janssen, T., Scha, R.: The visual grammar of information graphics. In: Narayanan, N.H., Damski, J. (eds.): Proceedings of the AID '96 workshop on Visual Representation, Reasoning and Interaction in Design, Stanford University (1996)
3. Engelhardt, Y.: The language of graphics: A framework for the analysis of syntax and meaning in maps, charts and diagrams. Unpublished Ph.D. thesis, Institute for Logic, Language and Computation, University of Amsterdam, The Netherlands (2002)

Defining Euler Diagrams: Simple or What?

Andrew Fish and Gem Stapleton

The Visual Modelling Group
University of Brighton, Brighton, UK
{andrew.fish, g.e.stapleton}@brighton.ac.uk
www.cmis.brighton.ac.uk/research/vmg

Many diagrammatic languages are based on closed curves, and various well-formedness conditions are often enforced (such as the curves are *simple*). We use the term **Euler diagram** in a very general sense, to mean any finite collection of closed curves which express information about intersection, containment or disjointness. Euler diagrams have many applications, including the visualization of statistical data [1], displaying the results of database queries [6] and logical reasoning [2, 4, 5]. Three important questions are: for any given piece of information can we draw a diagram representing that information, can we reliably interpret the diagrams and can we reason diagrammatically about that information? The desirable answer to all three questions is yes, but these desires can be conflicting. In this article we investigate the effects of enforcing the simplicity condition (as in [1, 2, 6]) or not enforcing it (as in [4, 5]).

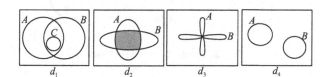

Fig. 1. Euler diagrams and reasoning problems

The diagram d_1 in figure 1 can be described by $\emptyset$, $\{A\}$, $\{A, B\}$, $\{B\}$, $\{A, B, C\}$. Each of the items in this list corresponds to a region in the diagram; for example $\{A\}$ corresponds to the region which is inside the curve A but outside the curves B and C. Any Euler diagram can be described in this manner. Whilst every Venn diagram description is drawable with simple closed curves, this is not the case for Euler diagrams [6].

When visualizing information, we may want to perform some logical reasoning or transform a diagram if information is updated and, ideally, these transformation rules will modify diagrams in such a way that users can easily identify the change. One such rule allows the removal of a minimal region (used when reasoning in [3, 5]); figure 1 shows that removing the shaded region from d_2 in a natural manner (by squashing it) gives rise to d_3 which contains non-simple curves. This can be redrawn using simple curves, giving d_4, but this has an obvious added cognitive load. Sometimes such redrawing is not possible: removing all minimal regions from Venn(9) except for those described by $\emptyset$, $\{A, B, C\}$,

D. Barker-Plummer et al. (Eds.): Diagrams 2006, LNAI 4045, pp. 109–111, 2006.
© Springer-Verlag Berlin Heidelberg 2006

$\{D, E, F\}$, $\{G, H, I\}$, $\{A, D, G\}$, $\{B, E, H\}$, $\{C, F, I\}$, leaves a diagram whose description is not drawable when simplicity is enforced [6].

The undrawability of some collections of set intersections can have profound effects on logical reasoning systems. Reasoning rules are usually defined in terms of a pre-condition and a post-condition, sometimes implicitly, (as in [3, 5]); the usual understanding of such a contract is that if the pre-condition holds for a well-formed diagram d_1 then there exists a well-formed diagram d_2 that satisfies the post-condition. However, specifying a strong enough pre-condition is diffi-cult because it is currently unknown which diagram descriptions are drawable under simplicity; since the application of any reasoning rule must return a well-formed diagram, this undrawability information must be incorporated into the pre-condition.

Not enforcing simplicity means that every description is drawable. An ex-ample of a diagram construction process can be seen in figure 2: starting with the description $\{A\}, \{C\}, \{B, C\}, \{A, B, C\}$, we draw four simple closed curves labelled appropriately which are glued together a single point, shown in the left-hand diagram; the required (middle) diagram consists of the three colour coded curves labelled by A, B and C; the rightmost diagram has the same description but better exploits the containment properties of Euler diagrams.

Fig. 2. Diagram construction

We can safely use definitions that rely on the interiors of simple closed curves because there is a well-defined (and intuitive) notion of what constitutes the interior. However, there are choices for the interior if we are presented with non-simple curves. The Euler/Venn diagram [5] d_1 in figure 3 is ambiguous: is Jean a software engineer? We hypothesize that many people would say Jean is a software engineer because Jean is seemingly placed inside Software Engineers.

The Euler diagram d_2 in figure 3 is also ambiguous: are all job seekers also lecturers or are none of them lecturers? The shaded part of d_3 indicates a likely 'interior' of the Lecturers curve. The interior of closed curves usually represents the set denoted by the curve's label (as in [2, 3, 4, 5]) and so identifying the interior is crucial for a formal interpretation of the diagram. For closed curves, there are various methods for identifying the interior and no such method has been given in any of the publications on Euler diagram reasoning where simplicity is not enforced. One such method is to take any bounded component of the plane minus the image of the curve to be interior. This agrees with our intuitive interpretation of d_1 but not d_2. There is a different method that agrees with the intuitive interpretation of d_2 but not d_1.

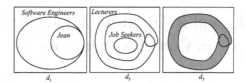

Fig. 3. Ambiguous diagrams

We have highlighted undesirable consequences of enforcing the simplicity condition in terms of undrawability. To overcome these consequences we can allow any self-intersecting curves but this is not sensible since it allows space filling curves or lines. We suggest that a refinement of the simplicity condition is used, such as allowing only a finite number of self-intersections, to provide a balance between conflicting desires. This will allow reasoning rules to be applied to more diagrams by enabling the pre-conditions to be relaxed. Such refinements may enhance usability in certain circumstances; for example, a minimal region can be removed using a natural process so that the resulting diagram looks similar to the original diagram. An understanding of the consequences of all well-formedness conditions on drawability, semantics and reasoning is needed.

Acknowledgement. Author Stapleton is supported a by Leverhulme Trust Early Career Fellowship.

References

1. S. Chow and F. Ruskey. Drawing area-proportional Venn and Euler diagrams. In *Proceedings of Graph Drawing 2003, Perugia, Italy*, volume 2912 of *LNCS*, pages 466–477. Springer-Verlag, September 2003.
2. E. Hammer. *Logic and Visual Information*. CSLI Publications, 1995.
3. J. Howse, G. Stapleton, and J. Taylor. Spider diagrams. *LMS Journal of Computation and Mathematics*, 8:145–194, 2005.
4. S.-J. Shin. *The Logical Status of Diagrams*. Cambridge University Press, 1994.
5. N. Swoboda and G. Allwein. Heterogeneous reasoning with Euler/Venn diagrams containing named constants and FOL. In *Proceedings of Euler Diagrams 2004*, volume 134 of *ENTCS*. Elsevier Science, 2005.
6. A. Verroust and M.-L. Viaud. Ensuring the drawability of Euler diagrams for up to eight sets. In *Proceedings of 3rd International Conference on the Theory and Application of Diagrams*, volume 2980 of *LNAI*, pages 128–141, Cambridge, UK, 2004. Springer.

Topological Relations of Arrow Symbols
in Complex Diagrams

Yohei Kurata and Max J. Egenhofer

National Center for Geographic Information and Analysis
and
Department of Spatial Information Science and Engineering
University of Maine
Boardman Hall, Orono, ME 04469-5711, USA
{yohei, max}@spatial.maine.edu

Abstract. Illustrating a dynamic process with an arrow-containing diagram is a widespread convention in people's daily communications. In order to build a basis for capturing the structure and semantics of such diagrams, this paper formalizes the topological relations between two arrow symbols and discusses the influence of these topological relations on the diagram's semantics. Topological relations of arrow symbols are established by two types of links, *intersections* and *common references*, which are further categorized into nine types based on the combination of the linked parts. The topological relations are captured by the existence/non-existence of these nine types of intersections and common references. Then, this paper demonstrates that arrow symbols with different types of intersections typically illustrate two actions with different interrelations, whereas the arrow symbols with common references illustrate a pair of semantics that may be mutually exclusive or synchronized.

1 Introduction

Illustrating a dynamic process with an arrow-containing diagram is a widespread convention in people's daily communications. Fig. 1a-c illustrate examples of diagrams for such dynamic processes as a workflow, an assembling procedure, and geographic propagations. If computers understand such arrow-containing diagrams, people can interact with computers more intuitively, for instance, by sketching a diagram on computer screens (Kurata and Egenhofer 2005a) to instruct the machines about the dynamic processes they will manage, support, or simulate.

Communication through an arrow-containing diagram requires the diagram readers to interpret the meaning of each arrow symbol in the diagram, because arrow symbols have a large variety of meanings (Horn 1998) and are used multi-purposely even in a single diagram without specification (Tversky, et al. in press). Such interpretations are not easy for computers (Kurata and Egenhofer 2005a), and sometimes even difficult for people without well-crafted context (Tversky, et al. in press). Kurata and Egenhofer (2005b) demonstrated that the interpretation of a diagram with a single arrow symbol can be partly derived from its syntactic pattern. People, however, often communicate using more complex diagrams with multiple arrow symbols (Figs. 1a-c). It, therefore,

D. Barker-Plummer et al. (Eds.): Diagrams 2006, LNAI 4045, pp. 112–126, 2006.
© Springer-Verlag Berlin Heidelberg 2006

remains a challenging problem to develop a formal method for interpreting such complex arrow-containing diagrams. As a first step toward this goal, this paper analyzes the spatial relations between arrow symbols in such complex diagrams and observes the influence of the spatial relations on the diagram's semantics. Among several types of spatial relations, this paper focuses on topological relations (i.e., spatial relations that are not affected by elastic deformations), because topological information is highly influential in people's conceptualizations of space (Egenhofer and Mark 1995).

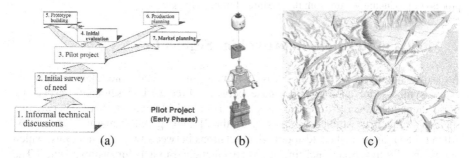

(a)   (b)   (c)

Fig. 1. Examples of complex arrow-containing diagrams that illustrate such dynamic processes as (a) a workflow (Horn 1998), (b) an assembling procedure (from a LEGO® manual), and (c) geographic propagation (Barraclough 2003)

In arrow-containing diagrams, arrow symbols are typically used together with other elements. A unit of an arrow symbol and the elements that the arrow symbol refers to (i.e., originates from, points to, or traverses) is considered a syntactic unit, called an *arrow diagram* (Kurata and Egenhofer 2005a). This paper extends this definition such that an arrow diagram is composed of one or more arrow symbols and the elements to which at least one of these arrow symbols refers. These arrow-related elements are called the *components* of the arrow diagram. Arrow diagrams are a subset of arrow-containing diagrams, since diagrams composed of arrow symbols alone are included in arrow-containing diagrams, but not in arrow diagrams.

An arrow diagram that contains n arrow symbols is called an *n-arrow diagram*. If $n>1$, the n-arrow diagram is also called a *multi-arrow diagram* (Fig. 2). The scope of this paper is to capture the meaningful structures embedded in such multi-arrow diagrams. Our premise is that such meaningful structures are sufficiently captured by a set of spatial relations between arrow symbol pairs. For simplifying the discussion, this paper deals with arrow symbols that neither intersect with themselves nor refer to the same component more than once.

The remainder of this paper is structured as follows: Section 2 reviews studies about line-line relations. Based on the formalization of topological line-line relations, Section 3 formalizes the topological relations between two arrow symbols, introducing two types of links that connect the arrow symbols directly or indirectly. Section 4 observes how such topological relations influence the semantics of multi-arrow diagrams. Section 5 demonstrates how this approach captures the structures and semantics of multi-arrow diagrams, using the example in Fig. 2. Section 6 concludes the discussion, pointing out some items for future research.

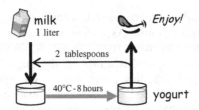

Fig. 2. An example of a multi-arrow diagram (4-arrow diagram), which illustrates the recursive process of producing yogurt from the mixture of milk and yogurt

2 Models of Relations Between Line Segments

An arrow symbol is essentially a directed line segment. Thus, the relations between arrow symbols are analogous to those between directed line segments, especially those embedded in a 2-dimensional space. Topological relations between two line segments have been studied extensively by the AI and spatial database communities. Allen (1983) distinguished 13 topological relations between two time intervals, which are essentially uni-directional line segments embedded in a 1-dimensional space. The *4-intersection model* (Egenhofer and Franzosa 1991) captured the topological relations between two spatial objects based on the existence/non-existence of geometric intersections of the objects' interiors and boundaries. The *9-intersection model* (Egenhofer and Herring 1991) extends the 4-intersection model by considering the intersections with respect to the objects' exteriors as well. In the 9-intersection model, the intersections between two spatial objects A and B are concisely represented by a 3×3 matrix (Eqn. 1), where $A°$, ∂A, and A^- are A's interior, boundary, and exterior, while $B°$, ∂B, B^- are B's interior, boundary, and exterior, respectively.

$$M(A,B)=\begin{pmatrix} A°\cap B° & A°\cap \partial B & A°\cap B^- \\ \partial A\cap B° & \partial A\cap \partial B & \partial A\cap B^- \\ A^-\cap B° & A^-\cap \partial B & A^-\cap B^- \end{pmatrix} \qquad (1)$$

Topological relations between A and B are captured by the existence/non-existence of these nine types of intersections. Thus, the matrices with different empty/non-empty entries correspond to different topological relations. Although the matrix distinguishes $2^9 = 512$ configurations, the configurations with geometric realizations are limited by some geometric constraints. Based on this model, Egenhofer (1994a) identified 33 topological relations between non-directed line segments embedded in a 2-dimensional space.

Another variation of the 4-intersection model distinguishes two boundaries (start and end points) of directed line segments (Hornsby, *et al.* 1999). In that model, topological relations between two directed line segments A and B are represented by a 3×3 matrix (Eqn. 2), where $\partial_s A$, $A°$, and $\partial_e A$ are A's start point, interior, and end point, while $\partial_s B$, $B°$, and $\partial_e B$ are B's start point, interior, and end point, respectively. Based on this model, Hornsby *et al.* (1999) identified 16 topological relations between two time intervals in a cyclic time (essentially uni-directional line segments embedded in a cyclic 1-dimensional space).

$$M(A,B) = \begin{pmatrix} \partial_s A \cap \partial_s B & \partial_s A^\circ \cap B^\circ & \partial_s A \cap \partial_e B \\ A^\circ \cap \partial_s B & A^\circ \cap B^\circ & A^\circ \cap \partial_e B \\ \partial_e A \cap \partial_s B & \partial_e A \cap B^\circ & \partial_e A \cap \partial_e B \end{pmatrix} \tag{2}$$

Clementini and di Felice (1998) introduced another model of topological relations between two directed line segments embedded in a 2-dimensional space, called *classifying invariants*. Their model captures more detailed topological relations than the 9-intersection model, yielding such distinctions as relations with different number of interior-interior intersections.

Some researchers explored the relations between two line segments other than topological relations. Schlieder (1995) defined point-set ordering in a 2-dimensional space, with which he identified 63 order relations between two straight directed line segments in a 2-dimensional space. Moraz *et al.* (2000), on the other hand, identified 14, 24, and 69 directional relations between two straight directed line segments in a 2-dimensional space, based on a set of relative positions of one segment's endpoints seen from the other segment at three different granularities. Rentz (2001) distinguished 26 order relations between two directed intervals (essentially two directed line segments) in a 1-dimensional space. Nedas *et al.* (in press) incorporated two metric measures, *splitting ratios* and *closeness measures*, into both the 9-intersection matrix and the classifying invariants as their metric refinements, following the premise "*topology defines, metric refines*" (Egenhofer and Mark 1995).

3   Topological Relations Between Two Arrow Symbols

Topological relations between two arrow symbols are established by two types of connections between these arrow symbols. One is *intersections*. Two arrow symbols may intersect with each other (Figs. 3a-b), as two line segments do. Thus, the topological relations between two arrow symbols are partly modeled in the same way as the relations between line segments are modeled based on their intersections. Another type of links is *common references*. A common reference of two arrow symbols is established when the arrow symbols refer to the same component (Figs. 3c-d). A common reference connects two arrow symbols indirectly through an intermediate component, while an intersection directly connects the arrow symbols. This analogy motivates us to model the topological relations between arrow symbols from a viewpoint of common references as well as that of intersections. In addition, since two arrow symbols are sometimes connected by both intersections and common references (Figs. 3e-f), the model for capturing both types of connections in a unified way is potentially useful. Sections 3.1 and 3.2, therefore, partly model the topological relations between arrow symbols based on intersections and common references, respectively, and Section 3.3 integrates the two models into a single hybrid model.

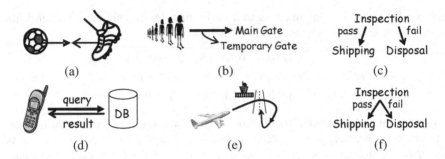

Fig. 3. Examples of 2-arrow diagrams where arrow symbols are connected by (a) a head-head intersection, (b) a body-tail intersection, (c) a common reference to the label "*Inspection*", (d) common references to the cell phone and database icons, (e) both a head-tail intersection and a common reference to the landing strip icon, and (f) both a tail-tail intersection and a common reference to the label "*Inspection*"

3.1 Topological Relations Established by Intersections

An arrow symbol consists of three different parts: *tail*, *body*, and *head*. The tail, body, and head are the rearmost point, interior, and headmost point of the arrow symbol, respectively. An arrow symbols corresponds to a time interval, since both are a kind of line segments with two qualitatively-different boundary points. Thus, following the 9-intersection model for time intervals (Hornsby, *et al.* 1999), this paper captures the topological relations between two arrow symbols A and B based on $3\times3 = 9$ types of their intersections: *tail-tail*, *tail-body*, *tail-head*, *body-tail*, *body-body*, *body-head*, *head-tail*, *head-body*, and *head-head intersections*. These nine types of intersections are concisely represented by a 3×3 matrix (Eqn. 3), where $\partial_{tail}A$, A°, and $\partial_{head}A$ are A's tail, body, and head, while $\partial_{tail}B$, B°, and $\partial_{head}B$ are B's tail, body, and head, respectively.

$$M_{I}(A,B) = \begin{pmatrix} \partial_{tail}A \cap \partial_{tail}B & \partial_{tail}A^{\circ} \cap B^{\circ} & \partial_{tail}A \cap \partial_{head}B \\ A^{\circ} \cap \partial_{tail}B & A^{\circ} \cap B^{\circ} & A^{\circ} \cap \partial_{head}B \\ \partial_{head}A \cap \partial_{tail}B & \partial_{head}A \cap B^{\circ} & \partial_{head}A \cap \partial_{head}B \end{pmatrix} \quad (3)$$

This matrix is called the *9-intersection matrix for arrow symbols*. The first, second, and third row correspond to A's tail, body, and head, while the first, second, and third column correspond to B's tail, body, and head, respectively. For example, since the two arrow symbols in Fig. 3b intersect only at one's body and another's tail, their 9-intersection matrix has only one non-empty element at $A^{\circ} \cap \partial_{tail}B$.

We first capture the topological relations between two arrow symbols by the existence/non-existence of these nine types of intersections alone. The existence/non-existence of each type of intersections is characterized by respective empty/non-empty entries in the 9-intersection matrix (Fig. 4).

$$\begin{pmatrix} \phi & \phi & \phi \\ \phi & \phi & \phi \\ \phi & \phi & \neg\phi \end{pmatrix} \quad \begin{pmatrix} \phi & \phi & \phi \\ \neg\phi & \phi & \phi \\ \phi & \phi & \phi \end{pmatrix} \quad \begin{pmatrix} \phi & \phi & \phi \\ \phi & \phi & \phi \\ \phi & \phi & \phi \end{pmatrix} \quad \begin{pmatrix} \phi & \phi & \phi \\ \phi & \phi & \phi \\ \phi & \phi & \phi \end{pmatrix} \quad \begin{pmatrix} \phi & \phi & \neg\phi \\ \phi & \phi & \phi \\ \phi & \phi & \phi \end{pmatrix} \quad \begin{pmatrix} \neg\phi & \phi & \phi \\ \phi & \phi & \phi \\ \phi & \phi & \phi \end{pmatrix}$$
$$\quad\text{(a)} \qquad\qquad \text{(b)} \qquad\qquad \text{(c)} \qquad\qquad \text{(d)} \qquad\qquad \text{(e)} \qquad\qquad \text{(f)}$$

Fig. 4. The 9-intersection matrices that capture the topological relations between two arrow symbols in Figs. 3a-f only in terms of intersections

Although the 9-intersection matrix distinguishes $2^9 = 512$ configurations with different empty/non-empty entries, not all configurations have geometric realizations. The head or tail of an arrow symbol is a point, which cannot intersect with more than one part of another arrow symbol that does not intersect with itself. Since we assumed that no arrow symbol intersect with itself, this condition leads to the following constraint on the 9-intersection matrix for two arrow symbols:

- *The first column, third column, first row, and third row have at most one non-empty element.*

On the other hand, the center cell (i.e., $A° \cap B°$) can freely be empty or non-empty. Among the 512 potential configurations of matrices, only 68 configurations satisfy this constraint (Table 1). Each of the 68 configurations corresponds to a different topological relation between arrow symbols in terms of intersections, some of which are shown in Fig. 5.

Table 1. Number of configurations of the 9-intersection matrices with geometric realizations

		Number of non-empty cells except $A° \cap B°$					
		0	1	2	3	4	
$A° \cap B°$	Empty	1	8	16	8	1	34
	non-empty	1	8	16	8	1	34
		2	16	32	16	2	68

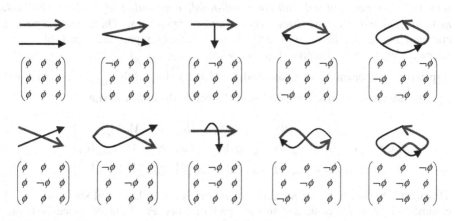

$$\begin{pmatrix} \phi & \phi & \phi \\ \phi & \phi & \phi \\ \phi & \phi & \phi \end{pmatrix} \quad \begin{pmatrix} \neg\phi & \phi & \phi \\ \phi & \phi & \phi \\ \phi & \phi & \phi \end{pmatrix} \quad \begin{pmatrix} \phi & \neg\phi & \phi \\ \phi & \phi & \phi \\ \phi & \phi & \phi \end{pmatrix} \quad \begin{pmatrix} \phi & \phi & \neg\phi \\ \phi & \phi & \phi \\ \neg\phi & \phi & \phi \end{pmatrix} \quad \begin{pmatrix} \phi & \phi & \neg\phi \\ \neg\phi & \phi & \phi \\ \phi & \neg\phi & \phi \end{pmatrix}$$

$$\begin{pmatrix} \phi & \phi & \phi \\ \phi & \neg\phi & \phi \\ \phi & \phi & \phi \end{pmatrix} \quad \begin{pmatrix} \neg\phi & \phi & \phi \\ \phi & \neg\phi & \phi \\ \phi & \phi & \phi \end{pmatrix} \quad \begin{pmatrix} \phi & \neg\phi & \phi \\ \phi & \neg\phi & \phi \\ \phi & \phi & \phi \end{pmatrix} \quad \begin{pmatrix} \phi & \phi & \neg\phi \\ \phi & \neg\phi & \phi \\ \neg\phi & \phi & \phi \end{pmatrix} \quad \begin{pmatrix} \phi & \phi & \neg\phi \\ \neg\phi & \neg\phi & \phi \\ \phi & \neg\phi & \phi \end{pmatrix}$$

Fig. 5. Examples of 9-intersection matrices that represent the topological relations between two arrow symbols in terms of intersections. Components referred by arrow symbols are not shown in this diagram.

Among the 68 topological relations, 20 relations are symmetric, while 48 relations have a converse relation. Depending on the existence/non-existence of a body-body intersection, these 68 relations are divided into two halves with one-to-one correspondences between them. For example, in Fig. 5, the upper five relations, each without a body-body intersection, have one-to-one correspondences to the lower five relations, each with a body-body intersection.

3.2 Topological Relations Established by Common References

In a 1-arrow diagram, components are located in front of, behind, or along the arrow symbol; therefore, an arrow symbol defines three different areas where its components can be located, called the *head slot*, *tail slot*, and *body slot* of the arrow diagram (Kurata and Egenhofer 2005a) (Fig. 6). This structure is extended for multi-arrow diagrams, such that each arrow symbol in a multi-arrow diagram individually defines its three slots. Consequently, an *n*-arrow diagram has $3n$ slots. Since we assumed that an arrow symbol does not refer to the same component more than once, the three slots of one arrow symbol cannot overlap with each other, whereas the slots of different arrow symbols may overlap and contain the same component.

Tail slot   *Body slot*   *Head slot*

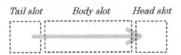

Fig. 6. Three component slots associated with each arrow symbol

A *common reference* is established when two arrow symbols refer to the same component, called the *common component*. If two arrow symbols, *A* and *B*, have a common reference, their common component is contained in *A*'s tail, body, or head slot, as well as in *B*'s tail, body, or head slot. Accordingly, based on the combinations of the slots that contain the common component, common references are classfied into $3 \times 3 = 9$ types: *tail-tail, tail-body, tail-head, body-tail, body-body, body-head, head-tail, head-body,* and *head-head common references*. These nine types of common references between *A* and *B* are concisely represented, just like their intersections, by a 3×3 matrix (Eqn. 4), where $c_{tail}(A)$, $c_{body}(A)$, and $c_{head}(A)$ are the respective components in *A*'s tail, body, and head slot, while $c_{tail}(B)$, $c_{body}(B)$, and $c_{head}(B)$ are the respective components in *B*'s tail, body, and head slot.

$$M_C(A,B) = \begin{pmatrix} c_{tail}(A) \cap c_{tail}(B) & c_{tail}(A) \cap c_{body}(B) & c_{tail}(A) \cap c_{head}(B) \\ c_{body}(A) \cap c_{tail}(B) & c_{body}(A) \cap c_{body}(B) & c_{body}(A) \cap c_{head}(B) \\ c_{head}(A) \cap c_{tail}(B) & c_{head}(A) \cap c_{body}(B) & c_{head}(A) \cap c_{head}(B) \end{pmatrix} \qquad (4)$$

This matrix is called the *9-CR matrix for arrow symbols*. Each cell shows the set of common components contained in each pair of slots. For instance, since both two arrow symbols in Fig. 3c contain *Inspection* in their tail slots, its 9-CR matrix has a non-empty element at $c_{tail}(A) \cap c_{tail}(B)$.

Since common references connect two arrow symbols in the same way intersections do, the topological relations between arrow symbols are captured in a similar way by the existence/non-existence of the nine types of common references. The existence/non-existence of the nine types of common references is characterized by respective empty/non-empty entries in the 9-CR matrix (Fig. 7). Although the 9-CR matrix distinguishes $2^9 = 512$ configurations with different empty/non-empty entries, not all these configurations have geometric realizations. The head or tail slot of an arrow symbol may contain more than one component (say, an icon and its caption), but it is unrealistic that these components belong to the different slots of another arrow symbol, since these components are located at the same (or undistinguishable) position pointed by the arrow symbol's head or tail. On the other hand, the components in the body slot of one arrow symbol can be contained in two different slots of another arrow symbol, since these components can be located at different positions. Since we assumed that no arrow symbol refers to the same component more than once, the following constraint on the 9-CR matrix is derived:

- *The first column, third column, first row, and third row have at most one non-empty element.*

This constraint is identical to that of the 9-intersection matrix for arrow symbols. Accordingly, among the 512 potential configurations of matrices, 68 configurations satisfy this condition. Each of these 68 configurations corresponds to different topological relation between arrow symbols in terms of common references.

$$
\begin{pmatrix} \phi & \phi & \phi \\ \phi & \phi & \phi \\ \phi & \phi & \phi \end{pmatrix}
\begin{pmatrix} \phi & \phi & \phi \\ \phi & \phi & \phi \\ \phi & \phi & \phi \end{pmatrix}
\begin{pmatrix} \neg\phi & \phi & \phi \\ \phi & \phi & \phi \\ \phi & \phi & \phi \end{pmatrix}
\begin{pmatrix} \phi & \phi & \neg\phi \\ \phi & \phi & \phi \\ \neg\phi & \phi & \phi \end{pmatrix}
\begin{pmatrix} \phi & \phi & \phi \\ \phi & \phi & \phi \\ \phi & \neg\phi & \phi \end{pmatrix}
\begin{pmatrix} \neg\phi & \phi & \phi \\ \phi & \phi & \phi \\ \phi & \phi & \phi \end{pmatrix}
$$

$$\qquad\text{(a)}\qquad\qquad\text{(b)}\qquad\qquad\text{(c)}\qquad\qquad\text{(d)}\qquad\qquad\text{(e)}\qquad\qquad\text{(f)}$$

Fig. 7. The 9-CR matrices that capture the topological relations between two arrow symbols in Figs. 3a-f in terms of common references

3.3 Integration of 9-Intersection Matrix and 9-CR Matrix

Intersections and common references are analogous in the sense that both associate two arrow symbols by connecting the tail, body, or head of one arrow symbol to the tail, body, or head of another arrow symbol. Intersections and common references are, therefore, generically called *links*. Based on the combination of the connected parts of two arrow symbols, links are classified into $3\times3 = 9$ types: *tail-tail, tail-body, tail-head, body-tail, body-body, body-head, head-tail, head-body,* and *head-head links*. Each type of link is further categorized into *direct links* (= intersections) and *indirect links* (= common references).

The intersections and common references of two arrow symbols A and B are represented by two matrices $M_I(A,B)$ and $M_C(A,B)$, respectively. These two matrices are easily integrated into a single 3×3 matrix (Eqn. 5), where $m_{I\,ij}$ and $m_{C\,ij}$ are the respective elements of $M_I(A,B)$ and $M_C(A,B)$ at (i,j) $(i,j \in \{1,2,3\})$.

$$M_L(A,B)=[m_{L_{ij}}]$$

$$m_{L_{ij}} = \begin{cases} \phi & \text{if } m_{I_{ij}}=\phi & \text{and} & m_{C_{ij}}=\phi \\ I & \text{if } m_{I_{ij}}=\neg\phi & \text{and} & m_{C_{ij}}=\phi \\ C & \text{if } m_{I_{ij}}=\phi & \text{and} & m_{C_{ij}}=\neg\phi \\ IC & \text{if } m_{I_{ij}}=\neg\phi & \text{and} & m_{C_{ij}}=\neg\phi \end{cases} \qquad (5)$$

$M_L(A,B)$ is called the *9-link matrix for arrow symbols*, because each cell of this matrix indicates the existence of links at each position pair and their types. For instance, the arrow symbol pairs in Figs. 3a-f correspond to the 9-link matrices in Figs. 8a-f. A merit of this hybrid matrix is that the topological relation between two arrow symbols is described by a single matrix even when two arrow symbols have both intersections and common references (Figs. 3e-f).

$$\begin{bmatrix} \phi & \phi & \phi \\ \phi & \phi & \phi \\ \phi & \phi & I \end{bmatrix} \quad \begin{bmatrix} \phi & \phi & \phi \\ I & \phi & \phi \\ \phi & \phi & \phi \end{bmatrix} \quad \begin{bmatrix} C & \phi & \phi \\ \phi & \phi & \phi \\ \phi & \phi & \phi \end{bmatrix} \quad \begin{bmatrix} \phi & \phi & C \\ \phi & \phi & \phi \\ C & \phi & \phi \end{bmatrix} \quad \begin{bmatrix} \phi & \phi & I \\ \phi & \phi & \phi \\ \phi & C & \phi \end{bmatrix} \quad \begin{bmatrix} IC & \phi & \phi \\ \phi & \phi & \phi \\ \phi & \phi & \phi \end{bmatrix}$$
$$\quad\text{(a)}\qquad\qquad\text{(b)}\qquad\qquad\text{(c)}\qquad\qquad\text{(d)}\qquad\qquad\text{(e)}\qquad\qquad\text{(f)}$$

Fig. 8. The 9-link matrices that capture the topological relations between two arrow symbols in Figs. 3a-f in terms of both intersections and common references

Since each cell is four-valued (ϕ, I, C, or IC), the 9-link matrix distinguishes $4^9 = 262,144$ configurations, although not all configurations have geometric realizations. The tail or head of an arrow symbol is a point and, therefore, cannot be linked to two different parts of another arrow symbol that neither intersect with itself nor refer to the same component more than once. The tail or head of an arrow symbol, however, may have both an intersection and a common reference with the same part of another arrow symbol (Fig. 3f). Thus, the following constraint on the 9-link matrix is derived:

- *The first column, third column, first row, and third row have at most one non-empty element.*

This constraint is, again, identical to that of the 9-intersection matrix for arrow symbols. Among the 262,144 potential configurations of matrices, 1,864 configurations satisfy this condition (Table 2), each of which corresponds to a different topological relation. Among the 1,864 topological relations, 184 relations are symmetric, while 1,680 relations have a converse relation.

Table 2. Number of configurations of the 9-link matrices with geometric realizations

		Number of non-empty cells except m_{L22}					
		0	1	2	3	4	
m_{L22}	empty	1	8×3^1	16×3^2	8×3^3	1×3^4	466
	non-empty	1×3^1	8×3^2	16×3^3	8×3^4	1×3^5	1,374
		4	96	576	864	324	1,864

In this way, we developed a model for capturing topological relations between arrow symbols in a unified way. This model is called the *9-link model for arrow symbols*. The 9-link model deals with both direct links (intersections) and indirect links (common references) between spatial objects, whereas the 9-intersection model deals with the direct links.

4 Topological Relations and Semantics

The meaning of multi-arrow diagrams is influenced by the arrow symbols' relations. Our premise is that in a multi-arrow diagram each arrow symbol illustrates *atomic semantics* together with its related components and links of the arrow symbols indicates interrelation between these atomic semantics. This section discusses what kinds of interrelations between atomic semantics are indicated by intersections, common references, and their combinations.

Before starting the discussion, we have to be careful about the *nested structure* in a multi-arrow diagram. An arrow symbol in a multi-arrow diagram sometimes refers to an inner arrow diagram instead of individual components, thereby forming a nested structure. In Fig. 9a, for example, the arrow symbol departing from *El Niño* refers not to *Fish catch*, but to the inner arrow diagram composed of *Fish catch* and a downward arrow symbol, which illustrates the decrease of fish catch. Accordingly, this diagram illustrates a dynamic process that El Niño triggers the decrease of fish catch. The use of such nested structures enriches the representation ability of multi-arrow diagrams. A problem is that the existence of a nested structure is not visually distinctive—in multi-arrow diagrams with a nested structure, arrow symbols apparently have a common reference (Fig. 9a) or an intersection (Fig. 9b). To focus the study on the fundamental aspects of arrow diagrams, the following discussion only deal with multi-arrow diagrams without nested structures.

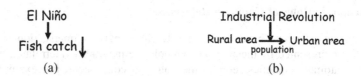

(a)   (b)

Fig. 9. Multi-arrow diagrams with a nested structure, which apparently have (a) a common reference to the label "*Fish catch*" or (b) an head-body intersection

4.1 Semantic Roles of Intersections

If two arrow symbols have intersections but no common references, each arrow symbol typically represents an *action* (i.e., movement of one component, sometimes triggered by or/and triggering an interaction with another component), and each intersection indicates an interrelation between such actions. In Fig. 3b, for example, each arrow symbol represents the movement to the main gate or temporal gate. Then, the body-tail intersection between these arrow symbols indicates that the movement to the temporal gate takes over a part of the mover to the main gate (i.e., a group of people). Like this example, each type of intersection indicates the following interrelations:

- A head-tail or tail-head intersection typically indicates that the tail-side arrow symbol completely takes over the mover of the head-side arrow symbol (Fig. 3e).
- A body-tail or tail-body intersection typically indicates that the tail-side arrow symbol partially takes over the mover of the body-side arrow symbol (Figs. 3b).
- A head-head intersection indicates that the movers of two arrow symbols meet and probably interact (Fig. 3a).
- A head-body or body-head intersection typically indicates that the mover of the head-side arrow symbol may merge with or influence the mover of the body-side arrow symbol.
- A tail-tail intersection typically indicates that the movers of two arrow symbols move away from each other, probably as a result of a certain event.
- A body-body intersection may indicate the interaction between the movers of two arrow symbols. Otherwise, the arrow symbols happened to cross at the body-body intersection.

These correspondences between intersections and interrelations of actions are summarized in a 3×3 matrix (Eqn. 9), following the structure of the 9-intersection matrix for arrow symbols. This correspondence, however, does not mean that the existence of each intersection always leads to the corresponding interpretation. For instance, body-tail intersection is occasionally used to indicate alternative scenarios of actions or events, without mentioning *partial takeover*.

$$\begin{pmatrix} \text{separation} & \text{parital takeover} & \text{complete takeover} \\ \text{partial takeover} & \text{(interaction)} & \text{merge or influence} \\ \text{complete takeover} & \text{merge or influence} & \text{meet (+interaction)} \end{pmatrix} \quad (6)$$

4.2 Semantic Roles of Common References

If two arrow symbols have common references but no intersections, the atomic semantics represented by these arrow symbols are inevitably interrelated in the sense that both atomic semantics refer to the common component. For example, the two atomic semantics of Fig. 10a—*the label "Mr. K" is assigned to the traveler* (which in turn is interpreted as *the traveler is Mr. K*) and *the person goes to Hawaii* (Fig. 10a')—are interrelated in the sense that they refer to the same traveler.

In addition to such weak interrelations, the atomic semantics may be strongly interrelated through mutual interference. For example, Fig. 10b illustrates *an exam results in pass or fail*, where its two atomic semantics, *an exam results in pass* and *an exam results in fail* (Fig. 10b'), are *mutually-exclusive* (i.e., they cannot be true at the same time). On the other hand, Fig. 10c illustrates *a cell phone sends a query to a database, which then returns a search result*, where its two atomic semantics, *a cell phone sends a query to a database* and *a database sends a query to a cell phone* (Fig. 10c'), are *synchronized* (i.e., whenever one is true, the other is also true). In this way, if arrow symbols have a common reference, the atomic semantics may be mutually-exclusive or synchronized. Which type of interference actually holds cannot be determined without background knowledge about, for example, whether the illustrated events may or must occur simultaneously. According to our observation, however, atomic semantics tend to interfere with each other when arrow symbols are symmetrically aligned (Figs. 10b-c). Consequently, symmetry of the 9-CR matrix may be useful for judging the possibility of such semantic interferences.

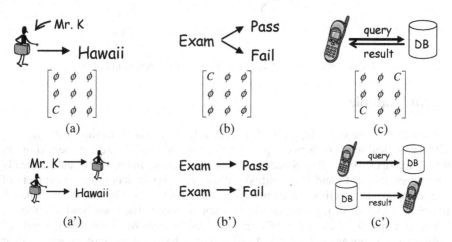

Fig. 10. (a-c) Examples of 2-arrow diagrams with common references and their 9-link matrices, and (a'-c') inner arrow diagrams of Fig. 11a-c, which illustrate the atomic semantics of the original 2-arrow diagrams

4.3 Semantic Roles of Coexisting Intersections and Common References

If two arrow symbols have both an intersection and a common reference, which connect exactly at the same positions of these arrow symbols, these two types of links work essentially as a single common reference (Fig. 3f). This is because two arrow symbols sometimes intersect unnecessarily when referring to a common component, and accordingly it appears that both an intersection and a common reference connect the same positions of two arrow symbols. Consequently, when the 9-link matrix contains IC in any cell except the center, the symbol IC can be replaced by I without changing the diagram's semantics (Fig. 8f).

On the other hand, if different positions of two arrow symbols are connected by an intersection and a common reference (i.e., the 9-link matrix has I and C in different cells, or possibly IC in the center), these two links should have individual semantic roles. Thus, the intersection of such arrow symbols implies that these arrow symbols illustrate the interrelated actions. For example, Fig. 3e illustrates *an airplane flies over a landing strip and then lands on it*, where its head-tail intersection indicates that the tail-side arrow symbol completely takes over the mover of the head-side arrow symbol (i.e., an airplane). On the other hand, the common reference of the arrow symbols always implies the *synchronization* of two actions, since the actions must occur simultaneously or continuously in order to have an interrelation. For example, the atomic semantics of Fig. 3e, *an airplane flies over a landing strip* and *something lands on a landing strip* (Fig. 11), must be synchronized such that the latter action takes over the mover of the former actions (i.e., the airplane). In this way, the combination of an intersection and a common reference at different positions indicate that the arrow diagram illustrates two interrelated and synchronized actions.

Fig. 11. Inner arrow diagrams of the 2-arrow diagram in Fig. 3e

5   An Example

Let the four arrow symbols in Fig. 2 be called *A*, *B*, *C*, and *D*, as shown in Fig. 12a. The topological relations between pairs of these arrow symbols are represented by six 9-link matrices (Fig. 12b). Since *D* intersects with *A* and *C*, it is automatically determined that *A*, *C*, and *D* individually illustrate a certain action (i.e., movement of one component sometimes accompanying an interaction). On the other hand, *B* illustrates a *change* of ingredients, which is not an action. The body-head intersection between *A* and *D* indicates that *D*'s mover (*2 tablespoons of yogurt*) merges with or influences A's mover (*milk*). Since *A* and *B* have a head-tail common reference, their semantics are interrelated in the sense that they refer to the same container, although these semantics are not mutually-exclusive or synchronized. Similarly, *B* and *C* have a head-tail common reference, which implies an interrelation between their semantics in terms of subject-sharing. Finally, the body-tail intersection between *C* and *D* indicates that *D* partially takes over *C*'s mover (*yogurt*). This interpretation is supported by the label on *D* (*2 tablespoons*).

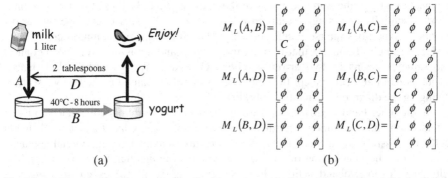

(a)   (b)

Fig. 12. (a) The 4-arrow diagram in Fig. 2 with arrow symbol identifiers *A-D* and (b) the 9-link matrices representing the topological relations between pairs of these arrow symbols

6   Conclusions

People often communicate dynamic information through multi-arrow diagrams. This paper formalized the topological relations of arrow symbols embedded in such multi-arrow diagrams from two viewpoints: intersections and common references. Then, we observed how these topological relations influence the semantics of multi-arrow diagrams. This work forms a basis for future research toward a computational model for interpreting complex arrow-containing diagrams. In order to achieve this goal, further extensions are needed to detect nested structures and to determine the type of semantic interference implied by common references. In addition, the correspondence between topological relations and semantics should be carefully examined with more examples of arrow diagrams and some systematic human subject experiments. Another item for future research is constructing an interpretation framework that may bind the relation between arrow symbols, instead of assigning a fixed semantics to each topological relation based on examples.

Since this paper limited the target to topological relations between arrow symbols, it should be meaningful to study the influence of other spatial relations, such as metric or directional relations, to the semantics.

Our discussions implicitly assumed that an arrow symbol always refers to the entire component. Arrow diagrams, however, sometimes have a *hierarchical common reference*, where one arrow symbol refers to a component while another arrow symbol refers to a part of the same component (Fig. 13). This setting leads to the possibility to extend our 9-link model, which will capture more detailed topological relations between arrow symbols.

Fig. 13. An arrow diagram with a hierarchical common reference. The thicker arrow symbol refers to a person carrying a bag, while the narrower arrow symbol refers only to the bag.

Another challenging problem is to analyze the conceptual neighborhoods of the topological relations between arrow symbols that we identified and to develop their composition table on these relations as a basis for qualitative spatial reasoning. Such problems are well-studied for regions (Egenhofer 1994b, 2005), but not yet for arrow symbols and even for directed line segments, which are often represented by arrow symbols. The study of these problems would probably leads to the findings of interesting properties of the topological relations between arrow symbols.

Acknowledgements

This work was partially supported by the National Geospatial-Intelligence Agency under grant number NMA201-01-1-2003. Max Egenhofer's research is further supported by the National Science Foundation under grant numbers EPS-9983432 and IIS-9970123; the National Geospatial-Intelligence Agency under grant numbers NMA201-00-1-2009, and NMA401-02-1-2009, and the National Institute of Environmental Health Sciences, NIH, under grant number 1 R 01 ES09816-01.

References

Allen, J. (1983) Maintaining Knowledge about Temporal Intervals. *Communications of the ACM*. 26(11):832-843.

Barraclough, G. (ed.) (2003) *Collins Atlas of World History*. 2nd revised edition. Borders Press: Ann Arbor, MI.

Clementini, E. and Di Felice, P. (1998) Topological Invariants for Lines. *IEEE Transactions on Knowledge and Data Engineering*. 10(1):38-54.

Egenhofer, M. (1994a) Definitions of Line-Line Relations for Geographic Databases. *IEEE Data Engineering Bulletin*. 16(3):40-45.

Egenhofer, M. (1994b) Deriving the Composition of Binary Topological Relations. *Journal of Visual Languages and Computing*. 5(2):133-149.

Egenhofer, M. (2005) Spherical Topological Relations. *Journal on Data Semantics III*. 25-49.

Egenhofer, M. and Franzosa, R. (1991) Point-Set Topological Spatial Relations. *International Journal of Geographical Information Systems*. 5(2):161-174.

Egenhofer, M. and Herring, J. (1991) Categorizing Binary Topological Relationships between Regions, Lines and Points in Geographic Databases. in: Egenhofer, M., Herring, J., Smith, T. and Park, K. (eds.) *A Framework for the Definitions of Topological Relationships and an Algebraic Approach to Spatial Reasoning within This Framework, NCGIA Technical Reports 91-7*. National Center for Geographic Information and Analysis: Santa Barbara, CA.

Egenhofer, M. and Mark, D. (1995) Naive Geography. in: Frank, A. and Kuhn, W. (eds.) *COSIT '95*, Semmering, Austria, Lecture Notes in Computer Science, 988, 1-15.

Horn, R. (1998) *Visual Language: Global Communication for the 21st Century*. MacroVu, Inc.: Bainbridge Island, WA.

Hornsby, K., Egenhofer, M. and Hayes, P. (1999) Modeling Cyclic Change. in: Chen, P., Embley, D., Kouloumdjian, J., Liddle, S. and Roddick, J. (eds.) *Advances in Conceptual Modeling*, Paris, France, Lecture Notes in Computer Science, 1227, 98-109.

Kurata, Y. and Egenhofer, M. (2005a) Semantics of Simple Arrow Diagrams. in: Barkowsky, T., Freksa, C., Hegarty, M. and Lowe, R. (eds.) *AAAI Spring Symposium on Reasoning with Mental and External Diagram: Computational Modeling and Spatial Assistance*, Menlo Park, CA, 101-104.

Kurata, Y. and Egenhofer, M. (2005b) Structure and Semantics of Arrow Diagrams. in: Cohn, A. and Mark, D. (eds.) *COSIT '05*, Ellicottville, NY, Lecture Notes in Computer Science, 3693, 232-250.

Moratz, R., Renz, J. and Wolter, D. (2000) Qualitative Spatial Reasoning about Line Segments. in: Horn, W. (ed.) *14th European Conference on Artificial Intelligence*, Berlin, 234-238.

Nedas, K., Egenhofer, M. and Wilmsen, D. (in press) Metric Details of Topological Line-Line Relations. *International Journal of Geographical Information Science*. http://www.spatial.maine.edu/~max/RJ53.html.

Renz, J. (2001) A Spatial Odyssey of the Interval Algebra: 1. Directed Intervals. in: Nebel, B. (ed.) *International Joint Conference on Artificial Intelligence 2001*, Seattle, WA, 51-56.

Schlieder, C. (1995) Reasoning about Ordering. in: Frank, A. and Kuhn, W. (eds.) *COSIT'95*, Lecture Notes in Computer Science, 341-349.

Tversky, B., Heiser, J., Lozano, S., MacKenzie, R. and Morrison, J. (in press) Enriching Animations. in: Lowe, R. and Schnotz, W. (eds.) *Learning with animation: Research and application*. New York: Cambridge University Press.

Extended Abstract of *Euclid and His Twentieth Century Rivals: Diagrams in the Logic of Euclidean Geometry*

Nathaniel Miller

University of Northern Colorado, Department of Mathematical Sciences
Ross Hall, Greeley, CO 80639
nat@alumni.princeton.edu

This extended abstract describes a forthcoming book which should be of interest to those attending the Diagrams 2006 conference and to others generally interested in diagrammatic reasoning in the context of Euclidean Geometry. The book, *Euclid and His Twentieth Century Rivals: Diagrams in the Logic of Euclidean Geometry*[4], is still in preparation, but will be published by CSLI press once it is completed.

In 1879, the English mathematician Charles Dodgson, better know to the world under his pen name of Lewis Carroll, published a little book entitled *Euclid and His Modern Rivals*. [1] Dodgson was concerned by the fact that quite a number of different nineteenth century authors had written their own treatments of planar geometry, most claiming to improve on Euclid, and each one slightly different in the order of its theorems, in which theorems it chose to include, in the proofs given of these theorems, in its treatment of the theory of parallel lines, and in other aspects. Dodgson's book was written "[i]n furtherance of the great cause which I have at heart—the vindication of Euclid's masterpiece...." It is written mostly in the form of a dream dialogue between a nineteenth century mathematician, Minos, and the ghost of Euclid. In it, they consider each of the modern rivals in turn, and conclude in each case that, while many of the rivals have interesting things to say, none of them are a more appropriate basis for the study of a beginning geometry student than Euclid's *Elements*.

At the time at which Dodgson wrote his book, the subjects of geometry and logic were both entering a period of rapid change after having remained relatively constant for two thousand years. There had been enough change already to make Dodgson feel that Euclid needed defending. In the hundred and twenty-five years since then, however, there have been much larger changes in these fields, and, as a result, rather than just undergoing some small changes, Euclidean geometry in general, and Euclid's proofs in particular, have mostly fallen out of the standard mathematics curriculum. This is at least in part because Euclid's *Elements*[2], which was viewed for most of its existence as being the gold standard of careful reasoning and mathematical rigor, has come to be viewed as being inherently and unsalvageably informal and unrigorous. One reason for this is certainly its essential use of geometric diagrams. Around the turn of the twentieth century, David Hilbert published his *Foundations of Geometry*[3], which fit more easily into the new framework of mathematical logic that was developing at that time, and did not use diagrams except as an explanatory tool. Since that time, most mathematicians have viewed Hilbert's *Foundations* rather than Euclid's *Elements* as the proper and more solid foundation of Euclidean Geometry.

Euclid and His Twentieth Century Rivals traces this history and tries to explain some of the reasons why this happened. It then brings twentieth century logical methods to

D. Barker-Plummer et al. (Eds.): Diagrams 2006, LNAI 4045, pp. 127–129, 2006.

bear on Euclid's proof methods, and shows how such methods can make Euclid's style of diagrammatic proofs as rigorous as Hilbert's proofs, which do not employ diagrams. A formal system, **FG**, is defined to manipulate geometric diagrams, and this system is used to study Euclid's informal methods of proof, methods that are still commonly used in informal treatments of geometry today. It is argued that, because they can be made rigorous in a formal system, these methods are in fact valid modes of argument.

Implicit in this method of evaluating an informal mode of argument by replacing it with an essentially similar formal system is what I call the *formality hypothesis*:

Hypothesis 1 (Formality Hypothesis). *An informal proof method is sound if and only if it is possible to give a formal system with the property that informal proofs using the informal methods can always be translated into equivalent correct proofs in the formal system.*

By a formal system, we mean a system of argument that is carefully enough specified that it can be implemented on a computer. Modern ideas of logic give us the means of saying, at least in principle, what it means for such a formal system to be sound—that is, whether or not it will only lead to correct conclusions. They don't give us, by themselves, the means to say what it means for an informal method of proof to be correct; but the formality hypothesis gives us a basis for evaluating such informal methods by translating them into corresponding formal methods.

Euclid and His Twentieth Century Rivals analyses Euclid's proof methods in this way, by translating them into a formal system, **FG**, which is strong enough and similar enough to what Euclid does to be able to formalize most of what is contained in the first four books of Euclid's *Elements*, which are the part of the *Elements* that deal with planar geometry. **FG** accomplishes this by giving a careful, formal definition of what constitutes a geometric diagram in this context, and then by giving careful rules for manipulating such diagrams. The definition of what constitutes a diagram is quite complicated, but it captures a comparatively simple and ancient idea: two diagrams are the same if they share the same topology. That is, if one can be stretched without tearing into the other; or, put another way, that all of the points and lines are arranged in the same way. This idea can already be found in the writings of Proclus, a fifth century commentator on Euclid, who writes that each case in a geometric proof "announces different ways of construction and alteration of positions due to the transposition of points or lines or planes or solids." (This is Sir Thomas Heath's translation, as given in [2].) A computer system, **CDEG**, which implements this formal system on a computer in order to demonstrate that it is, in fact, completely formal, is also discussed.

The book also discusses several other ways that we can use this formal system to better understand the history and practice of the use of diagrams in informal treatments of geometry. For example, commentators writing about Euclid's *Elements* often note that he uses the principle of superposition to prove the Side-Angle-Side rule and from then on uses this derived rule in place of the principle of superposition whenever possible, and conclude that Euclid must have thought that this principle was somehow suspect. However, *Euclid and His Twentieth Century Rivals* uses the formal system **FG** to show that there are very good reasons to use the Side-Angle-Side rule, once you have derived it, in place of the principle of superposition, even if there is nothing suspect about the principle. Furthermore, it is shown that by weakening the principle of superposition slightly, other weaker systems of geometry can be obtained in which the Side-Angle-Side rule is not derivable, and that these weaker geometries have a fascinating logical structure of their own.

Euclid and His Twentieth Century Rivals also discusses the significance of several technical results relating to the computational complexity of working with diagrams in the formal system **FG**, results published elsewhere in the *Journal of Complexity* [5]. The field of computational complexity measures how much time and other resources are necessary to do a given computation. Such results are highly relevant to the study of diagrammatic reasoning because once we know that arguments employing diagrams are logically sound, we would then like to know whether or not they require more time and resources than arguments that do not employ diagrams. In [5], it is shown that the ***diagram satisfaction problem*** of whether or not a diagram represents a physically possible arrangement of points and lines is solvable by computer in a finite amount of time, but that this problem is at least NP-hard, meaning that it can take an intractably large amount of time to solve in practice.[1] What does this mean? It means that, while diagram satisfiability can theoretically be decided by a computer, and case analysis in geometry can therefore be theoretically be done by a computer in a way that will never return unsatisfiable diagrams, it is nonetheless impractical to do so, and any real world computer system or formal system will therefore sometimes unavoidably return extra unsatisfiable cases. This shows us that the long tradition of considering and disposing of extra cases in geometry is unavoidable in a diagrammatic approach.

The formalization of Euclid's proof methods in the formal system **FG** is therefore useful for several different reasons: it lets us better understand his proofs; it allows us to prove metamathematical results about these kinds of proofs; and it shows that there is no inherent reason that the modern foundations of geometry must look completely different from the ancient foundations found in the *Elements*. Thus, the aims of *Euclid and His Twentieth Century Rivals* are not far removed from Dodgson's aims in 1879: to show that, while modern developments in logic and geometry may require changes in Euclid's development, his basic ideas are neither outdated nor obsolete.

References

1. Dodgson, Charles, *Euclid and His Modern Rivals*, 2nd edn, London: MacMillan and Co., 1885.
2. Euclid, *The Elements*, Translated with introduction and commentary by Thomas L. Heath, 2nd edn, New York: Dover, 1956.
3. Hilbert, David, *Foundations of Geometry*, 4th edn, Translated by Leo Unger, La Salle, Ill.: Open Court Publishing Co., 1971.
4. Miller, Nathaniel, *Euclid and His Twentieth Century Rivals: Diagrams in the Logic of Euclidean Geometry* (working title), Stanford, CA: CSLI Publications, to appear.
5. Miller, Nathaniel, Computational complexity of diagram satisfaction in Euclidean geometry. *Journal of Complexity* **22** (2006): 250–274.

---

[1] More specifically, it is shown in [5] that this problem, in addition to being at least NP-hard, is in the complexity class PSPACE, and in fact has the same complexity as the satisfaction problem for a particular fragment of the first order existential theory of the reals.

Flow Diagrams: Rise and Fall of the First Software Engineering Notation

S.J. Morris[1] and O.C.Z. Gotel[2]

[1] Department of Computing, City University, London
sjm@soi.city.ac.uk
[2] Department of Computer Science, Pace University, New York
ogotel@pace.edu

Abstract. Drawings of water are the earliest, least abstract forms of flow diagram. Representations of ideal or generalised sequences for manufacturing or actual paths for materials between machines came next. Subsequently documentation of production and information flow become subjects for graphical representation. A similar level of abstraction was necessary for representations of invisible flows such as electricity. After initial use to define control, flow diagrams became a general purpose tool for planning automated computation at all levels of composition. Proliferation of syntax variants and the need for a common language for documentation were the motivations behind standardisation efforts. Public communication of metalevel systems information superseded private comprehension of detailed algorithmic processes as a primary function. Changes to programming language structures and their associated processes caused the initial demise of flow diagrams in software engineering.

1 Introduction

This paper describes the origins of the flow diagram in engineering practice and charts the development of the underlying ideas as they became integral to early software engineering practice. It is the first in a two-part series examining the impact and legacy of flow conventions in contemporary software engineering diagrams.

In a paper written in 1946 Goldstine and von Neumann provided a succinct statement of a still not fully satisfied need in the field now know as software engineering, "An efficient and transparent logical terminology or symbolism for comprehending and expressing a problem, no matter how involved, in its entirety and in all its parts; and a simple and reliable step-by-step method to translate the problem ... into the code" [1]. A major part of their proposed approach [2] involved the drawing of 'flow diagrams' which came to form the work of all programmers until use of the first generation of higher level programming languages was widespread. The survival of this earliest form of special assistance into the time of manufacturing-like processes makes it one of the most long-lasting craft practices and suggests that its development might indicate typical characteristics for usefulness and longevity in software engineering notations.

D. Barker-Plummer et al. (Eds.): Diagrams 2006, LNAI 4045, pp. 130–144, 2006.

Graphical representations of material flows and, at a more abstract level, of process sequences were already commonplace and had fixed conventions by the time that the first electronic computers were being planned. Section 2 provides a selective view of these early manifestations. Section 3 sets out the contribution of von Neumann and Goldstine as described in their publications and uses archival material from the United Kingdom to outline developments in practice under their influence. Section 4 shows how the flow diagram notations were later both formalised in their syntax and altered in their application during the period when their role as a programming aid was diminishing. Section 5 outlines work examining the many progeny for which Goldstine and von Neumann's type of flow diagram is the sole ancestor.

2   Before Goldstine and von Neumann

2.1   Flow in Nature

Both as a hydraulic engineer and as a scientist fascinated by the workings of nature Leonardo da Vinci studied the flow of water throughout his life. A folio in the Codex Atlantico [3] shows details of a set of lock gates, part of Leonardo's Milanese work. On the lower part of one sheet there is a view of a weir and a cross-section of the sluice showing lines of motion past it. These drawings and their associated notes are certainly forms of engineering design, whilst others are clearly studies of water itself, always in motion and often impeded in some particular way.

A number of sheets now in the Royal Collection at Windsor Castle [4] all show a clear and rapidly moving body of water passing around an oblong obstacle, shown so that its position in relation to the flow is clear. Views from alternative positions also appear, in one case accompanied by notes comparing the movement to that of hair and describing the flows in terms of "whirling eddies, one part of which is due to the impetus of the principal current and the other to the incidental motion and return flow" [5].

The studies of water flow *per se* made by Leonardo emphasise the lines of movement and the shapes of vortices rather than phenomena such as spray or the play of light. They are illustrative figures which accompany texts and which, without representing exact appearance, give an outline or general scheme of what they represent. As such these drawings, concerned with the material that provides the first paradigm and metaphor for flow, are the earliest and least abstract forms of flow diagram. They are also early evidence of the heavy and continuous dependence of engineering on non-verbal learning and non-verbal understanding [6]. The developments that followed extended flow representations to other materials, man-made forces of movement and industrial processes, higher levels of abstraction and other rhetorical forms.

2.2   Course of Manufacture

The primary concern of Leonardo, shared by generations of engineers that followed him, was the harnessing of a natural force, the flow of water induced by gravitation, either as a source of power or in order to facilitate transport. The industrial revolution which began in the late eighteenth century added a broader interest in the motion of a wide variety of solid and liquid materials, in particular as manufacture transformed raw inputs into finished goods. In this context flow comes to mean, in the words of a textbook on flour manufacture, "the course through which any material travels whilst undergoing manufacture", and this course may be either some ideal and generalised sequence of manufacturing processes or an actual physical path from machine to machine "in their relative positions" or "giving the travel ... in the most direct manner irrespective of vertical or horizontal travel" [7].

Graphical representations of flows in this new manufacturing context normally have the title of 'flow sheet', a term that continued in general use, sometimes hyphenated, until the late 1940s and its supercession by 'flow diagram', by 'flow chart' and finally by 'flowchart'(as both noun and verb). The numbers of processes and machines involved often demanded a more general representation at some higher level of abstraction where the term 'block', or something similar term, came to be used. The same 1912 textbook refers to "care in 'block spacing' the main machines on the space available before filling in the flow lines. The skeleton or block flow-sheets ... emphasize the main masses into which each section of the several processes should be grouped ... the main operations are shown *en bloc*, and will serve as a frame upon which to hang the actual and of course more involved flow-sheet following ...". This notion of an hierarchy of representations, whether implicitly or explicitly defined and whether intended for illustration, or inscription of a design, has carried through to the most recent diagrammatic conventions, including those presented for software engineering.

2.3   Early Industrial Conventions

Early publications for chemical engineers show how the conventions used for material flows varied widely, particularly in the level of abstraction used in following engineering drawing conventions and in the deployment of arbitrary symbols to denote plant components or processes. Figure 1 shows a typical flow sheet published in 1943 [8] which incorporates standard symbols for a valve (similar to '><'), and for direction of flow (an arrow head), plus simple and complex graphical forms derived from engineering drawings in plan and cross section to denote pipes (single line) and evaporation equipment (labelled '1', '2' and '3').

Drawing derivatives persisted along with purely diagrammatic forms used to show abbreviated and abstracted descriptions. Figure 2 appeared in 1933 [9] to show the main operations within a process, described in detail in the surrounding text, for making pure potassium chlorine. This diagram uses the conventions which remain standard for 'top-down' sequence and layout, and for operations identified by title and isolated within rectangular boxes. These operations also have dual outputs and inputs, e.g. FILTRATION. At this level of abstraction the notion of a fixed sequence of processes becomes as important as any idea of the flow of material.

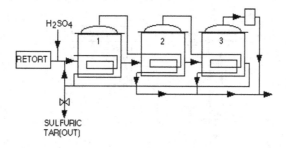

Fig. 1. 'Fig 16 Flow sheet of triple-effect evaporator for removing tars', 1943

Figure 3, originally also published in 1933 [10], is one of the earliest instances of the use of the term 'flow diagram' applied to a form akin to the current conventions. The text reads "Figure 7 represents preliminary flow sheets for a plant treating 2000 tons of raw polyhalite per day by processes 7 or 7A". In this diagram titles of operation are just underlined and boxes hold details of their outputs. Due to the number of inputs, products, by-products and residues, the top-down convention almost collapses and the direction of each flow arrow appears to be dictated merely by needs of layout. Such diagrams provide a means for presenting in an abbreviated graphical form the many stages in some complex chemical engineering.

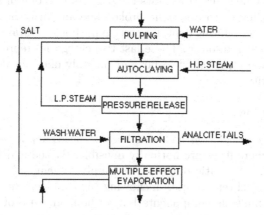

Fig. 2. Section from 'Figure 2. Flow sheet for Production of Potassium Chloride from Wyomingite', 1933

2.4   Abstraction of Flows

The OED attributes the first use of the term 'flow chart' to a book published in 1920 [11]. Although the figure to which the term is applied [12] is clearly what would now be called a Gantt chart, a diagram type invented at least two years earlier, this publication does illustrate another important source of concepts and

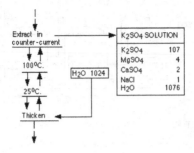

Fig. 3. Section from 'Flow diagram for Processes 7 and 7a', 1933

forms associated with the idea of flow. This source lies in the field of industrial management where the principal concern is the overall sequence of operations and hence the layout of machines on assembly floors in order to maximise use of machines, in particular by providing a steady supply of materials or partially completed products. Departmental organisation, documentation of production, movement of records, and hence flow of information, become subjects for graphical representation. The book by Knoeppel, as its title suggests, advocates a comprehensive use of graphical representations in production management.

The use of charts in the same book illustrates further abstractions of the concept of flow whereby it may refer additionally not only to the movement of physical documents about materials but also to the information contained in those documents. One figure [13] shows the organisation of a production control department including, with a specific broken line and arrowhead notation, the flows of 'records to accounting dept. for costing purposes', 'various shop controls', 'records', 'statistics' and 'data'. These last two categories represent the highest levels of abstraction represented so far and the only ones not directly denoting some physical entity.

2.5 Invisible Flows

A similar level of abstraction had also become necessary for another class of representations where flows are normally invisible. By the mid 1920s the terms 'block schematic', 'schematic diagram' or simply 'schematic' were in common use for any meta-level representation showing the arrangement of electrical systems comprising multiple components but without any use of detailed circuit notation. Figure 4 shows a typical example from 1935 [14]. The simple syntax of rectangular boxes with text labels, lines and arrowheads is the same as that for flow sheets and the arrangement of elements follows no two dimensional rules except the standard western preference for some form of left-to-right sequence. Such general representations of electrical systems would have been familiar to the engineers building and operating the first electronic computers and their use continued as a means for illustrating functional components and communication of early machines. Representations of hardware became an issue in themselves and are not considered here.

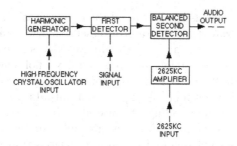

Fig. 4. 'Fig. 3 - Schematic of transmitting monitoring device', 1935

During the period prior to the appearance of flow diagrams in the field of computation, the notions of flow and the manner of their denotation became steadily more complex both in terms of the materials and processes involved and the manner of their representation. Regarding what was actually flowing, multiple, visible, synthetic materials, most clearly seen in the chemical industry, came to join simpler, single, natural or manmade materials as the subjects for examination and illustration. Enclosed manufacturing plant normally rendered flows as invisible as electric current. Multiple stage processes generating, manipulating or terminating flows, whether seen or not, steadily increased the level of complexity.

2.6   Symbolic Representation and Sequence

Although the schematised representations by Leonardo of manipulated water flows show a degree of abstraction from physical resemblance, they are as far from current notions of a flow diagram as are his drawings of machines from standard engineering drawings. In an intervening stage of the development of representations, conventional drawings such as pipework cross-sections derived from the scaled representations intended for construction, could be related to physical resemblance. In simplified forms these symbols acted as icons, in the Piercian sense derived directly from the appearance of the objects to which they referred. Such static representations required the addition of a classic Piercian index, the arrow or arrowhead, to indicate direction of flow and denote dynamic change.

Introduction of arbitrary symbolic elements in the early representations, most importantly the rectangular box, became necessary as the processes of flow became further subdivided and required some higher level description to record all significant components within one overall form. It is clear from the earliest examples that such boxes serve a graphical as much as a denotational purpose, circumscribing within an overall layout each of a number of individual textual labels which themselves do the referring to process definitions. However arrangement of the boxes themselves does contribute an essential element of meaning in so far as it follows some convention for sequence, most importantly top to bottom or left to right.

The notion of sequence is implicit in all the representations of flow considered so far. The later variants incorporate an essential new element of separation between two levels of an hierarchy. Individual elements in one diagram denote abbreviated or abstracted versions of more complex wholes at the level below in a number of other diagrams or forms of representation. In some variants there is also a means of showing multiple inputs and outputs to any component and thus the bifurcation and recycling of material. The notion of a flow diagram now incorporates a process broken down into individual sub-processes in a set sequence, each component denoted by an abbreviated title presented within a box and each stage in the sequence by at least one arrow or arrowed line between boxes representing succeeding and preceding component. This is essentially a meta-level form predicated on some lower and more detailed level of meaning into which the represented process will later be transformed by some means.

3   The GvN Type and After

3.1   Flow Diagram of Machine Control

The new flow diagrams of Goldstine and von Neumann (here abbreviated to 'GvN type') altered the basic definition in two fundamental ways. Their domain of operation is now abstract and mathematical rather than material, even at the sub-atom level, and their use is embedded within a staged process which makes use of an hierarchy of representations and is related to specified operational semantics of a computational machine.

Goldstine describes how "an exceedingly crude sort of geometrical drawing" developed from being a means to show the iterative nature of 'an induction' to being an essential means for expressing a mathematical problem and hence for programming [15]. Development of the notation took place incrementally. The 'assertion box' [16] was the last of a series of innovations which altered radically both the syntax and semantics of the flow diagram and defined its paradigmatic form.

The GvN type retained the single essential character from the syntax of its predecessors, the combination of at least two rectangular boxes representing

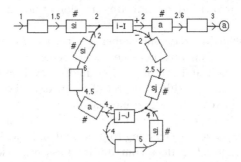

Fig. 5. Initial 'flow diagram' example given by Goldstine and von Neumann, 1947

operations of some kind and an arrowed line connecting them to define their sequence. Figure 5 shows the initial published example in which operations boxes are left unlabelled and have single connectors at opposite sides. Preceding types of flow diagram had incorporated operations with more than one input or output but without altering the generic operational meaning.

The most long-lasting innovation in the GvN type was the introduction an 'alternative box' with 1 input arrow and 2 output arrows. In the Figure 5 diagram the alternative boxes contain the 'induction variables', i and j, and the numbers at which induction ends, I and J. The outputs have plus and minus signs as labels indicating alternative courses if the test indicated within the box produces a positive or negative result. In combination with a simple junction notation to show where connectors converge, the operation and alternative boxes enable representation of single or multiple 'induction loops' and the corresponding iterative process.

3.2   Definition of New Semantics

It is the logic of such iterative mathematical processes rather than any flow of data that is the motivation for the GvN type. The diagram is "a schematic of the course of C(ontrol) through the sequence" of code that will be consequently defined. The accompanying text [17] gives precise semantics to the alternative outputs from alternative boxes by mapping them to specific conditional transfer orders (xCc and xCc' defined elsewhere [18]). Clarification of this course required the introduction of two additional types of boxes for 'substitution' and 'assertion', both labelled outside with what is now called a 'hash' (or 'pound') sign. In the simplest case a 'substitution' box provides the means to show that at least one variable, the induction variable, must change before control completes any loop, but specifically not to affect that change. In Figure 6 the loops conclude with boxes labelled 's i' and 's j'; boxes labelled 'a' appear at the exit from each loop after each 'alternative', making clear the 'assertion', for example 'i = I', which is a *de facto* restriction at this point.

The GvN type also incorporates the first specific semantics for the syncategorematic signs which form the connections between box symbols. These symbols dissect the flow diagram into 'constancy intervals', numbered with Arabic numerals, along which all changing variables and storage locations have constant values. A separate box connected by a broken line to the connector provided an optional space (not used in Figure 5 example) within which to record values. All successive contents of storage positions must also be shown in a separate 'storage table'. This notion of constancy within the interval between defined process elements is present in earlier and later types of flow diagram although neither made explicit nor discussed. Construction of a (usually partial) storage table comes in the second stage of the coding process of which the GvN type forms an integral part.

The drawing of flow diagrams came after mathematical preparations, the choice of suitable equations and conditions, clarification of the explicit procedures involved and estimation of precision. The authors emphasised that this first of four main stages had nothing to do with machines or mechanisation. To use contemporary expressions, the standard mathematical metalanguage already provided a suitable modelling framework.

In the second stage the drawing of flow diagrams provided the means to plan the course of the mechanical computation. This was a 'macroscopic' or 'dynamic' stage of coding [19], accompanied by the drawing of a (normally partial) storage table as the diagram evolved. The 'static' or 'microscopic' stage of coding [20] completing the contents of every operation and alternative box followed. Assignment of all storage positions and final order numbers comprised the fourth and final stage. Omitting the storage activities, these macro and micro stages correspond to levels now often called 'modelling' and 'implementation' in program development. The flow diagram proved to be an extremely successful syntactic form for computational 'models' related to the mathematical problems which were most of interest to many potential users. This remained the case until the market for commercial computers began to burgeon a decade after definition of the GvN type.

3.3   Standard Tool for Programming

Program documentation that survives from the immediately post-war period in the UK shows that the basic process defined by von Neumann and Goldstine provided the framework for programming the earliest versions of the machine designed initially by Turing at the National Physical Laboratory at Teddington (NPL). An example developed for the ACE Version 8A (the design immediately preceding the Pilot Ace that first ran in May 1950) is a program written by Mike Woodger for the solution of the differential equations for the motion of a dampened string [21]. The first three pages give a general description of the problem, its solution in principle and an elaboration of the method chosen. The fourth page defines 'Subsidiarys' (sub-routines for output and particular storage operations) and 'Storage Arrangements' for the fourteen delay line storage components. The next page consists of a flow diagram drawn on lined and columned paper using three columns on the left for boxes (reproduced in part as Figure 6). There follows a page of detailed code for the 'OUTLET' sub-routines and three pages of detailed code with the heading 'MAIN TABLE IN DETAIL'. The final two sheets show storage allocations in detail.

The intensive use of sub-routines began at an early stage in England [22]. This development did not effect the general function of the flow diagram; it merely made explicit the possibility of an hierarchy of generality in its use. In 1951 Turing used the much older term 'block schematic diagram' for an essential aid in defining subroutines. In Section 15 of his programmers' handbook for the Manchester machine [23], Turing defines 'Programming principles' which presuppose a two level combination of main program and sub-routines.

The principle steps that he recommends comprise 'plan making', 'breaking the problem down' and 'programming subroutines'. The plan should include the apportionment of storage to various duties as as well as decisions about mathematical approach and formulae to be used. The 'block schematic diagram'

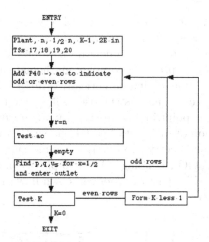

Fig. 6. Flow diagram for ACE 8A programme, 1947/48

provides a convenient aid to planning at the subsidiary level of the subroutine. It consists of "a number of operations described in English (or any private notation that the programmer prefers) and joined by arrows. Two arrows may leave a point where a test occurs, or more if a variable transfer number is used. Notes may also be made showing what is tested, or how many times a loop is to be traversed." [24]. After preparing the diagram "the operations appearing as blocks [within it] may be replaced by actual instructions." Finally there is the programming of the "main routine" about which there is simply the advice that it "follows principles similar" to those for the subroutines. The only flow diagram of any kind in this document (other than interesting contemporary additions inserted by Woodger on punched cards) is simple and shows a high level of generality (reproduced as Figure 7).

In all these diagrams the arrow symbol has lost the explicit and more restricted semantics assigned to it in the GvN type and the box symbols with enclosed labels now denote only operations or tests. The notion of flow has reverted to that of the controlled sequence. Flow diagrams have become a general purpose tool for planning an automated computation at all levels of composition or decomposition. The notion of flow has become synonymous with sequence as issues of computation and storage have been separated and although operations are dealt with one-by-one every machine has unique sets of commands and operations requiring its own sequence of operations to solve particular numerical problems.

Fig. 7. Untitled diagram from Manchester MkII Handbook, 1951

Although at a much later date a dedicated advocate of flow charts could write that "neither programmers nor analysts think in terms of flowcharts" and would justify their use by their "documentation value" [25], when the first attempts were being made to program complex mathematical procedures such as the Runge-Kutta method for solving numerically ordinary differential equations, "understanding the process demands the use of flow diagrams" [26]. This comment appears as the first point in a comparison made during 1959 between programs for the integration of differential equations written in the language used on DEUCE (the English Electric Company's commercial version of ACE) and in the newly defined ALGOL language. The complexity of the algorithms being constructed is illustrated by part of Woodger's own flow diagram reproduced in facsimile as Figure 8 [27].

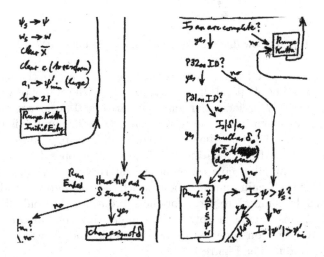

Fig. 8. Detail from flow diagram by M. Woodger for a differential equation algorithm, 1959. NPL Archive. Reproduced by courtesy of the National Museum of Science and Industry, London.

4   Formalisation, Diversification and Demise

4.1   Expanded Use by Manufacturers and Programmers

During the second half of the 1950s as computers became widely available for commercial use, their manufacturers prepared manuals clearly intended for operators expected to have little familiarity with any form of programming. In the UK Ferranti Limited produced commercial machines in collaboration with Manchester University and beginning with its second machine, the Pegasus first delivered in March 1956, prepared manuals for public use. The Pegasus manual presents an explanation of machine structure and an introduction to all aspects of its programming on the basis of principles easily recognised from earlier practice. It recommends the use of a block diagram as, "a less detailed flow-diagram

... in which whole groups of instructions are briefly described and not written down explicitly" when dealing with "a complicated problem" [28].

The flow-diagram itself "expresses the 'algebra' of the process". It is used as the means of expressing the results from an initial stage concerned with "converting the numerical method into a series of steps ... a small group of mathematical operations". Writing down the actual orders that the machine will have to obey is then "relatively simple once the flow-diagram has been obtained" [29]. Over the next two years the Ferranti literature incorporates a full range of uses for flow diagrams including a representation of wage processing at a high level of generality [30] and detailed examples of complete programs including full code [31].

A widening range of applications for flow diagrams and a rapid proliferation of symbols accompanied the increased commercial use of computers. The first ACM glossary of programming terminology published in 1954 defined 'flow chart' as "graphical representation of a sequence of operations using symbols to represent the operations such as compute, substitute, compare, jump, copy, read, write, etc." [32].

By the time of the most recent ISO document [33], now twenty years old, the number of symbols had increased from the basic post-GvN five (operation, test, flow arrow, stop, start) to forty including allowable variants. These standard symbols are to be combined to build up flowcharts at three levels of generality. Data flowcharts and program flowcharts show data flow and control flow at a basic level. System flowcharts show data flow at a higher level at which operations and their control are incorporated. At the same higher level the program network chart shows the path of program activations and interactions with related data. At the highest level the systems resource chart represents the configuration of 'data units' and 'process units'.

Proliferation of syntax variants, both by equipment makers and by major customers doing their own programming, and the need for a generally agreed language for documentation, were the motivations behind the standardisation efforts. Public communication of metalevel information about systems had come to supersede private comprehension of detailed algorithmic processes as the primary function of the flow diagram.

4.2   An Alternative Semantics

Provision of a means for public communication of numerical methods and procedures and for realizing a stated process on a variety of machines were the first two objectives for the definition of ALGOL [34]. This development prompted the only semantic definition of a flow diagram wholly separate from algorithm or system analysis. The syntactic flow chart for ALGOL [35] contains 328 symbols in all, labelled circles, ellipses and rectangles located within a space with positions defined vertically and horizontally. The sequence function represented by any arrowed line in the standard post-GvN form is now denoted only by the horizontal; the vertical denotes a definition of possible alternatives. This form of chart also incorporates recursion; a rectangular box indicates that the enclosed label refers to an element defined elsewhere at the given co-ordinates of the chart.

The definition of Algol came at the same time as a recognition that the problems then being addressed were those of "dealing not with computers but with general information transformers or symbol manipulators" [36]. Alternative data structures, firstly the 'linked lists' used in LISP and later the 'classes' of SIMULA67, were under development as the spaghetti-like characteristics of sequential programmes incorporating 'go to' statements were becoming a subject of controversy [37].

4.3   Conclusion

In the early days of programming with low-level machine and assembly languages, flow diagrams were obviously at a higher level of abstraction than program code and were routinely used to define the logic of algorithmic processes. The syntax and semantics of the notation had to accommodate the fact that data was stored and manipulated physically and processing was sequential. Flow diagrams emphasised flow of control, the fundamental requirement, and provided a comprehensive representation of a program. With changes in hardware, the development of modern operating systems and the introduction of higher-level languages, all converging during the 1960s, the complexity of the problems tackled could grow. Focus transferred away from the logical details of algorithmic processes and towards the broader concerns of software engineering, which emerged as a separate discipline following discussions at the NATO conference of 1968 [38].

As structured development gained favour in the 1970s and 1980s, flow diagrams came to be used to model the logic of workflow and structured flowcharts [39] were introduced to add discipline to the design of lower-level code. With the move to object-oriented analysis and design in the 1980s and into the 1990s there was even less need to express detailed algorithmic processing as part of routine practice. As software engineering tools flow diagrams of all varieties fell even further out of favour until the advent of the UML in the 1990s and the first introduction of its activity diagrams [40].

The GvN type flow diagram is the sole ancestor of an extended family of software engineering notations and diagrams. This continuing development of the concept of flow in software engineering and the many resulting notations and diagrams are the subject of further work.

References

1. Goldstine, H.H. and von Neumann, J. *On the principles of large scale computing machines.* In A.H.Traub (Ed.) John von Neumann, Collected Works Volume V, Design of computers, theory of automata and numerical analysis. Pergamon Press, Oxford, 1963, p30.
2. Goldstine, H.H. and von Neumann, J. *Planning and coding of problems for an electronic computing instrument*, Part II, Volume 1. Report prepared for US Army Ordnance Department, April 1947. In A.H. Traub (Ed.) 1963, pp 80-151.

3. *Il Codice Atlantico di Leonardo da Vinci nella Biblioteca Ambrosiana di Milano*, Editore Milano Hoepli, 1894-1904 (656a). Also http://www.museoscienza.org/english/leonardo/portellochiusa_d.jpg, January 2006.

4. Clark, K. and Pedretti, C. *The Drawings of Leonardo da Vinci in the Collection of H.M. The Queen at Windsor Castle* (2nd edition), London, Phaidon, 1968 (12579, 12659-12663) Also http://www.geocities.com/davincigateway/oldmanwater.14.jpg, January 2006.

5. Richter, J.P. *The Literary Works of Leonardo da Vinci* (2nd Edition), Oxford, 1939.

6. Ferguson, E.S. *Engineering and the Mind's Eye*. Cambridge, MIT Press, 1992, pxi.

7. Amos, P.A. *Processes of flour manufacture*. London, Longmans Green, 1912, pp171-178.

8. *Industrial and Chemical Engineering*, American Chemical Society, Volume 35, 1943, p295.

9. *Industrial and Chemical Engineering*, American Chemical Society, Volume 25, 1933, p375.

10. *ibid* p1159.

11. *Oxford English Dictionary*. Online edition at http://www.oed.com, August 2004.

12. Knoeppel, C.E. *Graphic Production Control*, The Engineering Magazine Company, New York, 1920, p135.

13. *ibid* p126.

14. *Proceedings of the Institute of Radio Engineers*, Volume 23, Number 7, p706.

15. Goldstine, H.H. *The Computer from Pascal to von Neumann*. Princeton, NJ, 1972, p266.

16. *ibid* p267.

17. *Planning and coding problems*, p87.

18. Burks, A.W., Goldstine, H.H. and von Neumann, J. *Preliminary discussion of the logical design of an electronic computing instrument*, Part I, Volume 1. Report prepared for US Army Ordnance Department, Second Edition, September 1947, p75. In A.H. Traub (Ed.)1963, pp34-79.

19. *Planning and coding problems*, p100.

20. *Planning and coding problems*, p101.

21. NPL Archive, National Museum of Science and Industry, London, N30/30 *ACE Version 8A Description, index, form of instruction word. Seven programs: 2 problems of Wilkinson, INORD, Laplace's equation of relaxation, forward integration, method of characteristics for vibrating string, interpolation 1947/48.*

22. Wilkinson, J.H. Turing's work at the NPL and the construction of Pilot Ace, DEUCE and ACE. In Metopolis et al., *A History of Computing in the Twentieth Century*. New York, Academic Press, 1980, p105.

23. NPL Archive N26/38 *Programmers' Handbook for Manchester Electronic Computer Mark II*, March 1951. Also in facsimile at http://www.turingarchive.org/browse.php/B/32, January 2006.

24. *ibid* p64.

25. Chapin, N. Flowcharting with the ANSI standard. A tutorial. *Computing Surveys*, Volume 2, Number 2, pp119-146, June 1970, p143.

26. NPL Archive N9 *December 1959 Points arising from the example of program 4532(6) in Algol.*

27. NPL Archive N9 *File: Simplified programming systems for ACE. M. Woodger work 1959. Incl. complete DEUCE prog. for shockwave boundary layer interaction, Ma4532, 1957-59.*

28. Ferranti Limited. *Ferranti Pegasus Computer Programming Manual Issue 1* List CS 50 September 1955, p1.7.

29. *ibid* p1.6.

30. Ferranti Limited. *Ferranti Pegasus Computer A description of a wage programme for a large dispersed organisation* List CS 169 September 1957.

31. For example Ferranti Limited. *Ferranti Pegasus Computer A simple programme 'Special Factorize'* List CS 153 June 1957.

32. Bright, H.S. Proposed standard flow chart symbols. *Communications of the ACM*, Volume 2, Number 10, October 1959, p17.

33. International Standards Organisation. *ISO 5807-1985 Information processing - Documentation symbols for data, program and system flowcharts, program network network charts and system resource charts.*

34. Backus, J. *The syntax and semantics of the proposed international algorithmic language of the Zurich ACM-GAMM conference*, p129. In Proceedings of the International Conference on Information Processing, UNESCO, Paris, June 1959, UNESCO, Paris, 1960.

35. Taylor, W., Turner, L. and Waychoff, R. A syntactical chart of Algol 60. *Communications of the ACM*, Volume 4, Number 9, p393, September 1961.

36. Gorn, S. *Common symbolic language for computers. Introductory speech.* In Proceedings of the International Conference on Information Processing, UNESCO, Paris, June 1959, UNESCO, Paris, 1960, p117.

37. Dykstra, E.W. Letters to the Editor: go to statement considered harmful. *Communications of the ACM*, Volume 11, Number 3, March 1968.

38. Naur, P. and Randell, B. (Eds) Software Engineering, Report on a Conference Sponsored by the NATO Science Committee, Garmisch, Germany, 7th to 11th October 1968, NATO, January 1969.

39. Nassi, I. and Shneiderman, B., Flowchart Techniques for Structured Programming, *SIGPLAN Notices*, Volume 8, Number 8, August 1973.

40. Booch, G., Jacobson, I., Rumbaugh, J. The Unified Modeling Language for Object-Oriented Development, Documentation Set Version 0.91 Addendum, Rational Software Corporation, Santa Clara CA, September 1996 pp27-30.

Reasoning by Intervals[*]

Benedek Nagy

University of Debrecen, Debrecen, Hungary
Rovira i Virgili University, Tarragona, Spain
nbenedek@inf.unideb.hu

Abstract. In this paper we present a way of reasoning by interval-values over [0,1]. Logical inference is visualized by interval-values. Boolean operators are extended to these values in a natural way. Some other useful operators are also defined. The way of reasoning by Euler/Venn diagrams works by interval-values as well. Moreover based on the length of the intervals probabilistic and fuzzy reasoning is possible.

1 Introduction

Reasoning is one of the oldest fields of philosophy. Ancient Greeks developed several possible ways of argumentation. In the 20th century formal models of reasoning appeared, based on (symbolic) logic. In logical reasoning, the task is to prove that a given formula is a logical consequence of a given set of formulae. Diagrammatic reasoning is an important concept to visualize inference. Venn and Euler diagrams ([Edwards]) are two dimensional figures based on closed curves. The notion of intervals is trivial. Drawing with a pencil on a paper we use points and (interval) lines. Restricting it to 1 dimension we get exactly the concept of (closed) intervals. In this paper a possible way of graphical inference is shown: reasoning by intervals over $[0, 1]$. In the literature there are some papers on intervals representing temporal relations; our model is essentially different.

2 Interval-Valued Logic

In classical logic each statement is either true (1) or false (0). The complex statements are constructed by connectives. In the Interval-valued logic ([Nagy]) the truth values are sets of points and/or disjoint intervals on [0,1]. Each point in [0,1] is either in the interval-value or not. We denote the empty set by $\perp$, and the whole set [0,1] by $\top$. An interval-value can be the union of open, closed and/or mixed intervals. Let A and B be the interval-values of two statements. Table 1 shows the results of their connectives. They are defined point-wise on $[0, 1]$ using the appropriate Boolean operator or by set theoretical operators on the interval-values. We present the operators Product and Divide for special type of interval-values. Given the interval-values A and B in the form $[0, a)$ and $[0, b)$ respectively. Let $\text{Product}(A,B) = [0, ab)$. If $b > a$, then let $\text{Divide}(B,A) = \top$

---

[*] The research is supported by grants OTKA F043090 and T049409.

D. Barker-Plummer et al. (Eds.): Diagrams 2006, LNAI 4045, pp. 145–147, 2006.
© Springer-Verlag Berlin Heidelberg 2006

Table 1. The basic logical operators of Interval-valued logic

Name	negation	conjunction	disjunction	implication
Sign	¬ A	A ∧ B	A ∨ B	A → B
Value	$|A| = [0,1] \setminus |A|$	$|A| \cap |B|$	$|A| \cup |B|$	$|A| \cup |B|$, $(|A| \setminus |B|)$
Set theoretically	complement of $|A|$	intersection	union	complement of difference

else let Divide(B,A) = $\left[0, \frac{b}{a}\right)$. There are some operators which connect the reals to interval-values and vice-versa. The operator ZeroStart assigns the interval-value $[0, x)$ to the real number x between 0 and 1. The operator OneEnd assigns $(1 - x, 1]$ to x. The Measure assigns the real value of the sum of the length of components to an interval-value. For an interval-value A FromZeroMeasure(A) is the real value x, where x is the maximal value such that $(0, x)$ is part of A.

3   Reasoning by Intervals

3.1   Simulating Euler (and Venn) Diagrams

Arguing method by Euler/Venn diagrams can be simulated using the set theoretical subset and equality relations among the interval-values. When two regions are the same, then they must have the same interval-value. If a region is inside of another one, then their interval-values must have the subset relation. If a point belongs to a region, then the correspondent point in the interval-value representation must belong to the given interval-value. (See Fig. 1.) Connectivity is not required in interval-values, moreover any of them can be split to any parts, therefore they can represent any possible relations.

3.2   Theorem Proving in Propositional Logic

Let n be the number of variables of a propositional formula. Let the value be assigned to the i-th variable $|A_i| = \bigcup_{j=0}^{2^{i-1}-1} \left[\frac{2j}{2^i}, \frac{2j+1}{2^i}\right)$. If the evaluation of the formula with A_i's has the value $[0, 1)$, then the formula is a logical law. If the result is ⊥, then the formula is unsatisfiable. A simple variation of this method gives the answer whether a formula is a logical consequence of some other formulae.

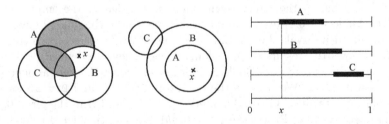

Fig. 1. Venn (left) and Euler (middle) diagrams and their simulation (right)

Table 2. The fuzzy logics and the probabilistic system use values between 0 and 1

	Gödelian logic	Probabilistic system	Łukasiewicz-logic
$\neg A$	1 for $A = 0$; 0 otherwise	$1 - A$	$1 - A$
$A \wedge B$	$\min\{A, B\}$	AB	$\max\{A + B - 1, 0\}$
$A \vee B$	$\max\{A, B\}$	$A + B - AB$	$\min\{A + B, 1\}$
$A \rightarrow B$	1, if $A \leq B$; B, otherwise	$\min\{B/A, 1\}$	$\min\{1 - A + B, 1\}$

3.3 Fuzzy and Probabilistic Reasoning (Table 2)

An interval-value has some measure, therefore we can do something more than the Venn-diagrams. In the simulation of Łukasiewicz logic we use ZeroStart for the (first) argument. At conjunction or disjunction we use OneEnd for the second argument, while at implication ZeroStart should be used. After the correspondent logical operator of interval-valued logic we use Measure to read the result.

To interpret the Gödelian logic we use ZeroStart for all arguments, the correspondent logical operator of interval-valued logic and FromZeroMeasure to read the result. In this system we loose the parts of the interval-values which are not connected to the point 0 after a logical operation (negation and implication).

Using the values as probabilities (P) $\neg A$ gets the probability that A is not true. Assuming that A and B represent two independent events, we get the result of their common occurrence: $P(A$ and $B) = A \wedge B, P(A$ or $B) = A \vee B$. The value of $A \rightarrow B$ is the maximal probability of B if A is true. With interval-values ZeroStart is used for all arguments. We use interval negation, Product or Divide for computing negation, conjunction or implication, respectively. Finally, Measure gives the result. The disjunction can be computed as $\neg(\neg A \wedge \neg B)$.

4   Conclusions

The presented reasoning system uses one dimensional intervals without the restriction of connectedness. It seems that the interval-values can work as models of predicate logic as well. It is a topic of future work to develop these systems.

References

[Edwards]   Anthony W.F. Edwards: *Cogwheels of the Mind: The Story of Venn Diagrams*, Johns Hopkins University Press, 2004.
[Hájek]   Petr Hájek: *Metamathematics of Fuzzy Logic*, Kluwer, Dordrecht, 1998.
[Nagy]   Benedek Nagy: A general fuzzy logic using intervals, in: Proc. 6th Int. Symp. of Hungarian Researchers on Comput. Intell., Budapest, Hungary 2005, pp. 613-624.

Generalizing Spiders

Gem Stapleton, John Howse, and Kate Toller

The Visual Modelling Group
University of Brighton, Brighton, UK
{g.e.stapleton, john.howse}@brighton.ac.uk
www.cmis.brighton.ac.uk/research/vmg

1 Introduction

Recent times have seen various formal diagrammatic logics and reasoning systems emerging [1, 4, 5, 7]. Many of these logics are based on the popular and intuitive Euler diagrams; see [6] for an overview. The diagrams in figure 1 are all based on Euler diagrams and are examples of *unitary* diagrams. *Compound* diagrams are formed by joining unitary diagrams using connectives such as 'or'.

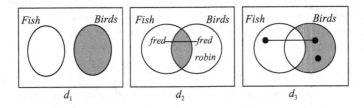

Fig. 1. Various extended Euler diagrams

We generalize the syntax of spider diagrams (of which d_3 in figure 1 is an example), increasing the expressiveness of the unitary system. These generalizations give rise to a more natural way of expressing some statements because there is an explicit mapping from the statement to a generalized diagram. Our theoretical motivation is to provide the necessary underpinning required to develop efficient automated theorem proving techniques: developing such techniques for compound systems is challenging and enhancing the expressiveness of unitary diagrams will enable more theorems to be proved efficiently.

2 Informal Syntax and Semantics of Spider Diagrams

Various systems of spider diagrams have been developed, for example [2, 3], and in this section we give an informal overview of their syntax and semantics. The unitary diagram d_1 in figure 2 contains three *contours*, labelled A, B and C. There is one *zone* inside A and two zones inside B. In total there are five zones (including the zone which is outside all three contours) of which two are *shaded*. This diagram also contains three *spiders*. The *constant spider* labelled s has a

D. Barker-Plummer et al. (Eds.): Diagrams 2006, LNAI 4045, pp. 148–150, 2006.

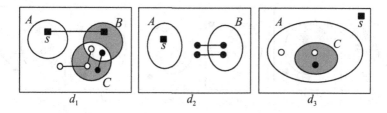

Fig. 2. Unitary spider diagrams

two zone *habitat*. The *Schrödinger spider* has a three zone habitat. The habitat of the *existential spider* is a subset of the habitat of the Schrödinger spider.

The diagram d_2 asserts that A and B are disjoint because the contours labelled A and B do not overlap. The diagram d_3 expresses $C \subseteq A$, since C is contained by A. Spiders denote elements in the sets represented by their habitats and the spider type affects the precise meaning. The diagram d_2 expresses that $s \in A$ whereas d_3 expresses that $s \in U - A$. The two existential spiders in d_2 denote the existence of two distinct elements in the set $U - A$. In d_3, the Schrödinger spider inside A but outside C indicates that there may be an element in the set $A - C$. Shading allows us to place upper bounds on set cardinality. For example, the shaded region inside C in d_3 represents a set with at least one element (because of the existential spider) and at most two elements (because of the Schrödinger spider): $1 \leq |C| \leq 2$. The Schrödinger spider inside A but outside C does not provide any information.

3   Generalizing Spiders

We generalize spiders as follows:

1. Different types of spider nodes can be connected with edges.
2. Each constant spider node is labelled.
3. We introduce *Schrödinger constant nodes*. Such nodes are unfilled and labelled in order to identify the individual that might be represented.
4. Spiders can have nodes placed in zones more than once.

Suppose we wish to make the following statement:

1. there is an element t in A or there is an element outside A, in which case it is either u or some unspecified element and
2. there is either another element t outside A or possibly another element s in A and
3. nothing else is in A.

In figure 3, $d_1 \vee d_2 \vee d_3 \vee d_4 \vee d_5$ and the generalized unitary diagram d_6 express this statement. However, d_6 naturally expresses the required statement whereas the compound diagram only implicitly makes this statement. Also d_6 is more concise than $d_1 \vee d_2 \vee d_3 \vee d_4 \vee d_5$. Without our generalizations, no unitary diagram is equivalent to d_6. We note that d_1, d_2 and d_4 are also generalized

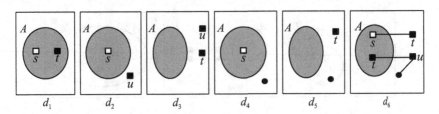

Fig. 3. Generalized spider diagrams

unitary diagrams, each being equivalent to a compound diagram containing two unitary parts. Thus, d_6 is equivalent to a non-generalized compound diagram containing eight unitary parts.

4   Conclusions

The presented generalizations increase the expressiveness of unitary spider diagrams. By allowing constant spiders' feet to be labelled, for example, we have provided a natural way of making some simple statements using unitary diagrams. Indeed, all of the extensions we have proposed enhance the spider diagram system by making it more flexible and notationally efficient. The development of reasoning rules is necessary before we can investigate the effect of our generalizations on the ability of theorem provers to find proofs. It will be interesting, and difficult, to provide a complete classification of the proof tasks where our generalizations are beneficial in terms of time taken to find a proof.

Acknowledgements. Author Stapleton is supported a by Leverhulme Trust Early Career Fellowship.

References

1. E. Hammer. *Logic and Visual Information.* CSLI Publications, 1995.
2. J. Howse, F. Molina, J. Taylor, S. Kent, and J. Gil. Spider diagrams: A diagrammatic reasoning system. *Journal of Visual Languages and Computing*, 12(3):299–324, June 2001.
3. J. Howse, G. Stapleton, and J. Taylor. Spider diagrams. *LMS Journal of Computation and Mathematics*, 8:145–194, 2005.
4. M. Jamnik. *Mathematical Reasoning with Diagrams.* CSLI Publications, 2001.
5. S.-J. Shin. *The Logical Status of Diagrams.* Cambridge University Press, 1994.
6. G. Stapleton. A survey of reasoning systems based on Euler diagrams. In *Proceedings of Euler Diagrams 2004*, ENTCS 134, pages 127–151, 2005.
7. N. Swoboda and G. Allwein. Using DAG transformations to verify Euler/Venn homogeneous and Euler/Venn FOL heterogeneous rules of inference. *Journal of Software and System Modeling*, 3(2):136–149, May 2004.

Diagrams in Second or Foreign Language Learning??!

Leonie Bosveld-de Smet

Rijksuniversiteit Groningen
9700 AS Groningen, The Netherlands
L.M.Bosveld@rug.nl

Abstract. Existing diagrams for tense use as presented in grammar surveys for foreign learners have been analyzed from a semantic theoretic point of view. The analysis reveals that the notational systems used in the diagrammatic representations of the tense rules dealt with are defective in that they can give rise to incorrect interpretations and ambiguities. Moreover, they do not allow the learner to get a coherent view of the tense system. The flaws of the existing diagrams are mainly caused by the lack of a clear delimitation and description of the critical information to be conveyed to the learner. In this paper, we try to avoid these defects by proposing a notational system that relies on a formal semantic approach of tensed discourse.

1 Introduction

Explicit grammar-based language instruction includes grammar points that are comparable to science problems in that they involve abstract concepts, properties and relations. Such problems are liable to get a diagrammatic representation. One such problem is the grammar of tense use. In this paper, diagrams representing the rules for tense use in English are discussed.

2 Semantic Analysis of Existing Tense Diagrams in Grammar Surveys

Three grammar surveys which address foreign learners of English are examined in detail. In each of the surveys consulted, the sections on verb tenses include rules how to use the tenses. The rules are represented verbally and graphically. The diagrams are based on different, though comparable, notational systems. Figure 1 shows examples of the diagrams representing the rule for the use of the Past Continuous as occurring in two different surveys.

For the semantic analysis of the existing diagrams, the theoretical framework proposed by Wang (see [2]) is used. In this framework, the link that associates the graphical domain (i.e. the pictures) with the application domain (i.e. the problem or subject matter visualized) is approached in two ways: by interpretation and by picture specification. The interpretation approach assigns meanings

D. Barker-Plummer et al. (Eds.): Diagrams 2006, LNAI 4045, pp. 151–153, 2006.

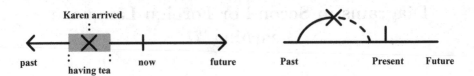

Fig. 1. Past Continuous

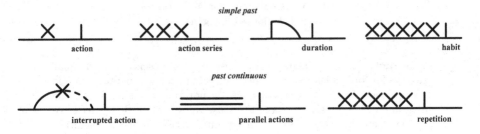

Fig. 2. Syntactic incorrectness

to pictures, the specification approach specifies picture classes for special objects in the application domain. For the interpretation of the diagrams, some assumptions are made. It is assumed that the terms used in the verbal rules (such as 'action', 'habit', 'duration') refer to different notions in the application domain. Moreover, the graphics involved in all diagrams of one grammar survey are taken to belong to the same notational system. The interpretation approach reveals that the diagrams fail to be syntactically correct, and the specification approach shows that the diagrams provide rather unconstrained picture specifications. Syntactic incorrectness is demonstrated in Fig. 2, which shows the diagrams given by one of the surveys for the various uses of the Simple Past, and of those for the Past Continuous. It can be observed that a series of crosses correspond to different application domain concepts, and that an 'action' is not always visualized as a cross.

3 Diagrams Based on a Well Delimited and Structured Application Domain

Natural language tense has received considerable attention in linguistics and logical semantics. The approach of Kamp and Reyle (see [1]) is adopted, in order to structure the application domain in an accurate way. In Kamp and Reyle's view, interpretations of sentences and text are constructed in the form of abstract structures, so-called Discourse Representation Structures (DRSs), obtained by construction rules. DRSs that represent tensed discourse account for the semantics of tensed sentences including the phenomenon that these sentences are temporally related to previous sentences. In the DRSs, events and states are adopted as irreducible semantic entities. Aspect, as indicated by Continuous and Perfect tenses, plays a role in the distinction of these two types of eventualities.

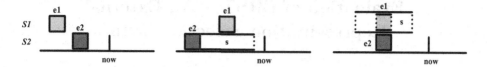

Fig. 3. Simple Past / Past Perfect / Past Continuous

Their temporal location is indicated by the tense of the verb or by other devices such as temporal adverbials. In the DRSs, events, states and/or times are related to each other by a precedence, overlap, abut, or inclusion relation. The proposal is to base diagrams on the temporal and aspectual information as represented in Kamp and Reyle's DRSs. Application domain concepts can be explicitly linked to graphical elements. Figure 3 shows possible diagrams visualizing the temporal and aspectual information of a sequence of two Past tense sentences S1 and S2. One sentence is in the Simple Past and the other either in the Simple Past, the Past Perfect, or the Past Continuous (e.g. S1:Henry ran downstairs. S2:A bomb exploded.— S1:Henry ran downstairs. S2:A bomb had exploded.— S1:Henry was running downstairs, S2:when a bomb exploded.). In the diagrams, e1 and e2 represent different events (e1:Henry-running event; e2:bomb-exploding event), s is the result state of e2 (Past Perfect), or the state generated by e1 seen from the inside (Past Continuous).

4   Conclusion and Future Work

Existing diagrams for tense grammar suffer from semantic defects. This might hamper learners' comprehension and correct use of the different tenses. Diagrams that rely on well delimited and described accounts of tense systems avoid the problems observed, and formal semantic descriptions such as the one put forward by Kamp and Reyle lend themselves well to this purpose. The results obtained so far call for empirical evidence. Further experimental research is required to find out whether diagrams have beneficial effects in explicit grammar-based second or foreign language instruction, and whether learning of tense grammar issues is facilitated more by the proposed diagrams than by the existing ones.

References

1. Kamp, H., Reyle, U.: From Discourse to Logic. Kluwer, Dordrecht (1993)
2. Wang, D.: Studies in the Formal Semantics of Pictures. Ph.D Thesis, ILLC-Publications, University of Amsterdam (1995)

Evaluation of ERST – An External Representation Selection Tutor

Beate Grawemeyer

Representation & Cognition Group
Department of Informatics, University of Sussex, Falmer, Brighton BN1 9QH, UK
b.grawemeyer@sussex.ac.uk

Abstract. This paper describes the evaluation of ERST, an adaptive system which is designed to improve its users' external representation (ER) selection accuracy on a range of database query tasks. The design of the system was informed by the results of experimental studies. Those studies examined the interactions between the participants' background knowledge-of-external representations, their preferences for selecting particular information display forms, and their performance across a range of tasks involving database queries. The paper describes how ERST's adaptation is based on predicting users' ER-to-task matching skills and performance at reasoning with ERs, via a Bayesian user model. The model drives ERST's adaptive interventions in two ways - by 1. hinting to the user that particular representations be used, and/or 2. by removing from the user the opportunity to select display forms which have been associated with prior poor performance for that user. The results show that ERST does improve an individual's ER reasoning performance. The system is able to successfully predict users' ER-to-task matching skills and their ER reasoning performance via its Bayesian user model.

1 Introduction

People vary in their knowledge of external representations (KER), in the range of representations that they can use effectively on reasoning tasks, and in their ability to match particular representations to tasks (i.e. in their knowledge of 'applicability conditions' ([19])).

Numerous factors are associated with ER-to-task matching skill. First, it is known that individuals differ widely in terms of their preferences for particular forms of external representation (ER) ([4],[6],[16]). Better reasoners organise their knowledge of ERs on a 'deeper' semantic basis than poorer reasoners, and are better at correctly naming various ER forms ([5],[7]).

Secondly, some types of tasks require a particular, specialised type of representation to solve whereas for other types of tasks, several different ER forms may be useful. The extent to which a problem is representationally-specific is determined by characteristics such as its degree of determinacy (extent to which it is possible to build a single, unique model of the information in the problem). ER selection skill requires, *inter alia*, knowledge of a range of ERs in terms of a) their semantic properties (e.g. *expressiveness*), b) their functional roles ([4],[23],[5],[2],[19])

D. Barker-Plummer et al. (Eds.): Diagrams 2006, LNAI 4045, pp. 154–167, 2006.

together with information about the 'applicability conditions' under which a representation is suitable for use on a particular task ([19]).

This paper describes the evaluation of ERST - a prototype External Representation Selection Tutor. ERST is an adaptive system designed to help users to choose effective representations for use across a varied range of tasks that involve answering queries using graphically displayed information from a database.

ERST's user model is being developed ([11],[12]) on the basis of empirical data gathered from a series of empirical studies. In these experiments a prototype automatic information visualization engine (AIVE) was used to present a series of questions about the information in a database. This approach is similar to that of [14], who used an empirical basis for the development of a user model for the READY system. That system models users' performance capacity under various cognitive load conditions.

This paper focusses upon an evaluation of ERST in which two versions of the system were compared. One version (used by participants in the evaluation group) used the adaptive ERST system with its user-modelling system turned on. A comparison group used another version of ERST - one with the user modeling subsystem turned off.

2   ERST

The aim of ERST is to enhance users' ER reasoning performance across a range of different types of database query tasks. The adaptive system is able to predict ER to task matching skills and ER reasoning performance, based on its user model. It drives ERST's adaptive interventions (by hinting or advising) or by 'hiding' inappropriate display forms.

ERST's user model and its user-adaptation mechanism have been developed on the basis of empirical data gathered from two experiments. The study [10] investigated the representation selection and reasoning behaviour of participants who were offered a choice of information-equivalent data representations (for example, tables, bar charts, etc.) for use on various database query tasks. Some tasks required the identification of unique entities, some required the detection of clusters of similar entities, and some involved the qualitative comparison of values.

A further study [11], investigated the degree to which some task types are more representation-specific[1] than others, with respect to reasoning performance and response latency.

The results showed that, display selection accuracy and database query answer performance were both significantly predicted by prior knowledge-of-external-representations (KER) pre-tests. Specifically, conceptual (classificatory) knowledge of ERs predicts success at appropriate information display selection on the AIVE tasks. In contrast, deeper, semantic (functional) knowledge of ERs was associated with success at *using* the selected ER i.e. reading-off information and using it to respond correctly to the database query.

---

[1] These are tasks for which only a few, specialised, representational forms are useful.

It was also found (unsurprisingly) that appropriate representation selection results in better query answering performance. Taken together the results suggested that for predicting query response accuracy, a participant's KER can be as powerful a predictor of question answering accuracy as display selection accuracy.

The selection latency results show that a speedy selection of a display type is associated with a good display-type choice. This could be interpreted to imply that users tend to do one of two things - either they recognise the 'right' representation and proceed with the task or they procrastinate and hesitate because of uncertainty about which display form to choose. Hence less time spent responding to the database query question is associated with a good display-type choice and correct query response. This suggested that the selection and database query latencies may be used in the system's user model as predictors of users' ER expertise.

However these effects differed somewhat across types of task. On highly representationally-specific tasks (e.g. correlate variables) high prior KER significantly predicted the user's ability to identify and select the optimal information display and hence to perform well on the database query task. In contrast, on less ER-specific tasks (eg 'locate'), on which several different types of information display are potentially equally effective, prior KER predicted performance less strongly.

The results of the experiments indicated that ERST (the adaptive version of AIVE), needs to take into account a) individual differences (like user's ER preferences), b) their level of prior ER knowledge and c) the domain task characteristics, in order to coach ER-to-task matching skills in an individualised way reflecting the individual's needs.

2.1   User Model

As described above, ERST's user model is derived from two experimental studies ([10],[11]) which examined the relationship between participants' background knowledge of external representations (KER) and their ability to select appropriate information displays in the course of responding to various types of database queries.

The experimental results show that particular types of data are crucial for modeling. A Bayesian network approach (e.g. [21],[17]) was chosen as a basis for ERST's user model. Bayesian networks have been applied successfully in ITS (e.g. [3]) and are suitable, *inter alia*, for recognizing and responding to individual users, and they can adapt to temporal changes.

The structure of a simple Bayesian network based on the experimental data can be seen in figure 1. The Bayesian network in ERST's user model has been 'seeded' with the empirical data from the experiments in order that it could, from the outset, usefully monitor and predict users' ER selection preference patterns within and across query types.

The aim was for ERST to be able to relate query response accuracy and latencies to particular display selections and contrive query/display option com-

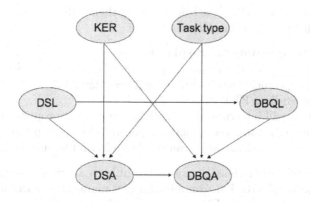

Fig. 1. Graph of a Bayesian network for ERST's user model. KER = knowledge-of-external-representations; DSL = display selection latency; DBQL = database query answer latency; DSA = display selection accuracy; DBQA = database query answer performance.

binations to 'probe' an individual users' prior knowledge of ERs. The empirical data was used to instantiate values in the relevant conditional probability tables (CPTs) at each node of the model. The bayesian network is sensitive to the representation specificity of different types of tasks. The network dynamically adjusts the CPT values and evolve individualised models for each of its participants in real time as they interact with the system. For example, for each ER selection and resulting database query performance score the corresponding CPT values will be updated and used from the system for an individual adaptation. The learned network is able to make the following inferences:

- **Predicting ER preferences and performance with uncertainty about background knowledge**
  If there is uncertainty about users' background knowledge of ERs, the system is able to make predictions about the dependent variables (e.g. background knowledge-of-external-representations (KER)), through a probability distribution of each these variables.
- **Learning about users' ER preferences and performance**
  Users' ER preferences and performance can be learned incrementally, through users' interaction with the system. The network can be updated with the individual characteristics and used to predict future actions and system decisions.

These inferences are used as a basis for ERST's adaptive interventions based on background knowledge, task type and ER preferences.

2.2 The Adaptation Process

As mentioned above, ERST's interventions consist of both overt hints or advice to users and also covert adaptations such as not offering less-appropriate display forms[2] in order to prevent users from selecting them. The system is able to adapt to the individual user in the following ways:

- **Hiding 'inappropriate' display forms**
  The system varies the range of 'permitted' displays as a function of each tasks' ER-specificity and the users' ER selection skill.
- **Recommending ERs**
  The system will interrupt and highlight the most appropriate ER (based on its user model) if too much time is spent on selecting a representation, after learning an individuals' selection display selection latency patterns.

Based on users' interactions the system is able to adapt the range of displays and/or recommend ERs. For example, if a user manifests a particularly high error rate for particular task/ER combinations, then the system will limit the ER selection choice if it believes that this task could be answered with an appropriate different ER. Additionally, after the system detected users average display selection latencies, ERST will recommend the most appropriate ER to the user, if the system believes that the user is unclear what kind of ER to choose and spends too much time in selecting a representation for a particular database query.

3   Evaluation of ERST

In order to evaluate ERST two version of the system will be compared. One version with the adaptive system turned on and the other version with the user modeling subsystem turned off, similar to the evaluation approach employed by [1] or [22]. It was hypothesised that ERST would improve an individuals ER reasoning performance across a range of different types of database query tasks.

3.1   Participants

Thirty two participants, 20 in the comparison and 12 in the evaluation group, were recruited for this evaluation. The twenty participants in the comparison group[3], which used non adaptive ERST, includes 5 software engineers, 1 graphic designer, 1 html programmer, 2 IT business managers, 7 postgraduate students, and 4 research officers/fellows (6 female/14 male). The evaluation group, which used the adaptive version of ERST consists of 2 software engineers, 1 IT business manager, 6 postgraduate students, and 3 research fellows (4 female/8 male).

3.2   Procedure

Participants in both groups were administered 4 pre-tasks, designed to assess their knowledge of external representations (KER) [8], before completing ERST's database query problem solving task.

---

[2] ERST having observed the user attempt to use such ERs unsuccessfully over several previous trials.

[3] The data from the second AIVE study is used in the control group.

3.3   Knowledge of External Representations (KER) Tasks

Four knowledge-of-external-representation tasks were employed. These consisted of a series of cognitive tasks designed to assess ER knowledge representation at the perceptual, semantic and output levels of the cognitive system [8]. A large corpus of external representations (ERs) was used as stimuli. The corpus contained a varied mix of 112 ER examples including many kinds of chart, graph, diagram, tables, notations, text examples, etc.

The first task was a decision task requiring decisions, for each ER in the corpus, about whether it was 'real' or 'fake'[4]. This was followed by a categorisation task designed to assess semantic knowledge. Participants categorised each representation as 'graph or chart', 'icon/logo', or 'map', etc. In the third (functional knowledge) task, participants were asked *'What is this ER's function'?* An example of one of the (12) multiple-choice response options for these items is *'Shows patterns and/or relationships of data at a point in time'*. In the final task, participants chose, for each ER in the corpus, a specific name from a list. Examples include 'venn diagram', 'timetable', 'scatterplot', 'Gantt chart', 'entity relation (ER) diagram'.

The 4 tasks were designed to assess ER knowledge representation using an approach informed by picture and object recognition and naming research [13]. The cognitive levels ranged from the perceptual level (real/fake decision task) to through production (ER naming) to deeper semantic knowledge (ER functional knowledge task).

3.4   ERST's Database Query Tasks

Following the KER tasks, participants' performed the ERST database query tasks. Participants were asked to make judgments and comparisons between cars and car features based on database information. The database contained information about 10 cars: manufacturer, model, purchase price, insurance group, CO_2 emission, engine size, horsepower, etc.

Each subject responded to 30 database questions, which were of 6 types: identify; correlate ; quantifier-set; locate; cluster (similarity); compare negative. For example, a typical correlate task was: 'Which of the following statements is true? A: Insurance group and engine size increase together. B: Insurance group increases and engine size decreases. C: Neither A nor B?'; or a typical locate task: 'Where would you place a Fiat Panda with an engine size of 1200 cc inside the display?'.

Participants were informed that to help them answer the questions, the system (ERST) would supply the appropriate data from the database. Further details of the tasks can be found in [10].

ERST also offered participants a choice of representations of the data. They could choose between various types of ERs, e.g. set diagram, scatter plot, bar chart, sector graph, pie chart and table. Only in the comparison group were all representations offered to participants, as in Figure 2.

---

[4] Some items in the corpus are invented or chimeric ERs.

Fig. 2. Representation selection interface

Fig. 3. Representation selection interface

For evaluation group participants (who used adaptive ERST), the list of representation choices might have been reduced or a single representation recommended (as discussed in 2.2) if the system believed that this intervention would enhance the individuals' ER reasoning performance on the basis of the current user's interaction and performance history. An example is provided in Figure 3.

Participants in both groups were told that they were free to choose any ER, but that they should select a form of display they thought was most likely to be helpful for answering the question. Participants then proceeded to the first question, read it and selected a representation.

The spatial layout of the representation selection buttons was randomized across the 30 query tasks in order to prevent participants from developing a set pattern of selection.

Based on the literature (e.g. [9]) a single 'optimal' ER for each task was identified (display selection accuracy scores were based on this - see results). However, each query type could *potentially* be answered with any of the representations offered by the system (except for set diagrams, which were only usable in quantifier-set tasks).

After the participant made his/her representation choice, ERST recorded the selection time (display selection latency) and then generated and displayed the representation instantiated with the data required for answering the question e.g. Figure 4).

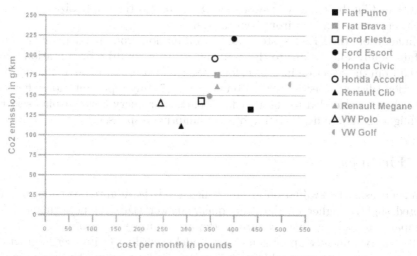

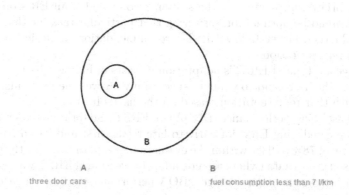

Fig. 4. Examples of ERST Euler's circles and plot representations

Participants then answered the question using the chosen visualization. ERST logged the time to answer the question using the chosen ER. Participants were not permitted to select a different representation following their initial selection. This constraint was imposed in order to encourage participants to carefully consider which representation was best matched to the task. Following a completed response, participants were presented with the next task and the sequence was repeated.

The following data was recorded: (1) the users' representation choices (display selection accuracy - DSA); (2) time to read question and select representation (display selection latency - DSL); (3) time to answer the question using chosen representation (database query latency - DBQL); (4) participants' responses to questions (database query answer - DBQA); and (5) the randomized position of each representation icon from trial to trial;

In addition, the ERST system used by evaluation group participants recorded details of any adaptations to the user that it made, ie: (6) representations eliminated from the selection interface, for a particular user, for a particular database question (display selection reduction - DSR); (7) any representations that were actively recommended to the user for a particular query task (display selection highlight -DSH); and (8) timing of such adaptive responses.

4   Findings

Over all tasks the evaluation group, which used the adaptive version of ERST scored slightly higher on database query answer (DBQA) performance (88% compared to 83% response accuracy) than the comparison group.

ERST dynamically updates its user model according to its user interactions. Not all adaptations were able to take place in early trials, because of sparse user data. After a few trials, enough data for ERST's user model was gathered and the system started to adapt. If the user spent too much time selecting an appropriate representation, the system recommended an ER - one which it believed would be optimal for answering that particular task for that user. The user could then either follow ERST's recommendation or decide to choose a different representation.

The second type of ERST's adaptation consists of hiding ERs in cases where it believes the user is not be able to successfully answer the particular database query with that representation, based on the user's history.

Database query performance (DBQA) for ERST's adaptations combined (highlighting and reducing ERs) for early to late trials (first and last 6 trials) shows an increase of 78% to 94% within the evaluation group. In contrast the equivalent comparison group data (where the non adaptive version of ERST was used) shows a similar increase from an average DBQA performance of 80% on early trials to 90% on late trials.

An evaluation method recommended by e.g. [25] is to break the adaptation down to its constituents. Figure 5 shows database query performance (DBQA) over ERST's different types of adaptation decisions for early and late trials.

Starting from early DBQA performance (78%) the highest increase can be seen in cases where users followed ERST's advice and selected the recommended representation (display selection highlight selection - DSHS). Here, every database query was answered correctly on late trials (DBQA = 100%). In contrast, *not following* ERST's advice resulted in poorer DBQA late-trial performance 75% (even lower than that on early trials).

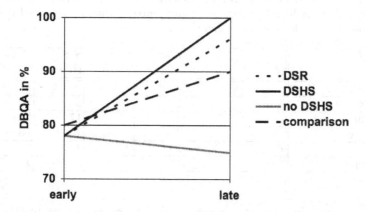

Fig. 5. Database query performance (DBQA) over ERST's adaptations in % for early and late trials. ERST's adaption for hiding a representations - DSR (display selection reduction); The user selected the recommended representation - DSHS (display selection highlight selection); User did not select ERST's highlighted representation - no DSHS; Comparison group performance (which used non adaptive ERST) - comparison.

ERST's other strategy (hiding particular ERs) resulted in an increase of DBQA performance to 96% on late trials. This is similar to trials where ERST did not adapt at all (within the evaluation group), DBQA on late trials increased to 96% as well. Hence 'recommending' seems to be a better intervention strategy than 'hiding'.

4.1 ERST's Adaptation in Relation to Task-Types and the Representational-Specificity of Different Tasks

The adaptations ERST conducted for each task type can be seen in table 1. The system was more active on its adaptation for some types of representations than others (like plot or bar charts), because these representations were used more often. For example, tables were selected 116 times, and scatterplots 114 times, whereas sector graphs were selected only 9 times (all of out of 360).

Some representations were highlighted by ERST on some tasks, but hidden for other tasks, such as the bar chart which was highlighted in the negative comparison task 22 times, whereas it was reduced 6 times in the cluster task, and 5 times in the identify task 1.

For highly representation-specific tasks (e.g. cluster or correlate), ERST was more robust in its adaptations. For example, on cluster tasks, the system recommended the plot chart 19 times but reduced the bar chart's availability 6 times (equivalent figures for sector graphs and set diagrams were 2 and 1 respectively).

Whereas for example, on the low representationally-specific quantifier-set task, ERST recommended set diagrams 17 times and tables 5 times. It reduced access to the table ER once and plot chart twice. It can be seen that in quantifier-set tasks,

Table 1. ERST's adaptations during evaluation according to task and representation type. Values for highlighting a particular representation in **bold** and reducing a representation from the selection menu in *italic*.

Task type	Bar chart	Pie chart	Plot chart	Sector graph	Set diagram	Table
Identify	5		*1*			**8**
Correlate	*3*		**17**		*2*	
Quantifier-set			*2*		**17**	**5**, *1*
Locate			*1*		*1*	**13**
Cluster	*6*		**19**	*2*	*1*	
Compare-neg	**22**	*1*			*1*	*1*

ERST recommending the set diagram to some users, and tables to other users, and sometimes reduced access to tables on some trials and encouraged their use on other occasions.

4.2   ERST's Adaptation Behaviour Related to Users' Prior Knowledge of ERs (KER)

Lower performance on the KER pre-tests was associated with more intervention by ERST during the database query trials. Low functional knowledge of maps, node and arrow/arc network and set diagrams was associated with high levels of ERST adaptation. Functional knowledge of maps is significantly negative correlated to ERST adaptation (r=-.70, p<.05), functional knowledge of node and arrow/arc networks is also significantly negative correlated (r=-.71, p<.01) as well as functional knowledge of set diagrams (r=.-64, p<.05).

It is interesting to note that these three KER sub-scores are derived among the most spatial types of ERs. Set diagrams use a spatial (containment) diagrammatic metaphor to represent set membership. In contrast to e.g. maps which are isomorphic and scaled representations of real-world space.

Exploratory regression analyses suggests that pre-test knowledge of highly spatial ER forms (set diagrams, maps etc.) are particularly predictive of ER selection accuracy and hence ERST's degree of adaptation. In future systems, this knowledge could be used to systematically test the level of individuals ER-to-task matching skills and knowledge of ERs - further research on this issue is planned.

5   Discussion

In the current study it was interesting to note that the 'best' representation was sometimes associated with poorer performance than other (suboptimal) representations, presumably due to individual differences in prior knowledge. This was observed in participants 1 and 9. Here, because of low prior knowledge scores, ERST recommended the tabular representation instead of the set diagram for quantifier-set tasks resulting in 100% success rate in database query answer performance. In this case adaptive ERST provides better reasoning support to such an individual than a system that recommends 'optimal' representations to all its users.

It was noted that subjects sometimes failed to follow the adaptive systems advice, which resulted in worse performance. The more representation-specific the task, the poorer the performance, if advice was not followed. For example, performance decreased dramatically on the highly representation-specific cluster task if ERST's advice was not followed (database query performance of 94% if participants followed the recommendation, in contrast to a performance of 0% if ERST recommendation was not followed). In contrast performance stayed the same whether following or ignoring the recommendation on the low-representation specific tasks (e.g. the locate task with database query performance of 100% for following/not following ERST's advice). A future system could indicate to its user the 'cost' associated with following/not following its advice.

Looking at individual recommendations, here some users refused to follow ERST advice on particular representations, which then were used successfully on other tasks. Participant 5, who refused to select the recommended ER, bar chart in negative comparison tasks, did select the bar chart on cluster and quantifier-set tasks with a surprisingly high (100%) database query response success. This can be contrasted with the recommendation of set diagrams. Here some participants refused to follow the recommendation and also never selected it on any other tasks. For example, participant 10 never selected the set diagram and did not follow ERST's advice to choose it on quantifier-set tasks. The findings suggest that participant 10 did not know how to use the recommendet ER. A future ERST-tutor version of the system could offer tutorial clarification about the functionality and cognitive and semantic properties of particular ER forms.

6   Conclusion

The current version of ERST is able to increase users' ER reasoning performance through recommending the most appropriate display. ERST's other adaptation strategy (varying the range of 'permitted' displays available to the user for selection) was not as successful in terms of increasing reasoning performance. This will be further investigated.

The next step will be to improve ERST's sophistication by enabling it to generate ER-to-task matching situations (contrived query trials) more proactively than it currently is capable of i.e. to systematically 'probe' an individual users' knowledge of ERs and his/her ER-to-task matching skills.

Then, in the case of a user manifesting high error rates for particular task/ER combinations, ERST could offer tutorial clarification, e.g. about the functionality of a particular ER, its cognitive and semantic properties, provide examples of good practice in its use, etc.

References

1. Brusilovsky, P., Eklund, J.: A study of user model based link annotation in educational hypermedia. Journal of Universal Computer Science, special issue on assessment issues for educational software **4** (1998) 429-448

2. Cheng, P.C.-H.: Functional roles for the cognitive analysis of diagrams in problem solving. In: Cottrell, G.W., eds.: Proceedings of the 18th Annual Conference of the Cognitive Science Society. Mahweh NJ, Lawrence Erlbaum Associates (1996) 207-212

3. C. Conati, A. Gertner and K. VanLehn: Using Bayesian networks to manage uncertainly in student modeling. User Modeling & User-Adapted Interaction **12** (2002) 371-417

4. Cox, R., Brna, P.: Supporting the use of external representations in problem solving: The need for flexible learning environments. Journal of Artificial Intelligence in Education **6** (1995) 239-302

5. Cox, R., Stenning, K., Oberlander, J.: The effect of graphical and sentential logic teaching on spontaneous external representation. Cognitive Studies: Bulletin of the Japanese Cognitive Science Society 2 (1995) 5-75

6. Cox, R.: Representation construction, externalised cognition and individual differences. Learning and Instruction **9** (1999) 343-363

7. Cox, R., Grawemeyer, B.: The mental organisation of external representations. European Cognitive Science Conference (EuroCogSci). Osnabrück (2003)

8. Cox, R., Romero, P., du Boulay, B., Lutz, R.: A cognitive processing perspective on student programmers' 'graphicacy'. In: Blackwell, A., Marriott, K., Shimojima, A., eds.: Diagrammatic Representation & Inference. Volume 2980 of Lecture Notes in Artificial Intelligence. Springer (2004) 344-346

9. Day, R.: Alternative representations. In: Bower, G., eds.: The Psychology of Learning and Motivation **22** (1988) 261-305

10. Grawemeyer, B., Cox, R.: The effects of knowledge of external representations and display selection upon database query performance. Proceedings of the Second International Workshop on Interactive Graphical Communication (IGC2003). Queen Mary, University of London (2003)

11. Grawemeyer, B., Cox, R.: A Bayesian approach to modelling user's information display preferences. In: Ardissono, L., Brna, P., Mitrovic, T., eds.: UM 2005: The Proceeding of the Tenth International Conference on User Modeling. Volume 3538 of Lecture Notes in Artificial Intelligence. Springer (2005) 233-238

12. Grawemeyer, B., Cox, R.: Graphical data displays and database queries: Helping users select the right display for the task. In: Butz, A., Fisher, B., Krüger, A. , Olivier, P., eds.: Smart Graphics, 5th International Symposium on Smart Graphics, SG 2005. Volume 3638 of Lecture Notes in Computer Science. Springer (2005) 53-64

13. Humphreys, G.W., Riddoch, M.J.: Visual object processing: A cognitive neuropsychological approach. Lawrence Erlbaum Associates, Hillsdale NJ (1987)

14. Jameson, A., Gromann-Hutter, B., March L., Rummer, R.: Creating an empirical basis for adaptation decisions. In: Lieberman, H. ,eds.: IUI 2000: International Conference on Intelligent User Interfaces. (2000)

15. Kirby, J.R., Moore, P.J., Schofield, N.J.: Verbal and visual learning styles. Contemporary educational psychology **13** (1988) 169-184

16. M. Kozhevnikov, M. Hegarty and R.E. Mayer: Visual/spatial abilities in problem solving in physics. In Anderson, M., Mayer, B. and Olivier, P., eds.: Diagrammatic representation and reasoning. Springer (2002)

17. Mitchell, T.M.: Machine learning. McGraw Hill, New York (1997)
18. Norman, D.A.: Things that make us smart. Addison-Wesley, MA (1993)
19. Novick, L.R., Hurley, S.M., Francis, M.: Evidence for abstract, schematic knowledge of three spatial diagram representations. Memory & Cognition **27** (1999) 288-308
20. Novick, L.R., Hurley, S.M.: To Matrix, Network, or Hierarchy: That Is the Question. Cognitive Psychology **42** (2001) 158-216
21. Pearl, J.: Probabilistic reasoning in intelligent systems: Networks of Plausible Inference. Morgan Kaufmann (1988)
22. Shute, V.J., Gawlick-Grendell, L.A., Young R.K., Burnham, C.A.: An experimental system for learning probability: Stat Lady description and evaluation. Instructional Science **24** (1996) 25-46
23. Stenning, K. , Cox, R., Oberlander, J.: Contrasting the cognitive effects of graphical and sentential logic teaching: Reasoning, representation and individual differences. Language and Cognitive Processes **10** (1995) 333-354
24. Vessey, I.: Cognitive fit: A theory-based analysis of the graphs versus tables literature. Decision Sciences **22** (1991) 219-241
25. Weibelzahl, S.: Problems and Pitfalls in Evaluating Adaptive Systems. In Chen, S. and Magoulas, G., eds.: Adaptable and Adaptive Hypermedia Systems. Heshey, PA: IRM Press (2005) 285-299

Changing Perceptions of Animated Diagrams

Richard Lowe

Department of Education
Curtin University, Australia
r.k.lowe@curtin.edu.au

Abstract. While the educational effectiveness of static diagrams is underpinned by a long heritage of manipulating visuospatial characteristics of the depicted content to improve explanatory power, no corresponding evolution has occurred regarding temporal manipulation of animated diagrams. When complex dynamic subject matter is presented using animations that depict temporal information in a behaviorally realistic manner, the perceptual properties of the display may be poorly matched to the learner's information processing capacities. An exploration of the effect on information extraction of manipulating the speed and presentation frequency of an animation depicting dynamically–complex subject matter suggests that such manipulations may be used to improve perception of thematically relevant information.

Static diagrams have long been used in education to simplify the way that complex subject matter is presented to learners. Recently, animated educational diagrams have emerged as a tool for teaching about dynamic systems. These animations appear better suited for presenting complex dynamic systems than do traditional static diagrams due to their capacity for explicit rather than implicit depiction of the referent's dynamics. Because the comprehension of such systems poses considerable challenges for learners, animations are increasingly popular for explaining this type of subject matter. Both *structural* and *behavioural* aspects of dynamic systems can contribute to their complexity. This means that in order to be educationally effective, their depiction must take account of the visuospatial and temporal demands likely to face learners when processing the information being presented. This paper reports an initial exploratory investigation of how temporal manipulation of such animated depictions may affect learners perceptual processing of different levels of dynamic information.

1 Animated Diagrams and Behavioural Realism

Educational benefits of static diagrams result from their manipulation of the referent's *visuospatial* properties in order to reduce problems learners face from structural complexity. However, this approach of manipulating properties has not been extended to encompass the *temporal* aspects of animated educational diagrams. Many current–generation animated diagrams fail to address the complexity of the referent's behaviour by depicting the situational dynamics realis-

D. Barker-Plummer et al. (Eds.): Diagrams 2006, LNAI 4045, pp. 168–172, 2006.

tically despite their extensive manipulation of the referent's visuospatial properties. In the domain of meteorology, learners performed poorly when required to extract thematically relevant information from *behaviorally realistic* animated diagrams even when given extensive user control over playback [5]. However, in the weather map animations used for that research, both structural and behavioural attributes contributed to the overall complexity. In contrast, Newton's Cradle is a dynamic system with a very simple structure but highly complex dynamics (especially if actual rather than idealized behaviour is depicted – see [1]). It therefore offers an excellent opportunity for teasing out the effect of temporally manipulating an animated depiction of a complex system whose complexity is largely due to its behaviour.

One of the reasons suggested for a lack of effectiveness of educational animations is that their presentation of information is poorly matched to the learner's processing capacities [6]. As a result of this poor matching, the learner's extraction of information relevant to the given learning task is inadequate. The most fundamental constraint on opportunities the learner has for such extraction is whether or not key aspects of the display required for the learning task are even resolvable at the playing speed used. However, even if there are no problems with the resolution of thematically relevant information, the learner may face other challenges in processing the animation [4]. Consider the case of a learner given the task of learning from a behaviorally realistic animation that provides a single presentation of dynamic subject matter shown at its actual speed. Various circumstances under which this situation could result in inadequate processing of the animation are suggested below:

– The aspects that need be extracted from the display are small, highly localised, and of very short duration
– The learning task requires that multiple graphic entities and events be extracted and related to each other.
– The multiple entities and events that need to be related are spread in space and time.
– The perceptual salience of the target information is low relative to that of its context.

These types of circumstances are likely to be most problematic for animations that present complex subject matter. The speed at which an animation is played and the number of times it is presented both have the potential to affect a learner's success in extracting target information from the display. Reducing the playing speed gives the learner more time to perform the necessary processing and alters the perceptibility profile of information presented so that changes at the micro level of the dynamic structure become more accessible. Further, repeating the animation several times gives a learner the opportunity to direct attention to different levels of analysis (micro to macro) on successive presentations and to partition limited attentional resources between spatial and temporal search across these occasions.

2 Newton's Cradle

Newton's Cradle consists of a number of identical balls freely suspended so that they are collinear and just in contact when at rest. This apparatus is widely used to demonstrate idealized abstract physics principles involved in collisions, such as conservation of energy and momentum (e.g. [2]). For example, when a ball at one end is pulled away from the other balls then released, its collision with the rest results in a ball at the opposite end being ejected from the row. This process of collision and ejection then occurs in reverse and a cyclic pattern is established. Recently, various animated diagrams of Newton's Cradle have appeared on the web and these typically depict the system behaving in an idealized manner involving cycling elastic collisions that continue unabated for as long as the animation is played. However, a *real* Newton's Cradle behaves in a far from ideal fashion. While its initial behavior bears a quite close resemblance to that shown in these idealized animated diagrams, over a period of a minute or so, this behavior gradually degrades and changes through the time taken for all movement to cease. Figure 1 summarises the main phases involved.

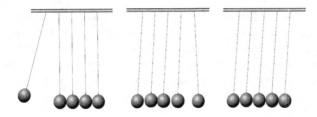

Fig. 1. Three Phases of Newton's Cradle dynamics: (1) 'end bounce', (2) 'middle jostle' (3) 'joint swing'. Phase two is much shorter and yet more complex than others.

This extended temporal change is far more complex than what is shown in an idealized version. It is hierarchically structured, varied in the types of motions involved, and subtle in the way causality operates. While some aspects of the sequence are visually simple and perceptually conspicuous (such as the prolonged synchronous swinging of all five balls in the 'joint swing' stage), other aspects (such as the interactions amongst the balls that redistribute energy and cause transitions between phases) are far less obvious. As a result, the dynamic events occurring across the whole sequence involve complexities and subtleties that pose considerable perceptual challenges to the viewer when observed just once at their actual speed.

3 Method

Eight undergraduate Teacher Education students volunteered to participate in this pilot investigation for course credit. The subject matter was a 1.5 minute behaviourally realistic portrayal of a five–ball Newton's Cradle showing the non–ideal dynamics of this system from the time a single ball was released until all motion ceased. Participants were asked to determine how this device worked and to think aloud while this sequence was played to them individually on three successive occasions. Verbalisations and eye tracking data superimposed on video records of the animation were collected during each of these episodes. Four of the participants were shown a normal speed version of the material and the other four were shown a half speed version. Participants' transcribed utterances were divided into idea units then coded in terms of visuospatial and temporal aspects of the mentioned information. These ranged from characterisations of the entities and their groupings to descriptions and explanations of various aspects of the system's behaviour. For example, the 'synchrony' category was used to code references to the balls moving to and fro in a coordinated manner (as occurs in the 'joint swing' stage). Video records were used to resolve coding ambiguities.

4   Results and Discussion

While verbalisations from participants in the normal condition tended to be more descriptive than explanatory overall, the reverse applied to slow condition participants. This trend is particularly pronounced for the synchrony, energy, and inference categories. The large disparity in the frequency with which synchrony was referred to (normal = 24, slow = 12) is a reflection that those who viewed the normal speed animation tended to describe the system more globally in terms of broad temporal patterns across groups of balls to the neglect of more detailed explanatory information. Conversely, participants in the slow group gave more explanatory accounts that referred to energy or made inferences more than twice as frequently as those in the normal group (energy: normal = 9, slow = 22. inference: normal = 7, slow = 16). Distinctive patterns also emerged within subject groups across the three playing occasions. Those in the normal group rarely produced any reference to the short second phase of the process, and for those who did, it was never mentioned during their first animation viewing. In contrast, all slow group participants identified three phases, as exemplified by the following example:

> One ball hits another ball and sends a reaction along to the other side causing it to move... the second two balls actually come off the middle ball and then as it slows down they stay connected to the middle ball and the two outer balls come off it... now all five of them together... they're all together and slowing down.

In this example, the first segment is an idea unit capturing the 'end bounce' phase while the second segment refers to the 'middle jostle' phase which is the transition to the ultimate 'joint swing' phase referred to in the final two segments. Eye fixation data suggested that participants in the slow group directed more of their attentional resources to the crucial micro level information within the animation than those in the normal speed condition.

5 Conclusions and Future Work

It seems that reducing the display speed affects the type and level of information being extracted. However, the individual contributions of speed manipulation and repeated presentation need to be isolated in subsequent work. We recently piloted the same task with a physics major in the slow condition and obtained very different results. Extensive physics background knowledge appeared to negatively affect the participant's capacity to perceive information that was not in accordance with the orthodox explanation of this device in terms of 'pure' physics principles. More extensive studies are currently underway to investigate the contributions of bottom up and top down processes in information extraction.

The results reported here suggest that manipulation of a referent's dynamic properties offers possibilities for improving learning from animated diagrams that parallel those found from hundreds of years of practical trial and error in manipulating the visuospatial properties of static diagrams. Since its arrival, the animated diagram has been widely adopted by educators who regard it as a major step forward over static diagrams for explaining dynamic content. Unfortunately, current use of animated diagrams in education is not underpinned by the same heritage of techniques for making them effective that we have for static diagrams. The challenges of using animated diagrams effectively as tools for learning should not be under–rated [3]. Given today's uptake of animated diagrams within learning resources, it is important for future research to explore temporal analogs of established visuospatial manipulations that have been so successful over the years in improving the explanatory power of their static precursors.

References

[1] Cheng, P.C.H: Thinking, expertise, and diagrams that encode laws. Technical Report 32, ESCR Centre for Research in Development, Instruction, and Training (1995), University of Nottingham
[2] Gavenda, J.D., Edington, J.R.: Newton's Cradle and Scientific Explanation, The Physics Teacher **35** (1997) 411–417
[3] Hegarty, M.: Dynamic visualizations and learning: getting to the difficult questions. Learning and Instruction, **14** (2004) 343–351
[4] Lowe, R.K.: Extracting information from an animation during complex visual learning. European Journal of Psychology of Education, **14** (1999) 225–244
[5] Lowe, R.K.: Multimedia learning of meteorology. In R.E. Mayer (Ed.), The Cambridge handbook of multimedia learning (2005) New York: Cambridge University Press
[6] Tversky, B., Bauer Morrison, J., Betrancourt, M.: Animation: Can it facilitate? International Journal of Human–Computer Studies, **57** (2002) 247–262.

The Visual and Verbal as Modes to Express Understanding of the Human Body

Sindhu Mathai and Jayashree Ramadas

Homi Bhabha Centre for Science Education, Tata Institute of Fundamental Research,
V.N. Purav Marg, Mankhurd, Mumbai—400088, India
{sindhu, jram}@hbcse.tifr.res.in

Abstract. In this study, students' expression of understanding of structure and function in three systems of the body through visual (drawn) and verbal (written and spoken) modes was probed. Those with good comprehension had high scores in both modes. Pedagogical practices must emphasise explicit use of drawings and words to link structure and function concepts. This can help students of lower ability to form an integrated mental model which will aid understanding and expression.

The study of living beings encompasses two unique but interconnected aspects, structure and function. The work reported here probed how students express understanding of structure and function in three systems of the human body through words and drawings. Understanding was inferred from expression through these two modes. We tested the following hypotheses:

- Structure-function scores are likely to be correlated among good students.
- More concepts are likely to be expressed through a verbal mode.
- More function than structure concepts are likely to be communicated.

Twelve mixed-ability students (5 girls, 7 boys) of Stds. 6, 7 and 8 (ages 10 to 13) from an English medium school in Mumbai, India were the subjects. They were asked to respond to three questionnaires on the digestive, respiratory and circulatory systems through drawings and words. The questionnaires required them to perform two tasks. The first was to draw the organs of that particular system (this question probed understanding of structure). The second required visualisation of the processes of digestion, respiration and circulation by forming a visual mental image (probed understanding of function) such as tracing the path of a favourite food through the body for the digestive system. This was followed by clinical interviews with each student.

Textual responses (written and spoken, broken into simple propositions) and drawings (schematic diagrams and exact depictions) were analysed for comprehension of structure and function. The scheme of analysis is shown (Table 1).

'Order' is a criterion for understanding both structure (location of organs) and function (order of action of organs in a system). 'Segmentation', specific to drawn responses, refers to the basic units in a drawing, in this case the organs. Hierarchy inherent in *understanding* function is 'functional hierarchy'. For the digestive system, there are two levels of hierarchy: passage of food through

D. Barker-Plummer et al. (Eds.): Diagrams 2006, LNAI 4045, pp. 173–175, 2006.

Table 1. Scheme of data analysis

Text/verbal responses		Drawings	
Comprehension of structure	Comprehension of function	Comprehension of structure	Comprehension of function
1 Organs of the system (names)	1 Order of action	1 Segmentation (organs drawn)	1 Order of action
2 Order (location in the system)	2 Hierarchy		2 Hierarchy

the alimentary canal and action of the liver and pancreas. Standard propositions from school textbooks provided guidelines to evaluate propositions from
students' verbal responses. For every system a student was assigned four scores
between 0 and 1 (for understanding of structure and function from verbal and
drawn responses). Pearson's correlation coefficient was determined for verbal-
drawing and structure-function scores across all systems (Table 2). Wilcoxon's
signed ranks test was used to check if structure-function and verbal-drawing
scores were significantly different.

Table 2. Pearson's correlation coefficient (r) for various scores

System of the body	Pearson's correlation co-efficient	
	Verbal with drawing	Structure with function
Digestive	0.24	-0.05
Respiratory	0.79**	0.82**
Circulatory	0.67*	0.39
Across all systems	0.62*	0.67*
*Significant correlation at $p<0.05$ level **Significant correlation at $p<0.01$ level		

For the digestive system it was found that there was no correlation between
structure-function and verbal-drawing scores. Most students were at roughly the
same level of understanding with scores at the higher end of the scale. A lack
of correlation could be attributed to the 'functional hierarchy' inherent in the
system. This led students to think that food goes into the liver and pancreas
just as it passes through organs of the alimentary canal. Another difficulty concerned the structural connection between the small and large intestine though
its functional connection was understood. This was perhaps a result of textbook
drawings which portray the small intestine as a separate organ enclosed by the
large intestine. Diagrams and content are presented separately with no clear linkages in the textbook and therefore learnt that way by students. For the respiratory system there is a direction in the structure of the system itself which helps
in understanding function. Hence structure-function and verbal-drawing scores
are correlated across written and drawn responses. The structural order of organs: nostrils-pharynx-trachea-bronchi-bronchioles-alveoli-bloodstream-body organs is also its functional order. The circulatory system has been treated quite
cursorily in textbooks at this level, and therefore there is no structure-function

correlation. Many students erroneously understood the capillary to be another term for arteries and veins.

Wilcoxon's test showed a greater expression of structure concepts through drawings. However, overall across both text and drawings, more function concepts were expressed through the familiar verbal mode. Since the working of a system is what is easily remembered and emphasised, function is spontaneously communicated. Also since there is no direct correspondence (at a macroscopic level) between structure and function in the human body, correlation of the two aspects becomes difficult. Expression through drawings, particularly *schematic diagrams* is not emphasised in schools. It is a common myth that exact depictions are indicators of good drawing and are the privilege of a few talented students. Diagrams in biology therefore are often used to convey structure alone, with function (which is represented schematically such as through arrows) to be inferred from it.

Subjects with high drawing scores were found to have high verbal scores too. Previous research has documented similar findings. Piaget (1966) emphasised the use of both imaginal abilities and logical reasoning to perform conservation tasks. Heiser and Tversky (in press) found that high ability students formed a 'unitary' mental model which incorporated both structure and function. This may have been the case for good students in this study, who made linkages despite an unfavourable learning environment. For the large majority of students, pedagogical practices must emphasise dual coding of content by explicitly linking text with drawings and structure with function. Mayer (2003) put forth certain principles to design effective multimedia explanations making use of both pictures and words. However, diagrams are open to interpretation and can reveal as well as conceal. Therefore, it is finally the teacher in a classroom environment open to receiving and asking questions, who should play an important role in facilitating picture-text and structure-function linkages.

References

Heiser, J., Tversky, B.: Arrows in comprehending and producing mechanical diagrams. Cognitive Science (in press)

Mayer, R.E.: The promise of multimedia learning: using the same instructional design methods across different media. Learning and Instruction 13 (2003) 125-139

Piaget, J., Inhelder, B.: Mental Imagery in the Child. Routledge, London (1966)

Interpreting Hierarchical Structure: Evidence from Cladograms in Biology

Laura R. Novick[1] and Kefyn M. Catley[2]

[1] Dept. of Psychology & Human Development
Peabody College #512
230 Appleton Place
Vanderbilt University
Nashville, TN 37203
Laura.Novick@vanderbilt.edu
[2] Dept. of Teaching & Learning
Peabody College #330
230 Appleton Place
Vanderbilt University
Nashville, TN 37203
Kefyn.Catley@vanderbilt.edu

Schematic diagrams (e.g., circuit diagrams, Euler circles), which depict abstract concepts, are important tools for thinking. Three interrelated diagrams that belong in this category—hierarchies, matrices, and networks—are common in both everyday and scientific contexts. For example, a hierarchy can be used to represent relations among members of the animal kingdom or the search space at a given point in a chess game; a matrix can be used to represent the grade sheet for a class or possible item pairings in a deductive-reasoning problem; and a network can be used to represent the food web for an ecosystem or the alliances among nations. Because these diagrams highlight structural similarities across situations that are superficially very different, by successfully constructing these diagrams, solvers would be led to see deep (i.e., structural) similarities among diverse situations that otherwise might not be salient (Novick, 2001). Structural understanding is a key factor underlying expertise. Thus, it is important to understand when hierarchies, matrices, and networks are each most appropriate to use, how they should be constructed, and what makes them easier or harder to understand (Hurley & Novick, 2006; Novick & Hurley, 2001).

Cladograms, a type of hierarchical branching structure, are one of the most important tools that contemporary biologists use to reason about evolutionary relationships. They depict the distribution of characters (i.e., physical, molecular, and behavioral characteristics) among taxa (see Figure 1 for alternative representations of the same cladogram). Cladograms are hypotheses about nested sets of taxa that are supported by shared evolutionary novelties called synapomorphies (e.g., one synapomorphy that supports the relationship between birds and crocodiles is the presence of a gizzard). Cladograms are general statements of relationships—i.e., that A and B are more closely related to each other than either is to C (e.g., in Figure 1, birds and crocodiles are more closely related to each other than either is to lizards); they make no attempt to define evolutionary ancestor/descendent relationships.

D. Barker-Plummer et al. (Eds.): Diagrams 2006, LNAI 4045, pp. 176–180, 2006.

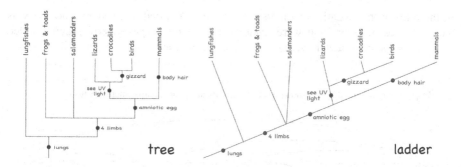

Fig. 1. Equivalent tree and ladder representations of a cladogram depicting evolutionary relationships among various vertebrates

As shown in Figure 1, cladograms can be represented in both a tree form and a ladder form. Over the past several years, these diagrams have begun to appear in middle school and high school biology textbooks (beginning at least in sixth grade). Interestingly, almost all of the textbook examples we have seen, from middle school life science through undergraduate zoology, are ladders.

1 Design

We investigated (1) students' understanding of the hierarchical relations presented in tree and ladder cladograms and (2) their ability to reason about the information presented in versions of these diagrams like those shown in Figure 1. Data from the second, reasoning, sub-study have been presented elsewhere (Catley, Novick, & Traynham, 2006). Here, we present data from the first sub-study. The subjects were Vanderbilt University undergraduates. We divided the students into two groups based on their most advanced biology class. The low knowledge group had taken either no biology courses in college (15 of 27 students) or one of the courses geared to non-science majors (e.g., human biology; 12 of 27 students). On average, this group had taken 0.8 biology (or relevant geology) courses in college. In contrast, the most advanced biology course for the higher knowledge group was either one or both semesters of the introductory biology sequence for science and pre-med majors (7 of 23 students) or a course that had the second semester of the former sequence as a prerequisite (16 of 23 students). On average, this group had taken 3.5 biology (or relevant geology) courses in college. The two groups did not differ in year in school (overall $M = 2.78$).

Subjects were given 10 problems with evolutionary relationships presented in one diagrammatic form, which they had to translate to a different (but structurally equivalent) diagrammatic form. For the first eight problems, subjects had to draw the indicated type of diagram. There were two problems of each of four types: (1) relationships presented in a circle diagram with a required translation to a tree diagram, henceforth referred to as circle to tree, (2) circle to ladder (e.g., see Figure 2), (3) ladder to tree, and (4) tree to ladder. To ensure that subjects focused on the structure of the presented relationships rather than their (possibly faulty) prior knowledge of the relevant taxa, Latin taxa names were used. We coded translation accuracy as well as types of errors that were made, with different types of errors being coded when subjects drew ladders versus when they drew trees.

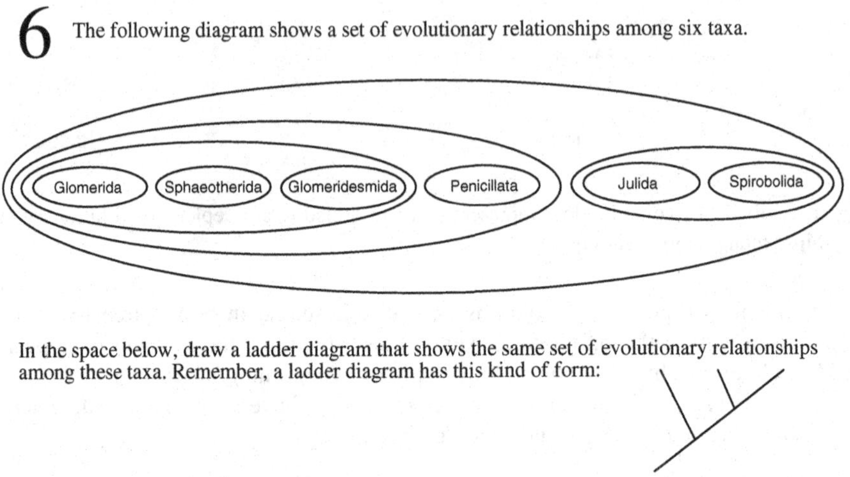

6 The following diagram shows a set of evolutionary relationships among six taxa.

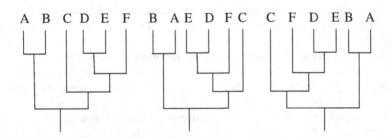

In the space below, draw a ladder diagram that shows the same set of evolutionary relationships among these taxa. Remember, a ladder diagram has this kind of form:

Fig. 2. A translation problem in which information was presented in a circle representation and students had to translate that set of relationships to an equivalent ladder representation

The last two problems, one tree to ladder and the other ladder to tree, were multiple choice; subjects had to choose the correct answer from among three alternatives. These problems tested subjects' knowledge of rotation because the correct answer involved a rotation of branches in the diagram. In both trees and ladders, parts of the diagrams can be rotated without the structure (i.e., the relationships) changing. For example, the three trees shown in Figure 3 are structurally (i.e., biologically) identical rotations of each other. The measure of performance on the multiple-choice problems was accuracy. We also obtained information about students' understanding of rotation by coding their drawings for the first eight problems in terms of whether their answer was a rotation of what was presented and whether it was correct given that it was or was not a rotation.

A B   C D E F   B A E   D F C   C F   D E B   A

Fig. 3. Examples of trees that are structurally-identical rotations of each other

2   Results and Discussion

Both subject groups found ladders more difficult to understand and interpret than trees. This was especially true for the low knowledge students, who were rarely correct in drawing translations whenever a ladder was involved (i.e., when the reference diagram was a ladder or when they had to draw a ladder). The nature of subjects' errors suggests that the difficulty in understanding ladders may reflect both perceptual and conceptual principles—the Gestalt principle of good continuation and a misleading analogy to real-world ladders, respectively.

Of the four problem types, the circle to ladder translation was most difficult (e.g., mean accuracy across all subjects for the problem in Figure 2 was 0.23). This may reflect the lesser visual similarity between circle and ladder versus tree and ladder diagrams. It may also reflect subjects' difficulty in rendering the circled group of three unresolved taxa (at the far left of the circle diagram in Figure 2) in a ladder. This particular structure happened only to be used for circle to ladder translations. All students had difficulty drawing this 3-item group (known as a bush) in a ladder, as reflected in their errors, but low knowledge students had greater difficulty.

A follow-up study, currently underway, is designed to disentangle the contributions of perceptual and conceptual factors to students' poor understanding of the ladder representation. A second goal is to determine the extent to which the greater difficulty of the circle to ladder compared with the tree to ladder translation is due to differences in visual similarity among these types of diagrams versus the presence of an unresolved 3-item bush.

From a psychological perspective, our results provide important insight into students' interpretations of hierarchical structure. In particular, our results indicate that hierarchies are easier to understand when the different levels are clearly visible. This is the case for the tree diagram but not the ladder diagram. Ladder diagrams appear to be harder to understand because (1) the Gestalt principle of good continuation makes it difficult to extract the key structural information about levels and (2) people's knowledge of real ladders is inappropriate for the task at hand.

From an educational perspective, our results are important in part because they demonstrate that the ladder representation is much more difficult for students to understand than the tree representation. Nevertheless, as previously noted, ladders are more commonly used in textbooks. In addition, by providing some insight into how students' interpret, and misinterpret, cladograms, our results begin to provide some guidance for improving evolution education. This is critical because students and teachers alike have a poor understanding of the processes that operate at the macro level, and no understanding at all of the history of life on our planet (e.g., Catley, 2002, in press). By using cladograms to visualize processes that occur over vast distances of time and space, evolution can become a tool for thinking and learning about biology, rather than simply a collection of facts about life.

References

Catley, K. M. (2002, April). Understanding Evolution: The state of teachers' knowledge. American Institute of Biological Sciences Annual Meeting, Washington, DC.

Catley, K. M. (in press) Darwin's missing link: A new paradigm for evolution education. Science Education.

Catley, K. M., Novick, L. R., & Traynham, B. (2006, April). Assessing Students' Understanding of Evolutionary Diagrams. 2006 Annual Meeting of the National Association for Research in Science Teaching, San Francisco, CA.

Hurley, S. M., & Novick, L. R. (2006). Examining the Conventions Associated with Constructing Matrix, Network, and Hierarchy Diagrams. Manuscript under revision.

Novick, L. R. (2001). Spatial diagrams: Key instruments in the toolbox for thought. In D. L. Medin (Ed.), The psychology of learning and motivation (Vol. 40, pp. 279-325). San Diego, CA: Academic Press.

Novick, L. R., & Hurley, S. M. (2001). To matrix, network, or hierarchy: That is the question. Cognitive Psychology, 42, 158-216.

Active Comparison as a Means of Promoting the Development of Abstract Conditional Knowledge and Appropriate Choice of Diagrams in Math Word Problem Solving

Yuri Uesaka[1] and Emmanuel Manalo[2]

[1] Department of Educational Psychology, Graduate School of Education
The University of Tokyo, Japan
yuesaka@educhan.p.u-tokyo.ac.jp
[2] The Student Learning Centre, The University of Auckland, New Zealand
e.manalo@auckland.ac.nz

Abstract. This study sought to address the problem of novices not being able to select the appropriate diagrams to suit given tasks. It investigated the usefulness of providing teaching sessions that involved active comparison of diagrams and review of lessons learnt following problem solving. Fifty-eight 8th grade participants were assigned to one of two instruction conditions. In both, traditional math classes were provided in which diagrams were used to explain how to solve math word problems. However, participants in the experimental group were additionally provided with sessions that required them to actively compare diagrams used, and consider and articulate the lessons they learnt from the problem solving exercises. The results showed that participants in the experimental condition subsequently constructed more appropriate diagrams in solving math word problems. In an assessment of conditional knowledge, these participants also provided more abstract and detailed descriptions about the uses of diagrams in problem solving.

1 Introduction

Using diagrams is a powerful strategy for problem solving. For example, Larkin and Simon [1] demonstrated that diagrammatic representations are superior to sentential representations because they diminish the memory, computational, and searching loads. Other studies have also empirically demonstrated the beneficial effects of providing some diagrams or visual representations in problem solving [e.g., 2, 3, 4].

Although there are some studies that have obtained results suggesting that self-constructed diagrams are not always effective [e.g., 5], recent studies have also shown that self-constructed diagrams are powerful heuristics in some problem solving situations [e.g., 6, 7, 8; see also a review by Cox, 9]. For example, Stern et al. [8] found that actively constructing and using linear graphs as reasoning tools while learning economics can better facilitate a transfer effect across subject content compared to a condition of passive diagram use.

D. Barker-Plummer et al. (Eds.): Diagrams 2006, LNAI 4045, pp. 181–195, 2006.
© Springer-Verlag Berlin Heidelberg 2006

However, school students might not always use diagrams effectively as teachers and researchers do. Numerous problems relating to students' use of diagrams have been identified in research involving educational practices and observations of students' actual problem solving: for example, there is a lack of spontaneity in using diagrams [10, 11, 12], poor choice of diagrams to use [11, 13], and failure to draw appropriate inferences when using diagrams [11, 13]. All of these problems need to be resolved if problem solving among school students is to be facilitated via the use of diagrams.

1.1 Inappropriate Choice of Diagrams Among Students

Among the problems involving the use of diagrams, one of the most critical is the inappropriate choice of diagrams to use. Cox [13] analyzed and categorized the notes and other workings that students produced during problem solving, and observed that one of the most common problems evident in the students' workings was their failure to choose the appropriate kinds of representations to use. Uesaka [11] also analyzed the strategies students used while attempting to solve algebra math problems, and found that some students were using diagrams that were totally inappropriate or too primitive. These studies suggest that the problem of inappropriate choice of diagram to use is relatively common among students.

Competence in selecting the most suitable diagram to use according to the task given can be considered an important aspect of graphic literacy for two reasons. Firstly, matching between task demands and types of representations is necessary to facilitate problem solving with diagrams [14]. Inappropriate diagrams may not only prove unhelpful but also cumbersome for problem solving. If a student consistently fails in selecting appropriate diagrams to use in problem solving, he or she is likely to become doubtful of the efficiency that could result from the use of diagrams. Secondly, the skills underlying the ability to choose the correct situational representations are not only helpful in constructing diagrams for problem solving but also in a host of other occasions such as when organizing materials for the purpose of communicating information to others.

In contrast to novices who experience problems in choosing appropriate diagrams, experts can be considered as having already acquired the necessary skills. For example, Novick and Hurley [4] demonstrated that many university students majoring in math education and computer science can be considered as experts in this area as they can, according to the task given, select the most appropriate type to use among three different kinds of diagrams - matrix, network, and hierarchy diagrams. Grawemeyer and Cox [15] also showed that students who possess more knowledge about external representations evidenced greater use of various kinds of diagrams and consistently achieved higher levels of performance in problem solving. These studies suggest that there is a gap between experts and novices as far as competence in choosing and constructing appropriate diagrams is concerned.

Previous studies in this area have revealed the deficits among novices and the skills that expert possess in choosing and constructing diagrams, but these studies have not shown how competence in constructing appropriate diagrams can be cultivated. It is therefore the main purpose of the present study to develop and evaluate a teaching method for cultivating students' competence in choosing and constructing appropriate diagrams, and hence examine the question of how it might be possible to bridge the gap between experts and novices as far as diagram use in problem solving is concerned.

1.2 Importance of Abstract Conditional Knowledge

Experts may be able to use appropriate diagrams because they possess abstract conditional knowledge about diagrams which can be considered as a kind of schema. Conditional knowledge of diagrams is knowledge about when and why certain kinds of diagrams are appropriate in certain situations. Dufour-Janiver, Bednarz, and Belanger [10] stated that "we need to be able to choose appropriate representations depending on the task, knowing why we make this choice" (p. 111). Novick, Hurley, and Francis [16] empirically demonstrated the existence of abstract knowledge relating to diagrams, and Novick and Hurley [4] revealed the content of such knowledge by collecting protocol data from undergraduate students in a science course in which students described the reasons why a certain type of diagram was chosen among three different options provided. Novick and Hurley [4] proposed that the structure of this abstract knowledge is schematic because it contains slots and values, and abstract knowledge is used when determining the appropriate diagram from the different kinds available.

Apart from those put forward by Novick and her colleagues, there are other theories suggesting that abstract and schematic knowledge mediate transfer of skills beyond the learning context. One of the key areas where abstract and schematic knowledge are considered as the mediating factors of problem solving skills transfer is research in analogical transfer. For example, Gick and Holyoak [17] put forward an argument for the importance of inducing schemas for generating analogical transfer by using "radiation problems" made by Duncker. Gick and Holyoak [17] demonstrated that showing more than one example and verbalizing the common features of the examples facilitated acquisition of the schemas and, as a consequence, promoted analogical transfer of problem solving. This process was referred to as "schema induction". Suzuki [18] also emphasized the importance of abstraction in analogical transfer when he reviewed relevant studies. He referred to this shared framework as "abstraction-based views".

1.3 Cultivation of Abstract Conditional Knowledge

Although previous studies relating to diagram use rarely investigated how it might be possible to cultivate abstract conditional knowledge, studies in other areas suggest that, through active comparison of cases in which diagrams are effectively used and review of lessons learnt after problem solving, it is possible to build this kind of knowledge. The evidence suggesting the importance of comparison comes from studies about analogical transfer. As described earlier, Gick and Holyoak [17] demonstrated that while schema induction did not occur in the condition where only one example was used, it did occur in the condition in which multiple examples were used prior to the administration of transfer problems and verbalized the common features of the examples. Catrambone and Holyoak [19] also found that comparing examples facilitated transfer of the necessary problem solving skills when dealing with math word problems immediately after learning, and suggested that a common structure among these problems might be acquired through the comparison undertaken.

While Gick and Holyoak [17] compared two structurally similar problems, Bransford, Franks, Vye, and Sherwood [20] pointed out the importance of comparing slightly different events and, via this comparison, making salient the features that are important. Their explanation, although focusing more on the identification of differences, appears to complement the explanations that postulate the acquisition of common structure. These findings therefore suggest that actively comparing several examples facilitate the formation of the necessary abstract rule to use – beyond each of the concrete events or examples dealt with.

In addition to the necessity to actively compare, as indicated in the psychological studies noted, educational contexts also suggest the importance of reflection after problem solving. Ichikawa [21, 22] proposed "lesson induction", a generalized strategy of schema induction, as a technique of personal tutoring and one of the heuristics of learning. He argued that it is helpful for teachers to urge students to draw out the lessons learnt following problem solving and provide explanations of why the students could or could not solve the problems given, focusing particularly on important aspects like the students' use of strategies.

Ichikawa also emphasized that activities like lesson induction make it possible to form more general and abstract rules that can be applied to a wider range of situations and may bring about transfer effects. The beneficial effect of lesson induction has been empirically shown in a study by Terao, Kusumi, and Ichikawa [23], in which a condition that involved drawing out the lessons learnt after working through an example problem evidenced a better performance in solving a structurally similar algebra word problem compared to a condition that did not involve this. A similar technique was described by Ellis and Davidi [24], which they referred to as "after-event reviews" (AERs). They emphasized the necessity of reviewing not only failure events but also successful events.

1.4 Hypothesis and Overview of the Experiment

The purpose of the experiments conducted as part of the present study was to test the hypothesis that active comparison of cases where diagrams have been effectively used in problem solving cultivates abstract conditional knowledge of diagrams; this in turn promotes the production of appropriate diagrams when solving math word problems.

To test the hypothesis, two conditions were prepared: the first was a condition involving active comparison of cases where diagrams have been effectively used, and the second was a condition without any active comparison – where diagrams were used in class, but passively provided by a teacher. In both conditions, a teacher used diagrams to explain how to solve problems, and students were provided adequate time to actively construct diagrams by themselves. The only difference between the experimental (first condition) and control (second condition) groups was that the experimental group was provided with the session comparing different kinds of diagrams and their suitability for helping solve particular kinds of problems.

It was predicted that the experimental participants would produce more appropriate diagrams when solving math word problems and, if transfer effect is mediated by abstract conditional knowledge, that the experimental participants would be able to verbalize more sophisticated conditional knowledge of diagrams.

2 Method

2.1 Participants

Fifty-eight 8th grade students voluntarily participated in the present study. They were recruited via an invitation letter sent to 8th grade students in two wards in Tokyo, Japan.

The participants were randomly allocated to one of the two conditions. The resulting proportions of students from private and public schools, and male and female students were confirmed not to be distorted.

2.2 Materials

Two types of mathematical word problems involving functions were used in the course (see Appendix 1 for the actual problems, and Figure 1 for examples of diagrams used during the teaching sessions). The first problem was a "mobile phone problem". A graph would have been the most helpful diagram to use in solving this problem because of the requirement to work out the changes in an amount. The authors checked and confirmed with three teachers, all with a certificate in mathematics education, that using a graph was the most appropriate for this problem.

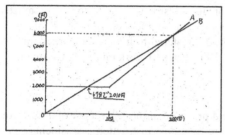 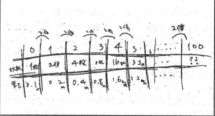

(a) Example for the Mobile Phone Problem   (b) Example for the Paper Folding Problem
(a) Example for the Mobile Phone Problem   (b) Example for the Paper Folding Problem

Fig. 1. Typical Diagrams Produced During the Instruction Sessions

The second problem was a "paper folding problem" (also shown in Appendix 1). In this problem, a table was deemed as the most appropriate diagram to use because the problem required finding out the rule of increment. The appropriateness of using a table was confirmed by the three teachers noted in the previous paragraph.

Pre-instruction and post-instruction assessments were held on the first and last days respectively. On these days, participants sat three assessments: in math word problem solving, conditional knowledge, and basic skills.

Math Word Problem Solving Assessment. This was administered at post-instruction to find out how well students can select the most appropriate diagram to use in solving math word problems. If the problems given in this assessment were too easy then students would not need to use diagrams at all to solve them, so more difficult problems were chosen. Three math word problems shown in Appendix 2 were prepared, and for each problem participants were given 6.5 minutes. Using a graph would have aided efficiency in solving the "water" and "travel" problems, while a table would have been appropriate for the "pentagon" problem (checked and confirmed with math teachers, as previously noted). The participants were provided with sheets of graph paper to use if they so wished.

Conditional Knowledge Assessment. This was also administered at post-instruction to examine what kind of conditional knowledge involving the use of diagrams had been acquired as a result of the courses provided. Participants were given 3 minutes to describe all of the merits of using each kind of representation (graph, table, formula), and to write out all of their ideas about when and why these three representations are effective. Only the descriptions provided for graph and table were analyzed and reported in this paper.

Basic Skills Assessment. The purpose of this assessment was to check the equivalence of the experimental and control groups at the start of the experiment as far as their ability to carry out transformations between different types of representations (equations, tables, graphs, and sentences). This assessment was administered in anticipation of avoiding a possible criticism that the experimental and control groups could have initially differed in their abilities to construct various kinds of representations. Seven short questions were given, requiring participants for example to construct a graph from a formula, or to construct a table from sentences representing a particular situation. Two versions of booklets containing all of the 7 questions were made, and each version was administered to half of the participants at pre-instruction. The participants were given 10 minutes to work on this assessment.

2.3 Procedure

The experimental classes were organized and provided at the University of Tokyo for five days during the summer of 2004. The pre- and post-instruction assessments were administered on the first and fifth days. In the three days in between, the instruction classes were provided, with each daily session lasting for 50 minutes.

In both conditions, the teacher presented two mathematical problems to students in the class and used basically the same teaching procedure for each problem during the instruction sessions. Firstly, the teacher asked participants to prepare some materials for explaining to other people. For example, in the mobile phone problem, participants were asked to make a pamphlet to explain to customers which mobile phone is cheaper depending on the calling time. Secondly, participants tried to work at solving the problems by themselves. At this time, the teacher provided as much help to all of the participants as required. For example, the teacher gathered in front of the board all of the students who wanted to receive hints, and taught the students how the problem could be solved using a graph and how an appropriate graph could be constructed.

After completing this problem solving task, some participants were asked to present the pamphlets they prepared in front of the class. For the presentations, not only pamphlets using a graph but also those using a table or using only equations were included. After the presentations, the participants were asked to choose the presentation that they considered easiest to understand and to write down the reasons for their choice. This final step of choosing a presentation and writing down the reasons for their choice was only included in dealing with the first problem. In the second problem, there was limited variation in the materials the students produced (with almost all using a table) – hence the final step was considered not viable for inclusion.

In the experimental condition, three additional manipulations were included to provide the participants with opportunities to compare the diagrammatic representations used after problem solving. Firstly, before starting to solve problems, students were asked to indicate their ideas about which kind of representation was most suitable for solving the problem given, although no feedback was provided at this stage. Secondly, at the end of the each problem, participants were asked to fill in the blanks on a sheet of paper that required them to write down the merits of each type of representation based on their most immediate experiences and observations in problem solving. Finally, after finishing solving both problems, participants were provided both of the problems used in the course and the participants' solutions considered most appropriate for each problem. The participants were then encouraged to consider when a particular kind of representations was most appropriate in problem solving, and what the merits are of each representation (graph, table, formula). Here, the teacher asked the participants to write down the rules they 'discovered' by filling in the parentheses of the following form: (a particular kind of representations) is effective for solving (a particular type of problem) because (the reason). Some of the students' ideas were shared in the class.

3 Results

When conducting analyses, participants who failed to take the pre-instruction or post-instruction assessments were excluded. The final numbers of participants included in the analyses were 26 in the experimental condition and 20 in the control condition.

3.1 Math Word Problem Solving Assessment Findings

To evaluate the kinds of diagrams participants constructed for solving problems, the first author and another scorer coded the types of representations that participants used. For the purposes of the present study, a diagram was defined as any representation of the problem except words (on their own), sentences, or numerical formulas. Tables were also counted as diagrams, and a table was defined as a depiction which contains at least a pair of values arrayed to represent two related variables. Drawings or illustrations considered unrelated to the problem were not categorized as diagrams. The inter-rater agreement was found to be 95.7%, which the present authors deemed as satisfactory. Examples of diagrams produced by participants are shown in Figures 2. When participants used more than one representation for a problem, the representation that was deemed more appropriate for the problem was counted.

The number of instances when appropriate diagrams were used was counted in each condition, and compared using a t-test. Here, responses using a graph were counted as appropriate for both the first "water problem" and the third "travel problem", and responses using a table were counted as appropriate for the second

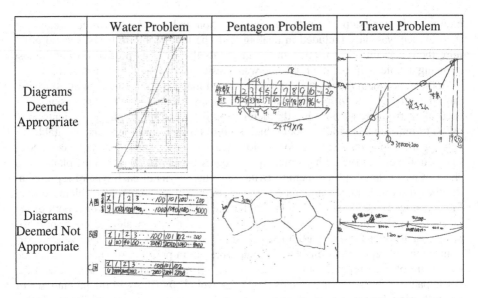

	Water Problem	Pentagon Problem	Travel Problem
Diagrams Deemed Appropriate			
Diagrams Deemed Not Appropriate			

Fig. 2. Examples of Diagrams Participants Constructed for Each of the Problems Given in the Math Word Problem Solving Assessment

"pentagon problem". The *t*-test analysis showed that instances of appropriate diagram use were higher in the experimental group compared to the control group ($t_{(44)} = 3.20$, $p < .01$; the experimental group's average for the three problems = 1.80, $SD = 1.10$; the control group's average = 0.85, $SD = 0.88$). The proportions of the kinds of representations participants used in each problem are shown in Figures 3.

The above findings indicate that participants in the experimental group were using diagrams according to the demands of the problems they were given (i.e., they were not simply using more of one kind of diagram). These results support the first prediction from the hypothesis.

3.2 Conditional Knowledge Assessment Findings

A two-stage classification process was used when analyzing the descriptions participants provided in this assessment. In the first stage, the descriptions provided were divided into four types: abstract statements, concrete statements, other descriptions, and no response. Descriptions were put in the first category ("abstract statements") if they referred to the application of a representation to many situations: for example, "A graph is useful for comprehending the overall trend", "When finding a rule about increasing values, a table is effective". Descriptions were put in the second category ("concrete statements") if they referred to the application of a representation in particular situations: for example, "A graph is useful when solving problems relating to time". All other descriptions provided were placed in the third category ("other descriptions"), including those which were unclear in meaning. Cases where the participants did not provide any descriptions were categorized as "no response". Both the

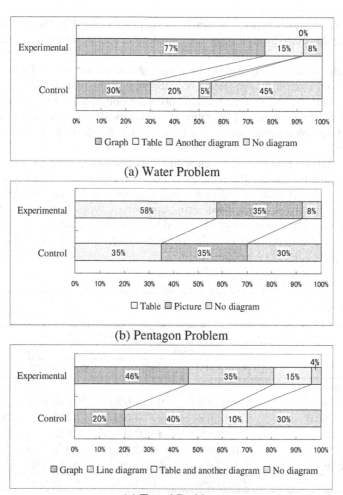

(a) Water Problem

(b) Pentagon Problem

(c) Travel Problem

Fig. 3. Distribution of Representation Types Used by Participants for Each Problem in the Math Word Problem Solving Assessment

first author and another scorer independently classified the descriptions the participants provided. The inter-rater agreement in this stage was found to be 92.7%, which the authors deemed as satisfactory.

For the second stage of classifying the participants' descriptions, responses under the "abstract statements" category were further divided into two categories: "elaborated statements" and "unelaborated statements". Descriptions were put in the elaborated statements category if they were more detailed descriptions and deemed more useful when making a decision as to which kind of diagram to use in a given situation (e.g., "A graph is useful for investigating the changes of two different things"). On the other hand, descriptions put in the unelaborated statements category were those more general and deemed not as useful when making decisions about which kind of diagram to use in a given situation (e.g., "A graph is easy to understand"). The inter-rater agreement during this classification stage was found to be 78.5%, which the present authors deemed as adequate for the purposes.

The results of these two classification procedures are shown in Table 1. The experimental and control groups' ratios of elaborated abstract statements, unelaborated abstract statements, and concrete statements were compared, and a chi-test analysis revealed that the differences of the ratios were statistically significant in both descriptions (for the graph descriptions, $\chi^2_{(2)} = 12.8$, $p < .05$; for the table descriptions, $\chi^2_{(2)} = 6.79$, $p < .05$).

Further analysis also revealed a clear link between conditional knowledge and appropriated diagram choice. Participants in both the experimental and control conditions who produced abstract and elaborated descriptions for both graphs and tables evidenced higher instances of constructing appropriate types of diagrams in the math word problem solving assessment compared to other participants ($t_{(44)} = 2.35$, $p < .05$; abstract and elaborated group's n = 22, average = 1.77, $SD = 1.11$; the other group's n = 24, average = 1.04, $SD = 1.00$).

Table 1. Classification of Participants' Descriptions Provided in the Conditional Knowledge Assessment

| | Description type | | | |
| | Graph | | Table | |
Category	Experimental	Control	Experimental	Control
Abstract rule				
Elaborated	23 (88.5%)	9 (45.0%)	20 (76.9%)	4 (20.0%)
Unelaborated	1 (3.8%)	2 (10.0%)	1 (3.8%)	2 (10.0%)
Concrete rule	1 (3.8%)	4 (20.0%)	1 (3.8%)	3 (15.0%)
Others	0 (0.0%)	0 (0.0%)	0 (0.0%)	1 (5.0%)
No response	1 (3.8%)	5 (25.0%)	4 (15.4%)	10 (50.0%)
Total	26 (100%)	20 (100%)	26 (100%)	20 (100%)

3.3 Basic Skills Assessment Findings

To confirm the equivalence of the experimental and control groups at the start of the study, the participants' performance on the Basic Skills Assessment administered at pre-instruction was analyzed. For each question in the test, a perfectly correct answer was given full credit (2 points), an answer that was deemed mostly correct but included a small mistake was given partial credit (1 point), and all others were given no credit (0 point). The participants' total scores from all seven short questions of the test were compared by using a two-way ANOVA in which the condition and the version of the test were used as independent variables. The result showed there were no main effects (effect of the condition, $F_{(1, 42)} = 0.06$, n.s.; effect of the version, $F_{(1, 42)} = 4.04$, n.s.), and no interaction between the two variables ($F_{(1, 42)} = 0.71$, n.s.). This result indicates that at the start of the study participants in the experimental and control conditions were equivalent as far as their skills in using representations were concerned.

4   Discussion

The results of this study suggest that active comparison of diagrams and reflection on lessons learnt after problem solving promote appropriate choice and construction of diagrams in math word problem solving. There are also indications that such activities promote the development of the appropriate abstract conditional knowledge, as the more abstract and elaborate statements about graphs and tables provided by the experimental participants suggest.

Through the Basic Skills Assessment administered at the start of the study, it was established that participants assigned to the experimental and control conditions were equivalent in their knowledge about, and abilities to depict and perform transformations between, the basic math representations of sentences, equations, tables, and graphs. This meant that any subsequent differences in use of diagrams between the experimental and control groups were not due to pre-existing differences in abilities/limitations of the participants to construct such representations.

After the instructions provided, subsequent assessments showed that participants in the experimental group made improvements in skills relating to diagram use that were greater than those made by participants in the control condition. The experimental group used significantly more appropriate diagrams in solving problems given in the Math Word Problem Solving Assessment. They also provided more abstract and elaborated descriptions in the Conditional Knowledge Assessment – suggesting that they had acquired a more generic understanding of the uses of tables and graphs, including details about when these might be most useful.

4.1   Contributions to Diagrams Research

One useful contribution of the present study to the area of diagrams research is that it identifies a possible teaching method for improving students' appropriate use of diagrams in math word problem solving. Previous research in this area [e.g., 11, 13] has pointed out the failure amongst student considered to be novices in selecting and using appropriate diagrams when solving problems, but no concrete suggestions have been put forward as to how this problem might be addressed through instruction. The findings of the present study provide encouraging indications that the method employed promotes appropriate diagram use and may be helpful in cultivating graphic literacy.

The present study was based on the notion that abstract conditional knowledge mediates students' appropriate use of diagrams. Although the role of abstract conditional knowledge has been discussed considerably in relation to areas such as analogical transfer in problem solving [e.g., 17, 18], and Novick and her colleagues have referred to it in their research concerning diagrams [4, 16], no investigations have previously been undertaken to find out if activities that are likely to promote such knowledge would result in more appropriate diagram use in problem solving situations. The present study has taken a first step in this direction and will hopefully stimulate further investigations of a similar nature. Future investigations should also more carefully examine the nature of this relationship between abstract conditional knowledge and appropriate use of diagrams. Although it is assumed in the present study that the development of knowledge leads to appropriate use, no evidence was gathered to confirm or refute this and it is equally possible that the direction of the relationship is the other way: in other words, that appropriate use of diagrams leads to the development of the relevant abstract conditional knowledge.

4.2 Implications for Math Education

In practical terms, the most important contribution of the present study is that it provides a concrete method that math teachers could use to promote the appropriate use of diagrams in math word problem solving. In general, math teachers use a lot of diagrams when explaining how to solve math word problems to students [10], and they tend to assume that through this demonstration students would not only develop the necessary skills but also be encouraged to likewise use diagrams when solving similar problems. However, the low spontaneous use of diagrams among students identified in previous research [11, 12] and noted earlier suggests an error in this assumption. The performance of participants in the control condition in the present study – a condition which provided instruction more similar to regular math classes – likewise suggests the shortcomings of simply demonstrating diagram use to students. The results of this study suggest that teachers need to additionally provide opportunities for students to actively compare diagrams used and to reflect on the merits of various diagrammatic representations following problem solving activities.

It would be useful in future investigations to examine more closely the relative benefits resulting from the additional activities provided to the experimental group in this study. To recap, in the experimental condition, the participants were additionally asked to (i) reflect on and indicate their ideas about which kind of diagrammatic representations might be most suitable prior to actual problem solving, (ii) explicitly compare cases where diagrams had been effectively used, and (iii) consider and describe the merits of diagrammatic representations in relation to different problem situations following their problem solving lessons. Which of these additional activities contributed most to the subsequently observed more appropriate use of diagrams was not explicitly looked at in the present study. The present study was also carried out in one group of 8th graders attending extra summer classes in Japan: future investigations will need to examine the generalizability of the findings here to students in other grades, and in classes in other countries.

Acknowledgement

The authors are grateful to Shin'ichi Ichikawa and Yoshio Kaburagi for their helpful comments during the planning of this study, and to Hayato Ishiwatari, Toshiyuki Kunimatsu, Masayoshi Yamazaki, Mikiko Seo, Takashi Kato, Yu Nakane, and Ken Hirakawa for assisting with the rating of participant responses and providing useful comments. This study was supported by a grant from the Center for Research of Core Academic Competences, School of Education, The University of Tokyo.

References

1. Larkin, J. H., Simon, H. A.: Why a diagram is (sometimes) worth ten thousand words. Cognitive Science. **11** (1987) 65-99
2. Cheng, P. C. H.: Why diagrams are (sometimes) six times easier than words: benefit beyond locational indexing. In: Blackwell, A., Marriott, K., Shimojima, A. (eds.): Diagrammatic representation and inference. LNAI, Vol. 2980. Springer, Berlin Heidelberg (2004) 242-254
3. Mayer, R. E.: The promise of multimedia learning: using the same instructional design methods across different media. Learning and Instruction. **13** (2003) 125-139
4. Novick, L. R., Hurley, S. M.: Improving cross-content transfer in text processing by means of active graphical representation. To matrix, network, or hierarchy: That is the question. Cognitive Psychology. **42** (2001) 158-216
5. De Bock, D., Verschaffel, L., Janssens, D., Van Dooren W., Claes, K.: Do realistic contexts and graphical representations always have a beneficial impact on students' performance? Negative evidence from a study on modelling non-linear geometry problems. Learning and Instruction. **13** (2003) 441-463
6. Cheng, P. C. H.: Electrifying diagrams for learning: principles for complex representational systems. Cognitive Science. **26** (2002) 685-736
7. Koedinger, K. R., Terao, A.: A cognitive task analysis of using pictures to support pre-algebraic reasoning. In: Schunn, C. D., Gray, W. (eds.): Proceedings of the Twenty-Fourth Annual Conference of the Cognitive Science Society. Lawrence Erlbaum Associates, Mahwah NJ (2002) 542-547
8. Stern, E., Aprea, C., Ebner, H. G.: Improving cross-content transfer in text processing by means of active graphical representation. Leaning and Instruction. **13** (2003) 191-203
9. Cox, R.: Representation construction, externalised cognition and individual differences. Learning and Instruction. **9** (1999) 343-363
10. Dufour-Janiver, B., Bednarz, N., Belanger, M.: Pedagogical considerations concerning the problem of representation. In: Janvier, C. (ed.): Problems of representation in the teaching and learning of mathematics. Erlbaum, Hillsdale NJ (1987) 110-120
11. Uesaka, Y.: Daisuubunsyoudainiokeru zuhyouno riyouwo unagasu youinto kainyuuhouhouno kentou [Investigating the factors and teaching method for promoting the spontaneous use of diagrams in mathematical word problems]. Unpublished Masters Thesis. The University of Tokyo, Japan (2003)
12. Uesaka, Y., Manalo, E., Ichikawa, S.: What kinds of perceptions and daily learning behaviors promote students' use of diagrams in mathematics problem solving? Submitted paper
13. Cox, R.: Analytical reasoning with multiple external representations. Unpublished PhD Thesis. The University of Edinburgh, United Kingdom (1996)
14. Norman, D. A.: Things that make us smart: Defending human attributes in the age of the machine. Perseus Books, Cambridge (1993)
15. Grawemeyer, B., Cox, R.: The effect of knowledge-of-external-representations upon performance and representational choice in a database query task. In: Blackwell, A., Marriott, K., Shimojima, A. (eds.): Diagrammatic representation and inference. LNAI, Vol. 2980. Springer, Berlin Heidelberg (2004) 351-354
16. Novick, L. R., Hurley, S. M., Francis, F.: Evidence for abstract, schematic knowledge of three spatial diagram representations. Memory and Cognition. **27** (1999) 288-308
17. Gick, M. L., Holyoak. K. J.: Schema induction and analogical transfer. Cognitive Psychology. **15** (1983) 1-38

18. Suzuki, H.: Justification of analogy by abstraction. In: Holyoak, K., Gentner, D., Kokinov, B. (eds.): Advances in Analogy Research: Integration of Theory and Data from the Cognitive, Computational, and Neural Sciences. New Bulgarian University, Sofia, Bulgaria (1998) 373-380

19. Catrambone, R., Holyoak, K. J.: Overcoming Contextual Limitations on Problem-Solving Transfer. Journal of Experimental Psychology: Learning, Memory, and Cognition. **15** (1989) 1147-1156

20. Bransford, J. D., Franks, J. J., Vye, N. J., Sherwood, R. D.: New approaches to instruction: because wisdom can't be told. In: Vosniadou, S., Ortony, A. (eds.): Similarity and analogical reasoning. Cambridge University Press, Cambridge (1989) 470-497

21. Ichikawa, S.: 'Kyoukunkinou'no tokutyouto sinrigakutekisyokenkyuutono kanren [The feature of 'Lesson Induction' and the relation to the psychological research]. In: Ichikawa, S. (Ed.): Gakusyuuwo sasaeru ninnti-kaunsering -Sinrigakuto kyouikuno aratana setten [Cognitive counseling that supports learning: A new approach bridging psychology and education]. Brain Press, Tokyo (1993) 52-61

22. Ichikawa, S.: Cognitive counseling to improve students' metacognition and cognitive skills. In: Shwalb, D. W., Nakazawa, J., Shwalb, B. J. (eds.): Applied developmental psychology: theory, practice, and research from Japan. Information Age Publishing, Greenwich Conn (2005) 67-87

23. Terao, A., Kusumi, T., Ichikawa, S.: Solution compression in mathematical problem solving: Acquiring abstract knowledge that promotes transfer. In: Shafto, M. G, Langley, P. (eds.): Proceedings of the 19th Annual Conference of the Cognitive Science Society. Lawrence Erlbaum Associates, Mahwah NJ (1997) 733-738

24. Ellis, S., Davidi, I.: After-event reviews: Drawing lessons from successful and failed experience. Journal of Applied Psychology. **90** (2005) 857-871

Appendix 1. Math Problems Used in the Instruction Sessions

Mobile Phone Problem

You are a clerk in a mobile phone shop. In this shop, two types of mobile phone plans are sold. When a customer comes in and says "I want to buy a mobile that will cost me the least", how will you explain the options to the customer? Please consider the best way of explaining and make a pamphlet that you can use in providing a clear explanation. When making a pamphlet, you can use not only words and sentences but also the graphs, tables and pictures as you may feel appropriate. Please make the pamphlet as easy to understand as possible.

> Plan A: A basic fee of 2000 yen including 100 minutes of free calling time. After 100 minutes, 40 yen per minute is charged.
> Plan B: There is no basic fee, and no free calling time. The cost of calls is 30 yen per minute.

Paper Folding Problem

An 8th grade student was asked to conduct an independent research as an assignment during the summer vacation. He decided to investigate the thickness of newspaper when he folds it 100 times, on the condition that a single sheet was 0.1 mm thick. He started to fold and found the rule that folding once produces a two-ply pile of 0.2 mm thickness, folding a second time produces a four-ply pile of 0.4 mm thickness, and so on. However, he could not fold the newspaper more than ten times so he could not find out the thickness that results from folding 100 times from actual trials. Please consider another way to find out the thickness of folding the newspaper 100 times, and provide an explanation that would be easy for other students to understand.

Appendix 2. Math Problems Administered at Post-instruction

Water Problem

The head of a company asked Taro to find out which of three countries – A, B, or C – would be best for establishing a factory. The factory uses water and water charges differed between the three countries. The different charging methods are described below. Please imagine you are Taro, and come up with an explanation that he could provide to the head of the company.

Country A: 1000 yen is charged as a basic fee, but you can use water without additional charge up to 100 litres. After 100 litres, 40 yen per litre is charged.

Country B: There is no basic fee. The cost of water is 20 yen per litre.

Country C: In addition to 2400 yen as a basic fee, there is a charge for every litre of water used. The charge is 4 yen per litre.

Pentagon Problem

There are many sheets of paper in the shape of a regular pentagon, with each side being 3 cm. These sheets are arranged one by one with the rule that a new sheet shares only one side with already arranged sheets. Find the circumference when arranging 20 sheets. (The case of how 2 sheets could be arranged was illustrated.)

Travel Problem

The distance between Hanako's house and the school she and her sister go to is 1200 meters. One morning, Hanako started to walk from her house to her school with a speed of 60 meters per minute. Three minutes after Hanako left, her sister also left for school by bicycle with a speed of 200 meters per minute. Her sister stopped when she was 800 meters away from their house and waited for her friend for 10 minutes. After that, she headed for school again at the same pace. Please answer the following four questions about this situation.

(1) How many times did Hanako and her sister meet before arriving at their school?

(2) Who arrived at the school earlier, Hanako or her sister?

(3) After Hanako left for school, how much time did it take until the distance between Hanako and her sister was the greatest?

(4) Who was nearer to school five minutes after Hanako started walking?

Synthesizing Visual and Action Routines Using Constraint Programming*

Bonny Banerjee and B. Chandrasekaran

Laboratory for AI Research, Department of Computer Science and Engineering
The Ohio State University, Columbus, OH 43210, USA
{Banerjee, Chandra}@cse.ohio-state.edu

Abstract. A diagrammatic problem-solver requires a library of visual routines (VRs) and action routines (ARs) - the VRs are used to obtain information of specified types from the diagram and ARs to modify the diagram in specified ways. The VRs/ARs required are unbounded - a new domain may call for new perceptions and actions. We report on progress on our research in building an automated VR/AR synthesis system that would take as input the definition of a new routine in terms of existing routines in the library and well-defined mathematical/logical constraints and synthesize the program for the desired routine using constraint programming. We illustrate the ideas by means of an example.

1 Introduction

Ullman [1] suggested that in order for the human visual system to work as efficiently as it does, there has to exist a finite set of elementary operations that can be combined to create an unbounded set of visual routines (VRs). Researchers have proposed different elementary operations and developed systems that can compose task-specific VRs from them, using a language of attention or genetic programming. Reasoning and problem-solving with diagrams requires a large repertoire of VRs and action routines (ARs) - VRs help perceive information from a diagram while ARs act on a diagram to modify it, where the diagram might be external, as on a piece of paper or on a computer screen, or internal, as a representation in the mind or in an artificial agent. A diagrammatic reasoning (DR) system [2] consists of a central problem-solver that calls VRs and ARs to interact with and modify the diagram, guided by the problem-solving goals. Researchers have proposed different VRs/ARs for different tasks, such as *left-of*, *inside, computing a path between two points avoiding obstacles, computing the point of intersection of two curves*, and so on (see [1, 2] for more examples).

Since DR requires essentially an unbounded set of VRs/ARs, it is not feasible to have a library of all VRs/ARs. Moreover, many of the VRs/ARs might be required only once, such as *inside a region and left of a point*, so there is no good reason to keep in store such VRs/ARs permanently. A feasible solution to this problem is

---

* This research was supported by participation in the Advanced Decision Architectures Collaborative Technology Alliance sponsored by the U.S. Army Research Laboratory under Cooperative Agreement DAAD19-01-2-0009.

D. Barker-Plummer et al. (Eds.): Diagrams 2006, LNAI 4045, pp. 196–198, 2006.

to devise a way to generate VRs/ARs automatically and store them, if desired, as programs for future use. Our research goals are three-fold - to develop a high-level language that supports the definition of a VR/AR as constraints on objects specified in terms of well-defined mathematical/logical functions and predefined VRs/ARs; to symbolically compute the desired VRs/ARs and diagrammatically represent their outcome with an appropriate instance; and to automatically synthesize their programs.

2   Solution Approach

Our framework for automatically computing VRs/ARs is based on spatial constraint satisfaction. While spatial constraint solving has been extensively used for computer-aided design (CAD) for the last few decades and more recently to origami construction, application of the same for automatically computing VRs/ARs for DR is novel. In our framework, a diagrammatic object is represented by its parametric equation where t is the parameter. An object o has associated with it a pair of polynomials in t for the x- and y-coordinates, denoted by $o.x(t)$ and $o.y(t)$ respectively. Given a finite set O of diagrammatic objects and a finite set C of spatial constraints defined on them, a VR returns True if O satisfies C, otherwise returns False. For example, the VR $Collinear(p, q, r)$ returns True iff the points p, q, r lie on a straight line., i.e.

$$Collinear(p, q, r) \Leftrightarrow (\forall t, \frac{p.x(t) - q.x(t)}{r.x(t) - q.x(t)} = \frac{p.y(t) - q.y(t)}{r.y(t) - q.y(t)})$$

Given a finite set C of spatial constraints defined on a diagrammatic object o, an AR determines whether there actually exists none, one or many objects o that satisfy C, and the spatial properties (extent and location) of the satisfying objects. For example, the AR $BehindCurve(p, c)$, where p is a point and c is a curve, is defined as the set of all points that lie behind c with respect to p. The VR $Behind(q, a, p)$ returns True iff the points q, a, p are collinear and a lies between q and p. Thus,

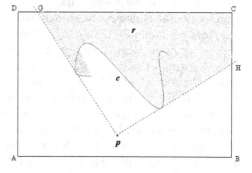

$BehindCurve(p, c) \rightarrow \{q : isaPoint(q) \land (\forall a, (isaPoint(a) \land On(a, c)) \Rightarrow Behind(q, a, p))\}$

$isaPoint(p) \Leftrightarrow (\forall t, p.x(t) = constant \land p.y(t) = constant)$

$On(a, c) \Leftrightarrow (\exists t, c.x(t) = a.x(t) \land c.y(t) = a.y(t))$

$Behind(q, a, p) \Leftrightarrow Collinear(q, a, p) \land Between(a, q, p)$

$Between(a, q, p) \Leftrightarrow (\forall t, 0 \le \frac{a.x(t) - q.x(t)}{p.x(t) - q.x(t)} \le 1 \lor 0 \le \frac{a.y(t) - q.y(t)}{p.y(t) - q.y(t)} \le 1)$

The VR/AR computing program searches the knowledge base (KB) for the definitions of the VRs/ARs used to define the desired routine in depth-first manner. If all definitions can be obtained from the KB, backward chaining is used to find all implications whose premises match with the routine and then replaces each occurrence of the routine definition by the consequent of the implications in the definition of the desired routine. If all definitions cannot be found, the program asks the user for the definitions of the particular routines that are undefined. We use a functional logic programming language in Mathematica for further reducing the constraints. Given the definition of the AR *BehindCurve* as above, the program synthesizes a program for *BehindCurve* and diagrammatically represents the region behind the curve (the shaded region in the figure).

2.1   Issues in Developing a Spatial Constraint Solver

Types of constraint-satisfaction problems. A problem might be well-constrained, over-constrained, or under-constrained. The solver outputs no solution if the constraints are inconsistent in the over-constrained case. In case of under-constrained problems, the system can incorporate appropriate constraints to produce a solution at the request of the user.

Types of constraint solvers. There are two basic strategies for solving the problem. An instance solver uses the numerical values of the given constraints while a generic solver uses symbols. For the purpose of synthesizing programs for VRs/ARs, a generic solver is more suitable. When only the solution to a specific instance is required, the instance solver is more useful. Synthesizing programs for VRs/ARs increases the library of the system which helps in defining more complicated VRs/ARs.

Approaches for spatial constraint solving. Our approach for automatically computing VRs/ARs exploits a fundamental principle - the existence and spatial properties of the desired diagrammatic objects can be captured as a logical combination of algebraic equations, with the constraints as parameters in the equations. A number of approaches have been proposed for constraint solving, such as algebraic, logic-based, graph-based, and theorem proving (see [3] for a review). A major bulk of this research has been dedicated to CAD, which deal with a limited set of objects with nice geometrical properties, such as points, lines, rays, line segments, circles and arcs; and also a limited set of constraints, such as distance, angle, parallelism, perpendicularity, incidence, and tangency between the objects. DR on the other hand requires a more general set of primitive objects and constraints, thereby demanding extension of the framework of spatial constraint satisfaction used for CAD and origami construction.

References

1. Ullman, S.: Visual routines. *Cognition.* (1984) 97–159
2. Chandrasekaran, B., Kurup, U., Banerjee, B., Josephson, J. R., and Winkler, R.: An Architecture for Problem Solving with Diagrams. In *Diagrammatic Representation and Inference*, A. Blackwell, K. Marriott, A. Shimojima (Eds), *Lecture Notes in AI* 2980, Berlin: Springer-Verlag. (2004) 151–165
3. Hoffmann, C. M. and Joan-Arinyo, R.: A Brief on Constraint Solving. *Computer-Aided Design and Applications*, Vol. 2, No. 5. (2005) 655–664

Deduction with Euler Circles: Diagrams That Hurt

Dustin P. Calvillo, Krista DeLeeuw, and Russell Revlin

Psychology Department, University of California, Santa Barbara, CA 93106
calvillo@psych.ucsb.edu
deleeuw@psych.ucsb.edu
revlin@psych.ucsb.edu

Abstract. Two studies evaluated the effectiveness of Euler circles in aiding participants in drawing conclusions to deductive reasoning problems. The problems were the ones that typically cause reasoners the most difficulty because their prior beliefs about conclusions interfere with their judgments of deductive validity. The use Euler circles reliably contributed to reasoners' inability to solve the problems. This pattern was shown for both young, university students and elderly retired people.

1 Introduction

Reasoning with syllogisms is not a natural act. If it were, students would not need detailed instructions on how to do it. Although reasoners seem to do quite well at deduction in some contexts, they are poor in others (e.g., Braine & O'Brian; Calvillo & Revlin, 2005; Evans, Newstead, & Byrne, 1993; Sloman, 1998). One source of students' difficulties with syllogistic reasoning is the interpretation of the quantified terms (e.g., Bucci, 1978; Chapman & Chapman, 1959; Revlin & Leirer, 1980; Wilkins, 1928). The use of diagrams, such as Euler (or Venn) circles, is the primary tool devised to help them with these terms. Anecdotal reports promote the use of such diagrams to help students solve deduction problems. The present study takes a closer look at reasoning with diagrams and systematically examines whether standard diagrams contribute to successful reasoning on the deduction problems that students ordinarily find to be the most difficult.

2 Errors in Reasoning

Two major sources of errors in deductive reasoning have been identified: one is concerned with how students represent the information to be reasoned about (e.g., Henle, 1962; Johnson-Laird & Byrne, 1991; Revlin & Leirer, 1978; Revlis, 1975;Wilkins, 1928) and the other source focuses on the mental operations by which reasoners evaluates tentative conclusions (e.g. Braine & O'Brien, 1998; Evans, 1989; Johnson-Laird & Byrne, 1991; Rips, 1994). These two factors naturally interact with each other because mental operations require a representation to operate upon. The Models Theory approach (e.g., Johnson-Laird, 1983) provides the most seamless description of both sources of error and has been successful in providing an account

D. Barker-Plummer et al. (Eds.): Diagrams 2006, LNAI 4045, pp. 199–203, 2006.
© Springer-Verlag Berlin Heidelberg 2006

of what on the face of it appears to be a non-rational influence on deductive decisions: the belief-bias effect (e.g., Klauer, Musch, & Naumer, 2000; Henle & Michael, 1956).

The belief bias effect shows that when reasoners are presented with premises and a possible conclusion for evaluation, they evaluate the conclusions based on their believability, even though reasoners are typically instructed to ignore "truth" in favor of validity. Participants in such studies typically construct one type of argument from the premises if the conclusion is believable and another type if it is not believable. In this way the representation and inferential processes are intertwined leading to errors in some cases, and correct deduction in others. These sorts of reasoning problems are largely impervious to simple instructions.

One early criticism of the Models Theory approach was that it was not substantively different from reasoning with Euler diagrams in the head (Newell, 1981). This raises the question whether the external use of Euler diagrams would be helpful to reasoners. Rizzo & Palmonari (2005) showed that reasoning accuracy was the poorest of a variety of conditions when reasoners use Euler circles. However, the deduction problems presented in that study were not ones in which a belief-bias effect occurs. It is possible that Euler circles might help reasoners sort through belief and validity. It is consistent with the historical use of diagrams in order to help student reasoners, it is sensible to anticipate that diagrams would be beneficial in just those cases where students do their worst: the ones where prior believability leads students to an incorrect conclusion.

3 Experiment 1

3.1 Method

Forty student volunteers solved 16 deductive reasoning problems. They read instructions that illustrated the problems and were admonished to imagine they were jurors who were only allowed to draw conclusions from the evidence and not from their personal knowledge. The problems were in the form of a paragraph followed by a single conclusion to which the reasoners had to indicate "follows" or "doesn't follow" (see procedures by Evans, Barston & Pollard, 1983). There were two within-subjects variables: Validity and Belief. Half of the problems contained conclusions that were deductively valid (EI syllogisms) and half were invalid (OI syllogisms). Half the problems contained conclusions that students would believe and half contained conclusions that students would disbelieve (Revlin, Leirer, Yopp, & Yopp, 1980).

To investigate the effect of Euler diagrams on the belief-bias effect, the students were randomly assigned to two groups. One group was given only standard instructions on deductive reasoning and the other group was given an additional sheet with examples of Euler diagrams. A single diagram corresponding to each quantifier that was present in the problems: *All, No, Some,* and *Some are not.* Although these diagrams did not exhaust the possible ones for each quantifier (e.g., there are five separate diagrams that describe the *Some* relations), they were just the ones necessary to accurately draw a conclusion to each of the syllogisms used in this experiment. After instructing the students on the diagrams' interpretations, the page of diagrams was left for the student to refer to while he or she worked each problem.

In addition, a measure of spatial abilities, the Paper Folding Test (Ekstrom, French & Harman, 1976), was administered after the students completed the reasoning problems. We included this measure in anticipation of possible individual differences in preferential reasoning strategies. It is possible that more highly spatial students would make better use of the diagrams, or that they would be more accurate on the syllogisms overall, if indeed they are using mental diagrams to solve them.

3.2   Results

If the belief-bias effect replicates with the present materials, then reasoners should endorse conclusions to which they agree and deny conclusions about which they disagree. For valid syllogisms, if they believe in the conclusion, they should be accurate in their judgment; if they disbelieve the conclusion, the students should reject it and be incorrect in their judgment. In contrast, on invalid syllogisms—problems where the reasoner should logically reject a conclusion—if the reasoner believed in the conclusion, he/she should accept it as true, and be incorrect. If the reasoner disbelieved in the conclusion, he/she should reject it and be correct. The results for the 320 judgments (20 students x 16 problems) are presented in Table 1, which shows that there is an overall belief-bias effect, $F(1,19) = 31.7$, p<.001.

Table 1. Percent of correctly accepted/rejected conclusions (chance = 50%) by undergraduate students

	Reasoning Accuracy of College Undergraduates			
	Valid Syllogisms		Invalid Syllogisms	
	Believed	Disbelieved	Believed	Disbelieved
No Euler	97.8	72.1	61.7	93.6
Euler	82.1	51.9	55.2	83.9

Table 1 also shows that the presence of Euler diagrams does not help students draw the correct conclusion. Indeed, there is a strong suggestion that these diagrams hurt performance, $F(1,25)=3.0$, p> .05. The diagrams do not selectively help or hurt performance on any particular type of problem (no reliable interactions were observed between diagrams and believability).

In the present study, there was no reliable correlation between spatial abilities and reasoning accuracy either overall or for specific subsets of problems. Nor was there an interaction between spatial abilities and the effectiveness of the Euler diagrams. Diagrams reliably hurt reasoning accuracy of all participants independent of spatial ability.

4   Experiment 2

The participants in Experiment 1 revealed that they were highly familiar with Euler circles and were quite comfortable using them (every student who was instructed on the diagrams used them). Experiment 2 considers the value of diagrams to help elderly reasoners who were not familiar with them. Spatial Ability declines with age (Salthouse & Mitchell, 1990). If the effectiveness of diagrams is independent of Spatial Ability, as shown in Experiment 1, we should see the same pattern of responding for elderly participants as we did for younger ones. We therefore obtained 26 volunteers from a local Adult Education class to participate in the study.

4.1   Method

The elderly volunteers (average age = 67 years) were asked to solve the same reasoning problems as the college undergraduates. They were given the same materials as described in Experiment 1, with the exception of the Paper Folding test.

4.2   Results

The reasoning accuracy of our elderly volunteers is summarized in Table 2, which shows the same belief-bias effect [$F(1,25)=14.8$, p<.01] as the undergraduates. In addition, diagrams reduced the overall accuracy of the elderly adults, $F(1,25)=2.3$, p=.1. An analysis of variance revealed there was no reliable difference in the pattern of responding between the two age groups.

Table 2. Percent of correctly accepted/rejected conclusions (chance = 50%) by elderly participants

	Reasoning Accuracy of the Elderly			
	Valid Syllogisms		Invalid Syllogisms	
	Believed	Disbelieved	Believed	Disbelieved
No Euler	89.6	77.9	70.6	88.2
Euler	87.5	60.8	40.0	90.0

5   Conclusion

The present findings challenge the prevailing notion that Euler diagrams help reasoners draw conclusions from deductive reasoning problems. Use of these diagrams was associated with diminished reasoning accuracy. We found that the use and effectiveness of these diagrams are independent of the users' Spatial Abilities. Although we selected the most challenging syllogisms in hopes that a floor effect might reveal any positive effect of the diagrams, the reasoners were reliably worse with Euler diagrams on all problems across two age groups.

References

Calvillo, D.P. & Revlin, R. (2005). The role of similarity in deductive categorical inference. *Psychonomic Bulletin & Review. Special Issue: Memory strength and recency judgments,* *12*, 938-944.

Ekstrom, R. B., French, J. W. and Harman, H. H. (1976) *Kit of factorreferencedcognitivetests,* Educational Testing Service, Princeton, NJ.

Evans, J. St.B T., Newstead, S. E., & Byrne, R. M. J. (1993). *Human Reasoning: The psychology of deduction*. Hove: Lawrence Erlbaum Associates.

Evans, J. St. B. T (1989). Bias in human reasoning: Causes and consequences. Hillsdale, NJ: Erlbaum.

Evans, J.St. B.T., Barston, J.L. & Pollard, P. (1983). On the conflict between logic and belief in syllogistic reasoning. *Memory & Cognition, 11*, 295-306

Henle, M. (1962). On the relation between logic and thinking, P*sychological Review,69*, 366-378.

Johnson-Laird, P.N. (1983). Mental models: Towards a cognitive science of lanuage,inference, and consciousness. Cambridge, UK: Cambridge University Press.

Johnson-Laird, P.N. & Byrne, R.M.J. (1991). *Deduction*. Hillsdale, NJ:Erlbaum.

Klauer, K.C. , Musch, J., & Naumer, B. (2000). On the belief bias in syllogistic reasoning. *Psychological Review, 107*, 852-884.

Newell, A. 1981. Reasoning, problem solving and decision processes: The problem space as a fundamental category. In R. Nickerson (Ed). *Attention and Performance. Vol.8*. Hillsdale, NJ: Erlbaum.

Revlin, R. & Leirer, V.O. (1978). The effects of personal biases on syllogistic reasoning: Rational decisions from personalized representations. In R. Revlin & R. E. Mayer (Eds.), *Human reasoning*. Washington, DC: Winston.

Revlin, R., Leirer, V.O, Yopp, H. & Yopp, R. (1980). The belief-bias effect in formal reasoning: The influence of knowledge on logic. *Memory & Cognition, 8*, 584-592.

Revlis, R. (1975). Two models of syllogistic reasoning: Feature selection and conversion. *Journal of Verbal Learning and Verbal Behavior, 14*, 180-195.

Rips, L. (1994). *The Psychology of Proof*. MIT Press, Cambridge, MA.

Rizzo, A. & Palmonari, M. (2005). The mediating role of artifacts in deductive reasoning. *Paper Presented at the 2005 Cognitive Science Society, Stresa, Italy*.

Salthouse, T.A. & Mitchell, D.R. (1990). Effects of age and naturally occurring experience on spatial visualization performance. *Developmental Psychology, 26*, 845-854.

Sloman, S. A. (1998). Categorical inference is not a tree: The myth of inheritance hierarchies. *Cognitive Psychology, 35*, 1-33.

Wilkins, M.C. (1928). The effect of changed material on the ability to do formal syllogistic reasoning. *Archives of Psychology, 102*, 1-83.

Diagrams as Physical Models[*]

B. Chandrasekaran

Laboratory for AI Research
Department of Computer Science & Engineering
The Ohio State University
Columbus, OH 43210 USA
chandra@cse.ohio-state.edu

Abstract. We discuss a variety of roles for diagrams in helping with reasoning, focusing in particular on their role as physical models of states of affairs, much like an architectural model of a building or a 3-D molecular model of a chemical compound. We discuss the concept of a physical model for a logical sentence, and the role played by the causal structure of the physical medium in making the given sentence as well as a set of implied sentences true. This role of a diagram is consistent with a widely-held intuition that diagrams exploit the fact that 2-D space is an analog of the domain of discourse. One line of research in diagrammatic reasoning is that diagrams, rather then being models, are formal representations with specialized rules of inference that generate new diagrams. We reconcile these contrasting views by relating the usefulness of diagrammatic systems as formal representations to the fact that their rewrite rules take advantage of the diagrams' model-like character. When the physical model is *prototypical*, it supports the inference of certain other sentences for which it provides a model as well. We also informally discuss a proposal that diagrams and similar physical models help to explicate a certain sense of *relevance* in inference, an intuition that so-called Relevance Logics attempt to capture.

1 Roles of Diagrams in Reasoning

Diagrams give many different types of assistance during problem solving. We identify five roles here: helping extend short term memory, helping organize problem solving by spatial organization of related information, as sentences in a 2-D language with specialized rules of inference, providing a model of the premises so that plausible subtasks may be hypothesized for formal inference, and providing a model

---

[*] This paper was prepared through participation in the Advanced Decision Architectures Collaborative Technology Alliance sponsored by the U.S. Army Research Laboratory under Cooperative Agreement DAAD19-01-2-0009, and by federal flow-through by the Department of Defense under contract FA8652-03-3-0005 (as a subcontract from Wright State University and Wright Brothers Institute). I am indebted to Peter Schroeder-Heister and Gerard Allwein for significant assistance in thinking about these ideas, to Neil Tennant and Stewart Shapiro for useful discussions, and to one of the reviewers who made useful suggestions for improvement.

D. Barker-Plummer et al. (Eds.): Diagrams 2006, LNAI 4045, pp. 204–217, 2006.

of the premises from which consequents can be inferred and asserted. The major concern of this paper is in the last role.

First, diagrams extend short term memory, by providing a spatially organized external location in which to note down information.   Second, they help *organize* problem solving. Simon and Larkin (1987) use the example of analyzing a pulley system – they show how the diagram of the pulley system helps the problem solver organize the sequence of equations to solve, or variables to assign values to. The problem solver can use his visual perception to locate the pulley that a strip of rope goes over, and thus to choose which tension variable to consider next.   Another example is the spatial organization of addends when we add two numbers: we line up the numbers such that the numerals in the ones, tens, etc., locations line up, and the locations and the spatial relations guide applications of the sequence of problem solving actions.

The third role is that of diagrams as two-dimensional syntax-controlled compositions of diagrammatic symbols[1]. Specialized rules of inference can be specified that can generate valid diagrammatic sentences. Allwein and Barwise (1996) contains a number of papers pursuing this perspective in productive ways. In this framework, e.g., theorems in set theory can be legitimately proved using an appropriate sequence of Venn or Euler diagrams.

The fourth and fifth roles both treat the diagram as a physical instance, a *model*, of a state of affairs of interest. That is, it depicts a situation that satisfies the premises. But the fourth and fifth roles deal with different ways of using the physical model. In the fourth role, the model suggests hypotheses to pursue in the formal proof. This is exemplified by the way diagrams are used in proving theorems in Euclid. In this kind of use, a diagram is *a* model of the premises. Not everything that is true in the model is necessarily true given the premises, but nevertheless a careful use of the model suggests possibly productive subtasks for theorem proving. For example, the fact that two angles are adjacent, and the theorem involves one of the two adjacent angles might suggest to the theorem prover that perhaps stored theorems involving adjacent angles may be useful in advancing the proof. Lindsay (1998) provides a review of the issues in the use of diagrams in geometry theorem proving. It has been estimated that the use of the diagram in this way reduces the search space by several orders of magnitude. In the traditional use of Venn or Euler Diagrams in proving theorems in Set Theory, the diagrams play a similar role. It is important to emphasize that the information from the model is not *asserted* as conclusion, but only used to find strategies for arriving at the general conclusion.

The fifth role for diagrams is also based on diagrams as physical models that satisfy the premises. The problem solver sees that the representation is also a model

---

[1] The Stanford Encyclopedia of Philosophy entry on Model Theory (http://plato.stanford. edu/entries/model-theory/) says, "…the overwhelming tendency of this work is to see pictures and diagrams as a form of language rather than as a form of structure. For example Eric Hammer and Norman Danner (Allwein and Barwise, 1986) describe a 'model theory of Venn diagrams'; the Venn diagrams themselves are the syntax, and the model theory is a set-theoretical explanation of their meaning."   This quote might overstate the case a bit ("overwhelming tendency"), but "diagrams as sentences or formal representations" is a common enough view. We comment later on how the views of diagrams as representations versus models might be reconciled.

for another assertion that is not explicitly part of the premises, and concludes that the assertion follows from the premises. Exactly when to generalize and by how much are issues for which answers differ from one diagrammatic application to another. Such a use of diagrams is much more common in applied rather than formal reasoning. This role of diagrams is my focus in this paper.

Consider two very simple examples. Given a simple addition problem in arithmetic, say to show $1 + 3 = 2 + 2$, suppose one draws four points (or arrange four stones on the ground) as below:

Fig. 1

Under the appropriate mappings, the situation is a model of $1 + 3$. But it is also a model of $2 + 2$. One can demonstrate to a child that $1 + 3 = 2 + 2$ by using the above diagram. Here the generalization issue is trivial: the child could, but typically wouldn't, say, "Maybe this is true when we add 1 star to 3 stars, but is it true when we add 1 slice of pizza to 3 slices of pizza?." Human intuitions about numbers seem sufficiently robust that this issue doesn't arise in a child or an adult. "Individuals that keep their distinct identity" seems to be the background intuition that is operational here, and using that people generalize from star marks on paper or stones on the ground to numbers in general.

Consider another example, one that will find frequent use in the rest of the paper. Given "If A is to the left of B, B is to the left of C, is A to the left of C?," people often draw a diagram as in Fig. 2:

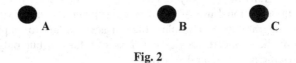

Fig. 2

There is a natural sense in which the physical diagram is a model of the problem situation[2]. The problem solver notices that indeed A is to the left of C, and declares that the inference is true[3]. Of course, the diagram only represents one specific way in which the points can be located to provide a model. Yet the problem solver makes

---

[2] More precisely, it a model of the conjunct of the given premise with axioms that capture the structure of space in terms of which the predicate Left is defined. For someone for whom the semantics of Left is that of spatially left in ordinary language, the axioms are implicit, and the Figure provides a model for the premise.

[3] In fact, the precise role played by the diagram in this inference is more complicated than appears at first, but for the current purposes, this partial account is adequate.

bold to assert that the conclusion is true for all the specific ways in which the points could be located.

Diagrams are just an example of physical models of this type. As mentioned earlier, architectural models and models of molecules constructed out of ping pong balls provide further examples, though their scopes in assisting in human reasoning are not as large as that of diagrams.

Regarding a main concern of logic – accounting for justifiable inferences – this style of reasoning based on a physical model needs to be part of any account of natural reasoning. It puzzles me that more has not been said in logic about the use of such physical models as aids to reasoning, given how prevalent diagrams are in everyday as well as professional reasoning, and the role played by architectural and molecular models in their respective disciplines. So this paper's goal is to raise the profile of physical models in logic. I raise a set of issues for deeper consideration by logicians.

Not everything about a diagram is model-like. It is important to mention that not all diagrams, or all aspects of a given diagram, are models. Actual diagrams have various notations in them, such as letters A and B in Fig. 2, or shadings as in Venn diagrams, that are not model-like. We return to this issue in a later section.

2 Physical Fragments Providing Models for Logical Sentences

A brief remark might be useful on the multiple, sometimes opposite ways, in which the term "model" has been used. In philosophy and practice of science, a description – a set of equations, e.g. – is called a model a domain if the description can be used to predict phenomena in the domain. Thus, Maxwell's Equations model electro-magnetic phenomena and physicists speak of the Newtonian model versus the Einsteinian model. In logic, the direction is from description to domain: a domain provides a model of a of axioms, e.g., arithmetic is a model of Peano's Axioms and plane geometry is a model of Euclidean Axioms. If the axioms are a description, the domain that fits the description is a model of the description. In related usage in logic, a model for a sentence is constructed by assigning truth values to the elements of the Herbrand Universe. In the rest of the paper, we use the term in the sense in which it has been used in logic.

It is useful to start with the standard definition of a model for a sentence in Logic. An *interpretation* for a sentence S consists of:

- A non-empty, possibly infinite, domain D of individuals
- Assignment of specific individuals in D to constant symbols in S
- Assignment to each n-ary function symbol in S of an n-ary function that maps from D^n to D.
- Assignment to each n-ary predicate symbol in S of an n-ary function that maps from D^n to {True, False}.

An interpretation for S is a *model* for it if S evaluates to True under the interpretation.

2.1 Modeling Physical Things

An informal description of how a physical entity may be used to provide an interpretation of a sentence might go as follows. Suppose one finds a physical entity, organizes it into parts, and models the parts in terms of particular sets of physical variables. A specific entity will have specific values for these variables. The causal structure of the physical entity will induce a set of causal constraints on these variables. Suppose, further, one is able to map the individuals in a given sentence S into the "parts" of the entity, and to map the predicates in S into relations between the variables of the parts. If the variable values of the entity are such that they satisfy S under this mapping, the entity could be said to be a physical model for S. What follows is a more formal rendering.

Domain of Individuals. Let Π be a fragment of physical world. Let $\Delta_\Pi: \{\pi_1, \pi_2, \ldots\}$ be a (possibly infinite) set of entities, each π_i a *part* – a subfragment – of Π. The entities need not be physically disjoint – one entity may be a physical part of another entity; nor it is necessary for Δ_Π to exhaust Π, i.e., the totality of physical fragments represented by the elements of Δ_Π to be equal to the matter represented by Π.

I will use two concrete examples to illustrate the ideas: Π_1, the set of points constituting a finite physical horizontal straight line, say drawn on a piece of paper; and Π_2, a physical object intended to be an architectural "model" of a house.

Examples in Π_1: The entire finite straight line is Π. Each point in it is a π, thus Π has an infinite number of parts in this model. Another model for the same physical object might subdivide the line into various segments, each providing a π.

Example in Π_2: The physical entity (the architectural "model") as a whole is Π, and the physical matter corresponding to various rooms, walls, doors, etc., are the π's.

Functions and Predicates. Let $\{\phi_i \mid 0 \le i \le k\}$ be a finite set of functions of various arities, such that if ϕ_i is n-ary, it is a function from Δ_Π^n to Δ_Π. Similarly, let $\{\rho_i \mid 0 \le i \le l\}$ be a set of functions of various arities, such that if ρ_i is n-ary, it is a function from Δ_Π^n to $\{True, False\}$. The ρ's are predicates defined on the physical variables, and thus the values that they take for their various arguments is determined by the causal structure of Π.

Examples in Π_1: The function, $\text{right}_1(\pi_i)$, defined as "the point that is exactly 2 inches to the right; if there is no such point, the right end point" is a unary function. Example of a binary predicate ρ is: $\text{Left}(\pi_i, \pi_j)$, with the obvious interpretation.

Examples in Π_2: Unary function Entrance-to(room_i), which takes values from the subset of parts of type "door." Thus, e.g., Entrance-to(room_5) = door_6. Example of ρ: Bigger-than(room_i, room_j) is a binary function which evaluates to *True* if the area of room_i is larger than that of room_j, and *False* otherwise.

Properties and Causal Structure of Π. For the purpose at hand, the physical structure is modeled in terms of a set of variables, selected *attributes* of the physical system. A specific physical instance will have specific values for these variables.   Let Θ_i: $\{ \theta_{i1}, \theta_{i2}, ... \theta_{ik_i} \}$ be a set of variables in terms of which entity π_i is modeled, and let Θ = $\bigcup_{i=1}^{n} \Theta_i$. The *causal structure* of Π, which constrains the values of the variables in Θ, determines the truth values of the various predicates for various values for their arguments, and thus the truth values of sentences composed out of these predicates.

Thus, part of modeling a physical fragment for the purpose of providing an interpretation for a sentence involves identifying a physical system with the right properties to provide an interpretation, and then setting its parameters – the values of the variables of the part – to those values for which the physical fragments provide a model for the sentence.

Examples in Π_1: Let part π_i be modeled in terms of a single variable x_i, the x-coordinate of π_i from some origin. Left(π_i, π_j) is defined by the values of x_i and x_j. Additionally, the constraints of the physical line result in constraints between predicates: if Left(π_i, π_j) and Left(π_j, π_k) are both *True*, then Left(π_i, π_k) is constrained to be *True*.

Examples in Π_2: The parts of the house may be modeled in terms of their length, width, height, area, etc. Color and material out of which a part is made may also be in the set of variables. Whether room$_5$ is larger than room$_3$ is fully determined by the physical dimensions of Π; there is no additional freedom to assign *True* or *False*. Further, in a physical architectural model, if room$_1$ is larger than room$_2$ which in turn is larger than room$_3$, the model will necessarily satisfy the predicate, larger-than(room$_1$, room$_3$).

In order to avoid confusion between different usages in science and engineering on one hand and in logic on the other, I use the term *P-model* to refer to a description of a physical entity as in the next definition – the description specifies a *point of view* to look at the physical entity. In order to take a gingerbread house as a possible model of some house, it needs to be viewed as a decomposition of matter into walls, rooms, etc., each as having lengths and heights, rather than as bits of sugar and ginger and flour.

Definition. A *P-model* of a physical fragment Π consists of the following specifications:

- Δ_Π, a set of individuals consisting of parts of Π

- a set $\{\phi_i\}$ of functions of various arities, such that an n-ary function is a mapping from Δ_Π^n to Δ_Π.

- a set of functions $\{ \rho_j \}$ of various arities, such that an n-ary function is a mapping from Δ_Π^n to $\{True, False\}$.

- a set of variables Θ in terms of which Π and elements of Δ_Π are modeled; a set Π_{ax} of causal constraints between the variables in Θ.

Remark. There is an infinity of P-models for a given physical entity.

2.2 A Physical Entity Supporting a Logical Model

Let a P-model M_Π of a physical fragment provide an interpretation for a sentence S.

Definition. If a sentence S evaluates to *True* under the interpretation provided by a P-model M_Π of a physical fragment Π, we say that Π provides a *physical model* for S.

Remark. What makes a predicate *True* or *False* in a physical model is that the variables take specific values in the physical fragment, and the causal structure Π_{ax} constrains values between variables.

Examples
Consider the following sentence S:

$$\forall x\, \forall y\, \forall z\ (L(x,y)\ \&\ L(y,z) \rightarrow L(x,z)) \tag{1}$$

Let Π be a physical 1-D spatial line fragment, and let the following be a P-model for Π. M_Π:

Δ_Π : the (infinite) set of points in the line fragment, $\{x_i\}$.

Θ: a single attribute, the co-ordinate of a point x_i with respect to some origin $\{\phi_i\}$: Null set

$\{\rho_j\}$: a single function, Less-than(x_i, x_j) = *True*, if the co-ordinate

of x_i is less than that of x_j; *False*, otherwise. $\tag{2}$

Under the interpretation M_Π, S is *True* in Π. A physical 1-dimensional line fragment is thus a physical model for the sentence (1).

As a more complex example, consider S': $[(\forall x\, \forall y\, \forall z\ (L(x,y)\ \&\ L(y,z) \rightarrow L(x,z))\ \&$

$(\forall x\, \forall y\ (L(x,y)\ \&\ L(y,x) \rightarrow Eq(x,y))]\ \&\ L(A,B)\ \&\ L(B,C) \tag{3}$

Consider the physical diagram in Figure 2 (where the little circles are to be taken as points), with the following M'_Π.

M'_Π: M_Π as defined in (2) plus the following assignments: Constants A, B and C assigned to the points in the 1-dimensional line fragment corresponding to the coordinates as in the Figure. Eq(x,y) assigned to function "Equal(x_i, x_j) = True iff x_i is the same as x_i, and False otherwise." $\tag{4}$

Under the interpretation M'_Π, S' evaluates to *True*, so the diagram in Fig. 2 is a physical model for S' in (3). Readers will recognize (3) as a simple axiomatization of left-ness plus the premises of the problem we stated at the beginning. M'_Π is also a model for the following, S'':

$[(\forall x \; \forall y \; \forall z \; (L(x,y) \; \& \; L(y,z) \rightarrow L(x,z)) \; \&$

$(\forall x \; \forall y \; (L(x,y) \; \& \; L(y,x) \rightarrow Eq(x,y))] \; \& \; L(A,C)$ (5)

Remark. In applied reasoning, the agent is reasoning in some domain of interest, D, and he is interested in making a model of a sentence, say S. Let D_{ax} be the set of axioms that describe the relevant aspects of the domain of interest. Thus, the agent is looking for a physical model of D_{ax} & S. Seeing Figure 2 as a model of Left(A,B) & Left(B,C) requires interpreting Left in the spatial meaning of the terms. This interpretation assumes D_{ax}. If instead S were Goo(A,B) & Goo(B,C), one wouldn't see Fig. 2 as its physical model. Successfully making a physical model of S when the agent is reasoning in D involves finding a physical medium such that its causal structure Π_{ax} has the right kind of homomorphism relation with D_{ax}.

There is no requirement that an arbitrary Π have a P-model that provides an interpretation for an arbitrary sentence S. In fact, it is a special situation where a physical model can be constructed so as to provide an interpretation for a sentence. The next section discusses how such physical models are often used.

2.3 Warrant for Generalization

Fig. 2 is a model for S', but it is just one model. There are infinitely many configurations of points for which the corresponding physical diagrams will provide a model for S'. Nevertheless, we generalize the inference to a class of situations. Fig. 2 also provides a model for "A is farther left of B than B is of C," but we know that this inference cannot be generalized.

This way of using physical models is quite common in applied reasoning. A chemist, who is considering whether S1 $\rightarrow$ S2, where S1 and S2 are sentences in his domain, might construct a chemical reaction which is a model of S1 (really a model of his domain axioms and S1), see if it is also provides a model for S2, and, though the specific chemicals in interaction model only instances of S1, generalize to the larger class. Of course, a good chemist would know just want sort of model to construct that would bear the generalization.

This style of proof might be called *physical-model-based proof.* The *Model-Based Rule of Inference* may be stated as follows:

> Given an inference problem, $S_1 \rightarrow S_2$, where S_2 is not a logical truth, in domain D with domain theory D_{ax}, and given a physical fragment Π such that it provides a P-model M_Π that satisfies D_{ax} & S_1, if M_Π also satisfies D_{ax} & S_2, and if M_Π *has warrant for generalization with respect to the inference* S_2, conclude $S_1 \rightarrow S_2$ in the general case in D.

One might use the term *prototypical* to describe a model that provides such a warrant for generalization. Asserting logical truths from the model is blocked for reasons related to relevance (see later section on Relevance Logics).

In many cases, the applied reasoner has limited or no access to D_{ax} is an explicit form. However, he has a body of intuitions and practices that help him construct prototypical models for classes of S's that provide warrant for generalization and help him scope them.

Let S be a sentence in a domain D characterized by axioms D_{ax}. Let Closure(D_{ax} & S) be the set of all inferences that are deducible from D_{ax} & S. If Π provides a model for S, it will also provide a model for all elements of Closure(D_{ax} & S). However, it will also provide a model for many other inferences that are not in Closure(D_{ax} & S). The reasoning agent needs to know how not to make the inferences that are not in Closure(D_{ax} & S), even though the model supports it. For example, he needs to know not to infer "A is farther left of B than B is of C" from Figure 2.

Different Types of Generalization involving Diagrams. Jamnik (2001) presents a system called Diamond that uses diagrams to prove results such as *1 + 3 + 5 ...+ (2n-1) = n^2*. The proof involves constructing, for a given *n*, an array of *n x n* dots arranged as a square. Because the proof is for a general n, the diagram uses ellipses to indicate the general case. Use of such a diagram involves two kinds of generalization, one of the sort we discuss in this paper, namely, use of *n* dots or stars (as in Fig. 1) to represent *n*. The use of ellipses in diagrams involves a very different type of generalization, inductive generalization over *n*. Jamnik uses an inference rule that she calls the ω–rule that is explicitly intended to help with the latter issue. However, the rule implicitly incorporates generalization over the specific icon (dot or star) used in the array to represent the number *n*. All the perceptions that the reasoning procedure calls for during reasoning involve treating a star or a dot as a singleton in the unary representation of a number. In the case of numbers, such representations and corresponding generalizations are so deeply rooted in our behaviors – we all know to treat the relevant icon as an integral entity and abstract it as a unit – that we don't stop a moment to think about them. However, in other uses of models, such as a chemist using chemical mixtures or use of unfamiliar diagrammatic schemes, learning the proper use of the scheme involves learning, implicitly or explicitly, the limits of generalization.

2.4 Diagrams as Models Versus Representations

Some of the well–known work in diagrammatic reasoning can be seen to be based on a diagram being a model, at least in parts. In Hyperproof (Barwise and Etchemendy, 1994), a computer-based system intended to help students learn logic, the left hand side, say, of the screen might contain certain premises and conclusions posed as predicate logic sentences in terms of visual and spatial properties of and spatial relations between objects such as cubes and tetrahedrons of different colors, sizes and locations. The right hand side might show a diagram where an area of space contains cubes and tetrahedrons of different sizes and shapes. The student learns to check if a certain sentence is true or false in the situation represented by the diagram, and learns to make or check inferences by building, or making use of, such diagrams. These parts of the diagram are physical models of the sentences on the left hand side. Similarly, a diagram showing a circle A inside another circle B is a model of the set theoretic assertion, $A \subset B$: the set of points in region A are indeed a subset of points in region B in the physical 2-D space. However, diagrams often have other elements in them whose role is more notational than model-providing, e.g., shading of regions in Venn diagrams to indicate emptiness of the corresponding set. These notational elements often obscure the role of the model parts of the diagrams, and encourage a view of them as simply formal representation, one that just happens to be different in kind from linear sentential ones. I think that these two views can be reconciled. Here's an outline of how.

First, a model itself is a *composition* of elements, the composition following some syntax. Venn and Peirce's diagrammatic systems for representing set-theoretic assertions can be described (Shin, 1994) as formal languages composed, following a syntax, of regions and various notational elements, and rewrite rules that permit replacement of diagrams satisfying certain conditions with other diagrams.

Suppose, in some sentential system, a rewrite rule supports writing S3, given S1 and S2. Suppose M1 and M2 are physical models for S1 and S2 respectively in a modeling system M, and suppose M3 is a composition of M1 and M2 such that it provides a model for S3. Suppose also that we treat the model system as a formal representation one of whose rewrite rules allows creating M3, given M1 and M2. If generalization from M3 to the assertion contained in S3 has warrant for generalization, we can treat the model system as an alternate formal representation for the assertions in the original sentential system. Thus, the sequence M1, M2 and M3 can be simultaneously viewed as models of S1 , S2, and S3, such that inferring S3 has the appropriate warrant for generalization, and as a proof of the content of S3 in an alternate formal representation language M. As I see it, the diagrammatic formal language that Shin develops works because the rewrite rules appropriately embody the generalization that the physical models allow.

Heterogeneous Proofs. When students use Hyperproof to make inferences, their proofs are *heterogeneous* (Barwise and Etchemendy, 1994) – the reasoning involves a sequence of steps some of which are sentential while others are inferences made from the diagrammatic part. As we just pointed out, what the students do with the diagrammatic component can be characterized as model-based inference or, equivalently, inference in a 2-D sentential system whose rewrite rules happen to allow just the kind of model-based inferences that support the needed generalization.

3 Prototypical Models

What makes a prototypical model? The specifics depend on the domain and the predicates of interest, but some general intuitions may be useful. The following ideas might help in the development of a more formal account.

The first idea is *minimality*. Let S be Left(A,B) & Left(B,C) (we are implicitly in the domain of 1-d space with a directed axis). Just as Fig. 2, Fig. 3 also provides a model for S. However, it provides a model for unrelated things such as Inside(D,E). Clearly, any inference based on this model, such as Left(A,B) & Left(B,C) → Inside(D,E), would be a mistake. Fig. 2 is in some sense minimal compared to Fig. 3 for Left(A,B) & Left(B,C).

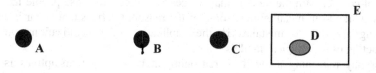

Fig. 3

The next idea is that of multiple prototype models. Let $S^{\wedge}$ be Left(A,B) & Left(A,C). While Fig. 2 provides a model for $S^{\wedge}$, it doesn't seem prototypical for another reason: it only accounts for a subset of instances. Fig. 4 provides another model.

Fig. 4

Fig. 2 provides a model for Left (B,C) and Fig. 4 provides a model for Left(C,B), neither of which follows from $S^{\wedge}$, thus neither of these inferences have a warrant for generalization. Applied reasoning in this case requires that two models be set up, each of which allows certain inferences, say Right(B,A), but not Left (B,C) or Left(C,B).

The third idea is a revisit of what we mentioned earlier, that Fig. 2 doesn't support the generalization of "A is farther left of B than B is of C," though the figure provides a model for it. Suppose a new predicate boogoo(x,y,z) is defined as "x is farther left of y than y is of z." Consider $S^{\#}$: Left(A,B) & Left(B,C) & Left(C,D), modeled by Fig. 5. Fig. 5 also provides a model for boogoo(A,C,D), which has a warrant for generalization. Even though boogoo(A,B,C) is True in the model in Fig. 2, the model does not provide a warrant for generalization.

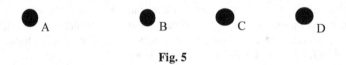

Fig. 5

It can be seen, from a metatheory of Left(x,y), that it provides an order between x and y. Thus any conjunctions of Left(x,y)'s would specify an order or alternate possible orders. Only information that follows from the order information has warrant for generalization. Such metatheories may be constructed in principle for every domain, but in practice, an agent performing applied reasoning in some domain usually has no access to such metatheories, at least in explicit forms. Even when he has access to the axioms for his domain, such as in the sciences, they are provisional and potentially revisable. So, reasoning in the practical world is aided by models constructed and interpreted with the aid of intuitions based on experience, training and partial theories. Such models play a large role in commonsense reasoning as well, as evidenced by the research of Johnson-Laird (1983) on how people solve syllogistic problems by constructing mental models. These mental models have many points of contact with the models that we describe here. Because people lack access to fully worked-out metatheories, some of the reasoning errors that occur in practical reasoning are due to mistakes in the application of generalization and in the construction of prototypical models.

Everyday reasoning as well as reasoning in professional disciplines is full of implicit and explicit guidelines about how to construct physical models – diagrams in particular – from which that desired information can be obtained perceptually, and about how to generalize. The ubiquity of such diagrams in human reasoning notwithstanding, it is important to note that discoveries of appropriate diagrams for classes of problems are hard-won. Such discoveries are prized – transmitted culturally for everyday reasoning, and made part of training in professional disciplines, with discoverers often honored with awards.

3.1 Relevance Logics and Prototypical Models

The model-based rule of inference blocks asserting a logical truth based on the physical model. This is because we wish the physical model to *play a role* in the assertion. Why? The intuition is the same as that which drives research on *Relevance Logics*, a summary of which is available in Mares (2006). Here's the main idea.

Relevance logics are a response to what some people take to be paradoxes of traditional implication. The paradox arises in that in some of the inferences authorized by the semantics of traditional implication, the antecedent doesn't seem *relevant* to the consequent. For example, $p \rightarrow (q \rightarrow p)$ and $p \rightarrow (q \rightarrow q)$; q in the first case and p in the second don't seem relevant to the conclusions. People performing applied reasoning, i.e., domain-specific every day argumentation, would object to someone bringing what appear to be irrelevant issues.

Now suppose a reasoning agent performing domain-specific reasoning uses prototypical physical models to assert consequences. That is, given p, he constructs a prototypical model in his domain for W that also models S, and that S has a warrant for generalization. In such cases, his assertion $p \rightarrow S$ will be a relevant inference *in W*. Technically, any model is also a model for logical truths, so we add a rule forbidding asserting a logical truth as a consequence of any sentence.

Consider as an example:

$$p \rightarrow (q \rightarrow p) \tag{6}$$

In this case p doesn't seem to play a relevant role in making $(q \rightarrow p)$ true. However, suppose there is a domain D in which q is a possible cause for p, and that q would definitely cause p. In that domain, it wouldn't be a surprise to assert that if p is true, then if q is known to be present, q caused p, and consequently that the truth of q would imply the truth of p. If D_{ax} denotes the axioms characterizing D, the following would not fail the test of relevance:

$$(D_{ax} \ \& \ p) \rightarrow (q \rightarrow p) \tag{7}$$

Suppose D describes a physical domain in which one constructs a physical fragment Π that provides a P-model for q. Π, under the same P-model mappings, would also provide a model for p. That is, there is no way to construct a model for q in D without it being a model for p as well. In this case, there is no failure of relevance to assert (6).

The discussion above leads to the suggestion that judgments of relevance in the case of implications arise in applied reasoning in specific domains, whose structures (causal structures if domains are physical) then can be judged to play or not play a role in the antecedent making the consequent true. Further, making such judgments is facilitated in the specific domains by constructing physical models when possible.

The foregoing is a précis of a slightly longer discussion that will appear in a forthcoming paper (Chandrasekaran, to appear). As far as I am aware, current approaches to Relevance Logics don't follow the above approach. My main reason to discuss this is to point to a potentially productive direction of research for the logicians interested in Relevance Logics.

4 Concluding Remarks

Diagrams are often-used reasoning aids in many situations. This paper views diagrams as just the most prominent example of larger class of reasoning aids that provide physical models to premises in some domain. Applied reasoning, where reasoning agents are concerned with inferences in specific domains rather than abstract notions of validity, has not drawn as much attention from logicians as it should. The use of such physical models in applied reasoning raises important issues in logic. I have attempted to formalize the notion of some piece of physical matter providing a model for a sentence. I identified a proof technique called physical model-based inference in which prototypical models in specific domains are constructed that support useful generalizations.

All of this is in accord with a central intuition of Johnson-Laird (1983) that people tend to reason *concretely*, by building small-scale "mental models" of concrete situations that satisfy the premises and generalizing from them, rather than by applying abstract rules of inference. Even though such mental models do not have to be diagrammatic, it is often the case that, when possible, people do construct diagrams of such concrete situations, which then serve as models in the sense that has been elaborated in the paper.

As can be seen from the example of pebbles for adding numbers, whose generalization properties even young children seem to understand, human intuitions on physical models and their prototypicality seem to draw on fairly deep structures in cognition. Developing reasoning skills in various domains includes acquiring or developing intuitions about how to construct prototypical models for specific reasoning situations.

I briefly described how the diagram as model view and the diagram as formal representation view can be reconciled by noting that if the diagrammatic rewrite rule happens to incorporate just those inferences that have the warrant for generalization in the specific model system, then the diagrammatic sequence can be seen as sequence of models or, alternatively, as a sequence of representations that result in the needed inference.

I also related such physical model-based reasoning to issues in Relevance Logics, where the goal is to identify when and how antecedents can be said to have a role in the consequent being true. Physical models, by incorporating the underlying causality of the domain, make it possible, under many conditions, to see whether or not the antecedents play a role in an implication being valid. One of my goals in this paper is to invite the attention of logicians more expert than I to look into what seems to be important lines of investigation.

The paper's account of how the causal constraints of the physical model result in entailments of assumptions also being satisfied by the model raises issues about mental images of diagrams. The degree to which the substrate of mental images mimics the causality of physical space is controversial. This issue requires a complex and nuanced discussion taking it beyond the scope of the current paper, but pointing out the existence of this issue is relevant to the goals of this paper.

Model-based reasoning of the kind I have discussed is not merely an issue in logic, but in artificial intelligence. AI has focused almost exclusively on what might be called linguistic representations, mirroring the logical form of natural language sentences. However, real reasoning in humans is multi-modal, with perceptual and kinesthetic modalities often contributing to problem solving. Diagrams provide an important window into such multi-modal representations. In (Chandrasekaran, et al, 2004), we describe a diagrammatic representation and reasoning architecture that integrates traditional symbolic reasoning with diagrammatic reasoning. This architecture can be viewed as a kind of generalization of heterogeneous reasoning of Barwise and Etchemendy (1994).

References

Allwein, G. and Barwise, J. eds. (1996), Logical Reasoning with Diagrams, New York, Oxford University Press.

Barwise, Jon, and John Etchemendy (1994), Hyperproof, Stanford: CSLI, and Cambridge: Cambridge University Press.

Chandrasekaran, B., Unmesh Kurup, Bonny Banerjee, John R. Josephson and Robert Winkler (2004), "An Architecture for Problem Solving with Diagrams," in Diagrammatic Reasoning and Inference, Alan Blackwell, Kim Marriott and Atsushi Shimojima, Editors, Lecture Notes in Artificial Intelligence 2980, Berlin: Springer-Verlag, pp. 151-165.

Chandrasekaran, B. (to appear), "Diagrams as Physical Models to Assist in Reasoning," in Model-Based Reasoning in Science and Engineering, L. Magnani, ed., King's College Publications, London.

Jamnik, M. (2001). Mathematical Reasoning with Diagrams: From Intuition to Automation. Stanford, CA: CSLI Press.

Johnson-Laird, P. (1983), Mental Models: Towards a cognitive science of language, inference, and consciousness, Cambridge: Cambridge University Press.

Larkin, J. & Simon, H. (1987), "Why a diagram is (sometimes) worth ten thousand words," Cognitive Science, 11:65-99.

Mares, Edwin "Relevance Logic," The Stanford Encyclopedia of Philosophy (Spring 2006 Edition), Edward N. Zalta (ed.), forthcoming, URL = <http://plato.stanford.edu/archives/spr2006/entries/logic-relevance/>.

Lindsay, Robert K. (1998), "Using Diagrams to Understand Geometry," Computational Intelligence, 14: 238-272.

Shin, S., 1994, The Logical Status of Diagrams. Cambridge: Cambridge University Press.

Visual Creative Design with the Assistance of Curious Agents

Ewa Grabska[1], Katarzyna Grzesiak-Kopeć[2], and Grażyna Ślusarczyk[1]

[1] Jagiellonian University, Institute of Computer Science, Nawojki 11,
30-072 Cracow, Poland
uigrabsk@cyf-kr.edu.pl, grazyna@ii.uj.edu.pl
[2] Jagiellonian University, The Faculty of Physics, Astronomy and Applied Computer
Science, Reymonta 4, 30-059 Cracow, Poland
katarzyna.grzesiak-kopec@uj.edu.pl

Abstract. This paper presents the new framework for visual computational design in which emergence is a key to creativity. Visual shape
grammar computations proceeded with the use of diagrams are situated
in a design context indirectly through the designer working with the
application or more explicitly through curious agents. Curiosity of intelligent agents is applied to search for novel and plausible design solutions.
The approach is illustrated by designing decorative patterns.

1   Introduction

This paper presents a new framework for visual kind of creative computational
design with the use of shape grammars and curious agent assistants. It takes
advantage of the diagrams not only in communication and cognition but in
creative thought as well. Visual shape grammar computations, emergent phenomena which enhance creativity and curiosity of intelligent agents are applied
to search for novel solutions. The approach is illustrated by the example of a
rosette design task.

2   Framework for Visual Computational Creative Design

Design knowledge and design results are often expressed graphically. A significant cognitive phenomenon of visual reasoning which enhances creativeness is
emergence. In design, these features which are not explicitly represented but
emerge from a design structure are called *emergent*. Emergence plays a key role
in the interaction between the designer and his/her drawing or sketch and has
a crucial influence on the dynamic context of design.

   The proposed framework [1] for visual kind of conceptual design focuses on
dynamic character of the environment. It enables the designer not only to shape
the design context on his/her own by modifying the design space but it makes the
use of the computational intelligence as well. Three crucial elements moulding
the design context are distinguished: the designer, curious agents and a shape

D. Barker-Plummer et al. (Eds.): Diagrams 2006, LNAI 4045, pp. 218–220, 2006.
© Springer-Verlag Berlin Heidelberg 2006

grammar. Adopting a shape grammar as a generative system enables to visualise on-line the ongoing design process, discover emergent shapes and introduce them into the process. Curious agents as the intelligent designer's assistants shape the design environment and are responsible for pointing the most promising design steps, which are likely to lead to creative design solutions. Their goal is to determine the novelty of the observed situation which means validating the originality of the current environmental state with respect to the history. They also have to identify unexpected consequences of future actions. Both novelty and surprise can be computed by comparing the current state of the agent with its previous experiences stored in memories. So, the degree of the agent's interest depends on the gained knowledge, while its curiosity can be defined as a form of motivation that promotes further exploration [3]. A single design task can be distributed among several curious agents communicating with one another. The very important aspect of the proposed solution is the interaction between agents and the designer. In this way, the designer can teach the assistants to take into account his/her individual preferences which are essential in design.

3   Curious Agent Assistants

The proposed framework has been verified by a dedicated application *DesignAnalizer2D* which deals with two different tasks: designing floor-layouts and 2D decorative patterns. The classified emergent elements are detected on-line by the system and new design rules containing them are defined [1]. By now, the first stage of the research on designing floor-layouts has been completed while the one on designing 2D decorative patterns is still in progress.

Let us consider the example of designing a 2D periodic pattern, namely a rosette. We inspect the environment with the help of two curious agent assistants: *inquisitive (i-agent)* and *sage (s-agent)*. Applying the initial design rules, the designer obtains an early rosette design with one emergent hole (Fig. 1(A,D1)).

The i-agent localises emergent elements which have come on the scene and determines the shapes which constitute them. Then, it generates a set of possible rules to represent emergent phenomena explicitly in the design structure and a set of rules to replace them. Finally, on the basis of its experience and knowledge, it proposes the most promising design rules to the s-agent(Fig. 1(B1,B2)).

The s-agent evaluates the rules proposed by the i-agent and rules which are already in the shape grammar by applying them to the current generation and verifying whether the obtained solution is valid. In our example, it checks whether the generated design is still a rosette. If the gained result passes the examination, the agent introduces some modifications to the emergent rules in such a way that they do not break the design constraints. While changing the rules the s-agent applies the same similarity transformation to every single shape which takes part in creating the emergent element(Fig. 1(C1,C2)). Finally, the s-agent advises the designer to select the sequence of rules which is the most interesting in its opinion as the next step of the present generation. Taking into account only a few new rules the s-agent is able to generate various creative solutions

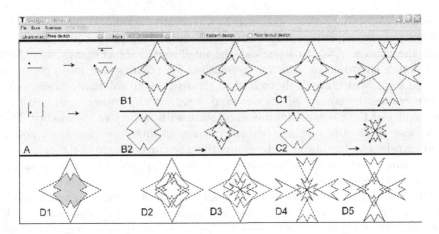

Fig. 1. (A) Initial grammar rules. (D1) Rosette design with one emergent hole marked. Emergent rules: (B1,C1) representing the hole, (B2,C2) replacing the hole. (D2-D5) Possible creative rosette designs.

which may be regarded interesting and plausible by the designer(Fig. 1(D2-D5)). None of them could be generated only with the use of the two rules provided at the beginning of the design task in a static design context.

4   Conclusions

The aim of the research is to build a computer system for visual kind of creative conceptual design. Visual computations proceeded with the use of diagrams are situated and connected to a context indirectly through the designer working with the application or more explicitly through curious agents. The intelligent agents perceive the changing environment and emergent phenomena that occur in it. Interacting with each other they help the designer to obtain interesting and valid solutions of a given design task.

References

1. Grzesiak-Kopeć, K.: Visual designing of graphical models with the use of emergent elements. PhD thesis (in PL). Faculty of Electrical Engineering, Automatics, Computer Science and Electronics. AGH University of Science and Technology (2005)
2. Knight, T.: Interaction in visual design computing. Visual and Spatial Reasoning in Design III. The invited paper (2004)
3. Saunders, R.: Curious design agents and artificial creativity. PhD thesis. Faculty of Architecture. The University of Sydney (2002)
4. Wooldridge, M. J.: Intelligent agents. Multiagent Systems: A Modern Approach to Distributed Artificial Intelligence. MIT Press. Cambridge. MA (1999)

The Logic of Geometric Proof

Ron Rood

Department of Philosophy, Vrije Universiteit Amsterdam
ron.rood@planet.nl

Logical studies of diagrammatic reasoning—indeed, mathematical reasoning in general—are typically oriented towards proof-theory. The underlying idea is that a reasoning agent computes diagrammatic objects during the execution of a reasoning task. These diagrammatic objects, in turn, are assumed to be very much like sentences. The logician accordingly attempts to specify these diagrams in terms of a recursive syntax. Subsequently, he defines a relation ⊢ between sets of diagrams in terms of several rules of inference (or between sets of sentences and/or diagrams in case of so-called heterogeneous logics). Thus, diagrammatic reasoning is seen as being essentially a form of logical derivation. This proof-theoretic approach towards diagrammatic reasoning has been worked out in some detail, but only in a limited number of cases. For example, in case of reasoning with Venn diagrams and Euler diagrams (Shin [5] and Hammer [2]).

Nevertheless, when taken as a general approach towards the development of a logic of geometric proof, several problems can be pointed out (see below). These problems become manifest especially when we require that our logic should take account of certain cognitive and methodological features of geometric proof. In order to meet with these issues, we propose an alternative approach. The main difference is that, instead of proof-theory, ours is entirely oriented towards model-theory (cf. Barwise and Feferman [1]). We provide the first steps towards a mathematical formulation of our logic.

In order to bring to light the issues we have in mind, consider the following well-known theorem of elementary Euclidean plane geometry: the sum of the internal angles of a triangle is equal to two right angles. The following proof of this theorem is equally well-known. First, one constructs a triangle. Second, the base side of the triangle is extended. One sees that there arises an external angle adjacent to an internal one; together, these two angles are equal to two right angles. Third, the external angle is divided by constructing a line parallel to the opposite side. One sees that the two adjacent external angles that arise accordingly are respectively equal to the two opposite internal angles. This completes the proof.

The first object computed in the course of the proof described in the previous paragraph is referred to as a triangle. The second object can be referred to as a triangle *cum* extended base side, and so on. Thus, one obtains a series of objects such that any object in the series (except from the first) is always *built on* previous objects. In this sense, the proof proceeds constructively. The series of objects constructed approaches a "limit object." The informational content of this limit object can be described in terms of a statement of the theorem proved.

D. Barker-Plummer et al. (Eds.): Diagrams 2006, LNAI 4045, pp. 221–225, 2006.

As suggested earlier, it is often assumed that a reasoning agent computes diagrammatic objects in the course of a geometric proof. Upon closer inspection, however, the evidence that can be provided is rather shaky and often not very transparent. For example, a geometric proof typically goes together with the construction of a written diagram. However, on itself this can be hardly counted as evidence for the claim that the corresponding internal representations are also diagrammatic in nature. To this end, we need subtler methods. We will not enter into this vast and controversial territory here.

Larkin and Simon [4], p.106, have speculated that external diagrams play the same functional role as their corresponding internal representations. This makes us wonder about the precise role of external diagrams. It has been pointed out that in case of mechanical reasoning tasks, for example, an external diagram is often used as a form of external memory. One could then infer that the same function is fulfilled by the internal representation corresponding to it.

Note, however, that a memory resource in general not only serves to store information. Typically, one retrieves information from it as well. However, as is suggested by our example, geometric proofs proceed only by carrying out construction processes. Indeed, in case information would have been retrieved from an external diagram, then a geometric proof would certainly be considered as incorrect. In pure mathematics generally, "reading off" information from a diagram is not permissible when one's aim is to prove a theorem. As a result, when concrete diagrams are produced in the course of a geometric proof, their function as a medium of external memory is only limited. Though they may be used to store information, one does not subsequently retrieve this information in order to develop the proof, at least when this proof is to be carried out correctly. It would therefore seem that no information is retrieved from the corresponding internal representations as well.

What, then, *is* the function of the external diagrams that often accompany geometric proofs? Without elaborating deeply upon the issue, Larkin and Simon have also noted in passing that external diagrams sometimes represent information with substantially more detail than their corresponding internal representations (*ibid.*). We submit that this points towards an important cognitive function of external diagrams: they improve the quality of a geometric proof in that they make one see the truth of a theorem more clearly.

By way of summary, we provide two reasons why we cannot think of geometric proof as a logical derivation of a representation from others. First, as our example shows, the representations are constructively built on one another. Accordingly, a geometric proof is characterized by information *growth*. A logical derivation, in contrast, is always characterized by information *decrement*: the content of any derived representation (sentence or diagram) never goes beyond the content of the representation it is derived from. Second, as is suggested by our example, all representations are constructed. A logical derivation, in contrast, always starts from given representations (axioms or hypotheses).

When restricted to geometric proof, our considerations suggest the following (informal) semantic interpretation of "a proves φ": (i) a constructs a series of objects approaching a limit object $\mathfrak{A}$, (ii) the informational content of $\mathfrak{A}$ can be described in terms of φ. From an abstract logical point of view, what is computed is a series of first-order relational structures (possibly with constants) such that any structure in the series is "built on" previous structures. We can view this in terms of a productive directed system of relational structures. The direct limit of such a system satisfies the theorem.

Let I be a directed set, i.e., a set endowed with a pre-order $\leq$ such that $\forall i, j \in I : \exists k \in I : i, j \leq k$. For any $i \in I$, let Σ_i be a signature and let $\mathfrak{A}_i$ be a relational Σ_i-structure. Denote the universe of $\mathfrak{A}_i$ as A_i (which is nonempty). Whenever $i \leq j$, we assume that we have a mapping $\varphi_{ij} : A_i \to A_j$. We call the family $\{\mathfrak{A}_i, \varphi_{ij}\}$ a *productive direct system* provided that:

(1) $\varphi_{ii} = 1$ $(i \in I)$ and $\varphi_{jk} \circ \varphi_{ij} = \varphi_{ik}, i \leq j \leq k$;
(2) for any relation symbol $R \in \bigcup \Sigma_i$ there exists $i \in I$ such that $R \in \Sigma_j$ whenever $i \leq j$; similarly, for any constant $c \in \bigcup \Sigma_i$ there exists $i \in I$ such that $c \in \Sigma_j$ whenever $i \leq j$;
(3) given $\varphi_{ij} : A_i \to A_j$ $(i \leq j)$, then for every n-ary $R \in \Sigma_i \cap \Sigma_j$,

$$\{\varphi_{ij}(x_1, \ldots, x_n) : (x_1, \ldots x_n) \in R^{\mathfrak{A}_i}\} \subseteq R^{\mathfrak{A}_j}$$

and for every constant $c \in \Sigma_i \cap \Sigma_j$ we have $\varphi_{ij}(c^{\mathfrak{A}_i}) = c^{\mathfrak{A}_j}$.

It is because of condition (3) that the system is called productive.

Let $A = \prod A_i$. Where $a, b \in A$, let $a \equiv b$ iff $\exists k \in I : \forall i \geq k : a(i) = b(i)$. Then $\equiv$ is an equivalence relation on A. Denote the equivalence class of $a \in A$ in $A/\equiv$ as $\bar{a}$. Let $p : A \to A/\equiv$ be the projection. The restriction $\pi_i = p|A_i$ is called the canonical map of A_i into $A/\equiv$. Clearly, $\bigcup \pi_i(A_i) = A/\equiv$. It follows that $A/\equiv$ is nonempty.

Define a model $\mathfrak{A}$ with universe $A/\equiv$ as follows. For every n-ary relation symbol $R \in \bigcup \Sigma_i$ there exists $i \in I$ such that $(\bar{a}_1, \ldots, \bar{a}_n) \in R^{\mathfrak{A}}$ iff for every $j \geq i, (a_1(j), \ldots a_n(j)) \in R^{\mathfrak{A}_j}$. Similarly, for every constant $c \in \bigcup \Sigma_i$ there exists $i \in I$ such that $c^{\mathfrak{A}} = \bar{a}$ iff for every $j \geq i, c^{\mathfrak{A}_j} = a(j)$. This interpretation is well-defined. The model $\mathfrak{A}$ is called the *direct limit* of the system $\{\mathfrak{A}_i, \varphi_{ij}\}$ and is denoted as $\varinjlim \mathfrak{A}_i$.

In our definition of a productive direct system we could have opted for a fixed signature for all the structures instead of letting each structure have its own signature. However, from a computational point of view, it is preferable to have a relation symbol or a constant available only when one needs it. Furthermore, once introduced, a relation symbol or constant should remain available from then on. This is what motivates condition (2).

Often, Σ_i is a subset of Σ_j and φ_{ij} is injective whenever $i \leq j$. In such a case, it follows that $\mathfrak{A}_i$ is embedded in the Σ_i-reduct of $\mathfrak{A}_j$. From the assumption that the φ_{ij} are injective, it follows that the π_i are also injective. Hence, each model $\mathfrak{A}_i$ embeds in $\mathfrak{A}$.

Suppose we also assume that the index set I is well-ordered, so that $\{\mathfrak{A}_i\}$ is a chain. Recall that $a \equiv b$ iff a and b are both identical to a constant value after some point. Denote this constant value as z. Then $f : A \to \bigcup A_i$ given by

$f(\overline{a}) = z$ is well-defined. Furthermore, f is an isomorphism between $\mathfrak{A}$ and the union of the chain $\{\mathfrak{A}_i\}$.

In case of our earlier example, the limit object was the last object in the series. Note, however, that not every geometric proof proceeds by constructing a series of objects having a last object. For example, some proofs appear to proceed by the construction of an infinite series of objects, e.g., when one (geometrically) proves the existence of space filling curves by way of an appropriate infinite sequence of curves uniformly converging to a limit curve. Upon closer inspection, however, not all the terms in the sequence are actually produced. For example, one often constructs the first two or three terms in the sequence and accordingly sees the possibility of an infinite sequence of objects (e.g., Kuratowski [3], p.222-4). Thus, our approach describes possible proofs, and these can be infinite. Alternatively, one can specify the sequence analytically (e.g., Sierpiński [6]). In this case, however the proof seems to lose its peculiar geometric character and therefore becomes considerably less relevant to consider in the present context.

Logically speaking, an agent computes a model by employing axioms, definitions and theorems of an underlying theory. For example, a triangle is computed by employing the definition of a triangle (defining a triangle as, for example, a certain system of line segments). One typically computes a triangle by constructing a *minimal* model of the definition of a triangle. Intuitively, the model is minimal in the sense that one ignores any relation (or constant) not cited in this definition, e.g., the length of the sides (see also below). Subsequently, one extends the base side of the triangle by employing further axioms, theorems, and/or definitions, and so on. Accordingly, our attention is drawn to a language. The sentences (or sets thereof) of this language are interpreted as procedures to the effect of constructing certain models, generally by building on other models.

Considering our running example, the construction of a model may in very broad outlines proceed as follows. First note that one typically announces the construction of a triangle verbally by "let ABC be a triangle." We may imagine that a signature is introduced consisting of the constants AB, BC and AC. We can then consider the Herbrand universe corresponding to this signature and interpret its elements as the respective sides of the triangle. In such a way, one builds a model of the definition of a triangle. One subsequently employs other sentences in order to carry out further constructions. For example, one announces the extension of the base side of the triangle by saying "extend the base side AB to a point D" and then uses the relevant axioms and/or theorems. We may imagine that the constant BD is added to the signature and that the universe of the model is extended accordingly so as to obtain a Herbrand universe corresponding to the new signature. Thus, a second model is built. Note that the first model is a submodel of a reduct of the second. A detailed specification of the entire computation requires some work. Unfortunately, we cannot treat the matter here.

As said, mathematician proving the theorem typically writes down the abstract models in terms of concrete diagrams. The object in its written form has a lot more determinations when compared with the corresponding model, which is abstract in turn. For example, the sides of the diagram will have a determinate length, the angles will have a determinate size etc. Be that as it may, in order to prove the theorem one merely has to run a number of procedures associated with sentences of an underlying theory. Since one only takes account of these processes, the irrelevant properties and relations are disregarded accordingly. When the proof is taken as a process, the objects computed are partial objects.

We believe that our model-theoretic approach can be applied to proofs in algebra too. In this case, however, the models considered are algebraic structures instead of relational ones.

References

[1] Barwise, J., S. Feferman (eds.). *Model-theoretic logics.* New York: Springer-Verlag (1985)
[2] Hammer, E. *Logic and visual information.* Stanford: CSLI Publications, Center for the Study of Language and information, Stanford University (1995)
[3] Kuratowski, K.*Introduction to set theory and topology.* Oxford: Pergamon (1961)
[4] Larkin, J., H. Simon. Why a diagram is (sometimes) worth ten thousand words. *Cognitive science* 11 (1987), 65-99. Repr. in: *Diagrammatic reasoning: cognitive and computational perspectives,* J. Glasgow, N.H. Narayanan, B. Chandrasekaran (eds.), Cambridge: MIT Press and Menlo Park: AAAI Press (1995), 69-109.
[5] Shin, S.-J. *The logical status of diagrams.* Cambridge: Cambridge University Press (1994)
[6] Sierpiński, W. Sur une nouvelle courbe continue qui remplit tout une aire plane. *Bulletin international de l'Academie Polonaise des Sciences et des Lettres, Cracovie, serie A, sciences mathématiques* (1912), 462-78.

Exploring the Effect of Animation and Progressive Revealing on Diagrammatic Problem Solving

Daesub Yoon[1], N. Hari Narayanan[2], SooCheol Lee[1], and Oh-Cheon Kwon[1]

[1] Telematics·USN Research Division,
Electronics and Telecommunications Research Institute, Korea
{eyetracker, juin, ockwon}@etri.re.kr
[2] Department of Computer Science & Software Engineering
Auburn University, AL 36849, USA
naraynh@auburn.edu

Abstract. We conducted eye-tracking studies of subjects solving the problem of finding shortest paths in a graph using a known procedure (Dijkstra's algorithm). The goal of these studies was to investigate how people reason about and solve graphically presented problems. First, we compared performance when the graphical display was animated to when the display was static. Second, we compared performance when the display was initially sparse, with detailed information being progressively revealed, to when the display presented all information simultaneously. Results suggest that while animation of the procedure or algorithm does not improve accuracy, animation coupled with progressively revealing objects of interest on the display does improve accuracy and other process measures of problem solving.

1 Introduction

There is increasing research interest in the intelligent user interfaces community in attentive interfaces that track and respond to the user's attention. To develop such interfaces, we have to understand how people view, comprehend and respond to visual displays of information. One aspect of this that our research has addressed is how people solve problems drawn from domains with the following five characteristics: (1) objects of the domain are spatially distributed; (2) the domain is dynamic, i.e. objects and their properties change over time; (3) objects interact with each other; (4) such interactions can be traced along chains of dependency relationships that branch and merge in spatial and temporal dimensions; and (5) predicting the future evolution of a system or object configuration requires reasoning from a given set of initial conditions and inferring chains of events along paths of dependency.

Understanding the cognitive processes underlying such reasoning can provide insights into the design of information displays that actively aid the problem solver and enhance his or her performance. Given the increasing use of large format interactive displays for tasks such as weather forecasting, emergency management and military planning, results from such research may find significant

D. Barker-Plummer et al. (Eds.): Diagrams 2006, LNAI 4045, pp. 226–240, 2006.

practical application. In this context, we report on two experiments that investigated how people reason about graphs. In addition to determining the accuracy of their answers, we measured their response times and collected data on their eye movements. Our goal was to understand the relations between accuracy and patterns of visual attention allocation across the display as well as response times.

The rest of this paper is organized as follows. First, we summarize earlier work on diagrammatic reasoning that has a bearing on the present research. The next section explains Dijkstra's algorithm or procedure for finding shortest paths in an undirected weighted graph using a simple graph as an example. This section also illustrates the operation of animated and progressively revealing displays of this procedure. The following section (4) describes the experiments. Section 5 presents the results and a concluding discussion.

2   Prior Research

Diagrams are often used in problem solving as external representations. Problem solving with such a static pictorial representation sometimes requires mental simulation of the behavior of the system that the diagram depicts, which can require the reasoner to mentally manipulate the configuration or situation depicted in the diagram during reasoning. In this case, diagrammatic reasoning may involve several cognitive processes. Prior research on diagram-based problem solving may be classified as theoretical analyses, empirically-based cognitive modeling attempts, and empirical investigations of the problem solving process.

Larkin and Simon undertook a theoretical analysis of diagrammatic versus sentential representations and described features of diagrams that aid reasoning [1]. Extending this line of inquiry, Cheng proposed twelve functional roles of diagrams in problem solving [2]. These analyses suggest that diagrams of mechanical devices can aid a problem solver by explicating the structure, components, relations, states and causal dependencies in the devices.

Based on several experiments, Narayanan and Hegarty developed a cognitive model of multimodal comprehension and suggested guidelines for designing information displays [3]. They describe the process of how people construct a mental model during multimodal comprehension. First, a static mental model is constructed by diagram decomposition, making representational connections, making referential connections, applying knowledge about basic laws, and hypothesizing causality. Then, people construct a dynamic mental model by mental animation and inference. Based on several experiments, they showed that multimedia systems designed according to their cognitive model were better in terms of facilitating comprehension and learning. Based on this result, they suggest six principles for information display design: decomposition principle, prior knowledge principle, co-reference principle, basic laws principle, lines of action principle, and mental simulation principle.

Several studies have focused on uncovering characteristics of the process of diagram-based problem solving. [4] employed the protocol analysis approach to discover characteristics of subjects' behaviors during diagrammatic reasoning. To explain how diagrams are used in reasoning, they proposed a process model of how people might solve mechanical reasoning problems from diagrams. Then they interpreted experimental results within the framework of their process model. They found that diagrams aid the indexing and recall of relevant knowledge in two different ways. First, when a subject attended to a diagram, related information is recalled from long-term memory or retrieved from the diagram. Second, configurational and shape information extracted from the diagram and combined with prior knowledge about components and behaviors facilitated this indexing and recall of inferential knowledge.

Their research suggests that diagrams of systems are decomposed into components by reasoners and that this decomposition may be guided by spatial adjacency, causality or both. When attention shifts across multiple components, their model predicts that these will either be connected/contacting components or components related by dependencies. For instance, their results indicated that diagrammatic reasoning about mechanical devices proceeded along the direction of causality.

In another experiment [5], these researchers further explored the role of the diagram in guiding the reasoning process. They used the diagram of an impossible or contradictory mechanical device. In that problem, the causal chain of events in the operation of the device split and merged at certain points. They postulated that reasoning trajectories of diagram-based problem solving were influenced by device structure, inferred causation, search strategy, verification goals, and the need to replenish short-term memory. They did indeed find that search strategy, verification goals and replenishing short-term memory were factors that influenced focus shifts.

One limitation of their experiments is that they did not collect eye movement data by using an eye tracker to infer focus shifts. Instead, they used verbal protocols and gestures of subjects to infer the components that subjects were focusing on. This is an indirect measure. Eye movement data can provide direct information on focus shifts when subjects solve diagrammatic problems.

Rozenblit and his colleagues [6] conducted three experiments to determine if eye movement data could provide crucial information about moment-by-moment cognitive processes when subjects were reasoning about diagrammatic problems. They found that independent raters who did not participate in the experiments were able to actually predict principal axes and principal directions of the visual stimuli presented to subjects simply by considering only subjects' scan paths without seeing the stimuli. In case of mechanical reasoning problems, raters predicted principal axes and principal direction more accurately than for non-mechanical reasoning problems. In another experiment, independent raters tried to predict subjects' accuracy by observing their eye positions overlaid on the diagram of the device; and more than 75% of the time, the raters correctly predicted subjects' accuracy. This suggests that eye movement patterns contain crucial information about how people are reasoning with a diagrammatic problem.

Based on the result of [6], Grant and Spivey conducted two experiments to find out how people look at a diagram and how their eye movement patterns correlate with inference making [7]. In the first experiment, they discovered that when people solve a diagrammatically presented medical problem, those who had the correct solution looked at a particular area of the diagram for a long time. This area contained critical information to solve the problem. They looked only at the fixation time (i.e. how long someone looked at a specific area of the diagram), not at gaze patterns. Based on this result, they did a second experiment with three different conditions, one of which was designed to attract subjects' attention to this area using a blinking technique on the display. During the second experiment, they did not track subjects' eye fixations. Subjects who saw the blinking diagram performed better than those who saw a static diagram. This result tells us that attention guiding may help diagrammatic reasoning. But the problem is that they did not know if the subjects actually looked at the blinking area during the experiment, because they did not use an eye tracker.

Yoon and Narayanan conducted an eye tracking study of diagrammatic problem solving [8] in the domain of mechanical devices. Their study revealed eye movements that suggested subjects used mental imagery to solve mechanical reasoning problems presented as diagrams. The researchers found that subjects who employed mental imagery during problem solving exhibited more eye fixations, looked at more components of the problem, spent more time looking at important components, and gazed across the diagram in a more systematic fashion. This suggests that attentive interfaces that help such users better guide their visual attention may improve their problem solving performance.

3   The Shortest Path Problem and Its Solution Procedure

Many problem solving situations admit well-defined procedures that specify components or elements of the problem that a successful problem solver must pay attention to. So it is reasonable to hypothesize that a display that guides the attention of the problem solver to these components, in a particular order if so specified by a problem solving procedure that is available, can positively influence the user's problem solving process and its results.

The problem we chose to investigate this hypothesis is the Shortest Path (SP) problem: given a graph G with edge costs and a starting node s, find the shortest cost paths from s to every other node in G. Many practical problems, such as route planning and project planning, can be represented as a SP problem. An approach to solving such problems is given by Dijkstra's algorithm. This algorithm, invented by Edgar Dijkstra, solves the problem in stages. In the first stage, it considers edges to all nodes adjacent to s one at a time. The cost of the path to each adjacent node (from s), which is nothing but the cost of the edge connecting it with s, is then attached to that node. In the second and following stages, the algorithm picks one node, say j, from among all nodes with a finite cost attached to them - j will be the node with the smallest cost. Then the algorithm declares that the shortest path from s to j is now known, and

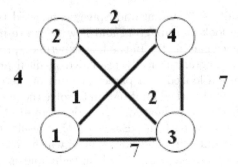

Fig. 1. A Simple undirected weighted graph

proceeds to consider, in turn, each node adjacent to j to see if a smaller (than its current path cost) cost path from s to it (that passes through j) can be found. If so, the attached cost of that node is appropriately updated. Then, as before, the algorithm picks the next node j to declare that the shortest path to it is now known. When this declaration has been made for all nodes, the algorithm terminates. This is an efficient and systematic way to find all shortest paths from s to all other nodes. It can be shown that n to the power 3 is an upper bound for the number of steps needed to solve this problem for a graph with n nodes.

It is obvious that the algorithm considers elements of the problem (i.e. nodes and edges of a graph) in a systematic fashion, and produces and updates relevant information (changing path costs from s to various nodes), during the course of problem solving. What we investigated was the question of whether users trained on applying (i.e. mentally simulating) Dijkstra's algorithm on graphs are helped to different extents by different kinds of information displays that present SP problems to them.

One possible display to present a SP problem to a user is a static one - the picture of the graph with all relevant node and edge information included. A second possible display is one that graphically animates the various steps of the algorithm. Figure 2 shows various snapshots of an animated display that illustrates the operation of Dijkstra's algorithm on an undirected weighted graph (shown in Figure 1).

Read Figure 2 column by column (left to right) and each column top to bottom. It shows a 4 node graph with node 1 being the start node s. Each edge has a cost attached to it. Each node has an information triple attached to it. The first item of this triple is either 0 (indicating that the shortest path to it is not yet known) or 1 (indicating that the algorithm has declared that the shortest path to this node is known). This item is 1 for node 1 since it is the start node, and therefore the shortest path from it to itself is already known. This node is also colored blue to indicate this (in the black and white figures we use a horizontal pattern to indicate a blue colored node). The second item of the triple is the cost of the shortest path to this node found thus far. This item is 0 for node 1 since it is the start node. The third item is the previous node in this shortest path. This item is S for node 1 since it is the start node.

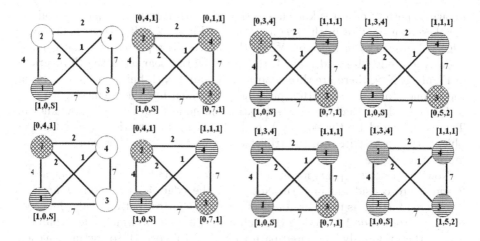

Fig. 2. Illustration of Dijkstra's algorithm operating on a simple graph, showing stages of an animated display; discussed in text column-wise left-to-right and top-to-bottom in each column

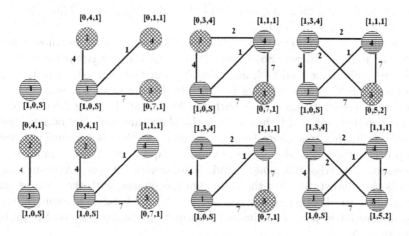

Fig. 3. Snapshots of a progressively revealing animated display of Dijkstra's algorithm operating on a simple graph; discussed in text column-wise left-to-right and top-to-bottom in each column

In the first stage, the algorithm considers edges to all nodes adjacent to node 1 one at a time. So it first considers the edge 1-2, which is indicated by that edge turning red from its default black color. Next, the algorithm computes the cost of the path from node 1 to node 2 (which, in this case, is simply 4). This is indicated by node 2 turning red (in the black and white figures we use a diamond pattern to indicate a red colored node). Now the cost of this path is attached to node 2. This is indicated by the information triplet [0,4,1] appearing beside node 2, indicating that while the shortest path to node 2 is not yet known, a path of cost 4 is known and the previous node in that path is node 1. All these can be seen in the second graph in Figure 2. After these operations, edge 1-2 reverts to black color. When this process is repeated for edge 1-3 and node 3, and edge 1-4 and node 4, the graph will appear as in the third graph of Figure 2. Now there are three nodes, 2, 3, and 4 with finite path costs. So in the second stage the algorithm picks node 4, which has the smallest path cost of 1, and declares that the shortest path from node 1 to node 4 is now known. This is indicated by the edge 1-4, node 4 and its information triplet turning blue, with the first item of the triplet changing from 0 to 1, as can be seen in graph 4 of Figure 2.

Now the process repeats for node 4. Its edges 4-2 and 4-3 and adjacent nodes 2 and 3 are considered in turn. Note that the edge 4-1 and node 1 are not considered further as these are parts of a known shortest path already. The cost of a path from node 1 to node 2 through node 4 is 1+2=3, which is less than the cost of the previously computed path to node 2 (of cost 4), so the information triplet of node 2 is updated from [0,4,1] to [0,3,4]. The cost of a path from node 1 to node 3 through node 4 is 1+7=8, which is more than the cost of the previously computed path to node 3 (of cost 7), so its information triplet is not updated. These operations can again be illustrated by the appropriate edges, nodes and information triplets changing color and the values changing. This situation is depicted by graph 5 in Figure 2.

Now there are two nodes, 2 and 3, with finite path costs and whose shortest paths are not yet known. So in the third stage the algorithm picks the one with the smallest path cost, 2, and declares that the shortest path from node 1 to node 2 is now known. This is indicated by the edge 4-2, node 2 and its information triplet turning blue, with the first item changing from 0 to 1, as can be seen in graph 6 of Figure 2. Now the algorithm considers nodes adjacent to node 2. In this case there is only one node, 4, that is left and the rest of this process, therefore, should be obvious. Graph 7 in Figure 2 indicates the consideration of edge 2-3, node 3 and its information triplet updating. Graph 8 shows the final state, after the algorithm terminates, with the shortest paths from node 1 to all other nodes (i.e. 1-4, 1-4-2 and 1-4-2-3) in blue. What we have illustrated is how an animated information display for the SP problem might operate.

One might conceive of a display that not only shows this animation of the shortest path finding procedure but also takes a "just in time" approach by only revealing components of the graph that the animation is illustrating at any moment. So, initially, only node 1 and its state are displayed (Figure 3; read this figure also left to right column-wise and each column top to bottom). Then, when the procedure considers node 2, it and the edge to it is revealed with appropriate color codings as described above. Similarly, other nodes and edges are also revealed only when they are considered for the first time by the procedure. This will result in a progressive revealing of information as shown in Figure 3.

4   Experiments

4.1   Comparing an Animated Display to a Static Display

Cognitive and computational modeling in the mechanical domain [9, 4, 5] suggest that construction a dynamic mental model of a system is often accomplished by considering components individually, inferring their behaviors due to influences from other connected or causally related components, and then inferring how these behaviors will in turn affect other components. This can involve both rule-based inferences that utilize prior conceptual knowledge and visualization processes for mentally simulating component behaviors [4, 5, 10, 11]. In the domain of machines, a spatial visualization process called mental animation [9] is involved in the simulation of component behaviors. Mental animation appears to be constrained by working memory capacity such that people are only able to mentally animate one or two component motions at a given time [9]. Working memory demands are imposed when several mechanical components constrain each other's motions, so that the motion of components cannot be inferred one by one [12] or if imagining the motion of a component changes the configuration of components so that it no longer corresponds to the external display [4]. Furthermore, this type of mechanical reasoning is particularly demanding of spatial working memory processes [11]. This evidence suggests that an adaptive display that can provide local animation or other kinds of visualizations showing the behaviors of individual components of a system is likely to improve problem solving performance by reducing working memory demands and freeing up more mental resources to meet processing demands of the problem at hand. This prediction was tested.

In this study, we have two conditions. Subjects in both conditions were given training on Dijkstra's algorithm prior to start of the experiment. The experimental condition shows subjects an animation of how to find shortest path in a weighted undirected graph according to a Dijkstra's algorithm as explained in Section 3. Figure 4 shows the stimulus used. The animation did not run to completion. Instead, it was stopped in the middle and subjects were asked to predict the next stop (the problem to be solved appears in Figure 4). The control condition showed a static diagram of the state of the graph at the point at which animation stopped (Figure 4 shows the state of the graph at this point).

4.2   Comparing a Progressively Revealing Animated Display to an Animated Display That Presents All Information Simultaneously

The cognitive process model of Narayanan and Hegarty predicts that successful decomposition of a system into its constituents is a necessary precursor to accurate mental model construction. This process can be hindered if the display is dense, with many variables shown, or if the problem solver lacks the necessary background knowledge of the domain and representational conventions to successfully parse the display. For example, Hegarty and Shimozawa [13] reported the following findings from the meteorology domain: (1) it takes longer to verify a fact

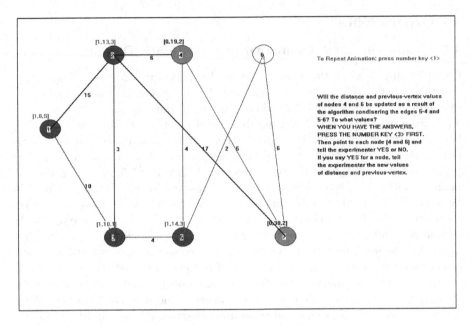

Fig. 4. Graph used in the experiments

about a specific variable from a weather map if the map shows more variables; (2) performance is somewhat faster and more accurate for integrated maps than separate maps; (3) but performance is somewhat slower and less accurate for maps that included additional irrelevant variables; (4) the number of variables displayed on a weather map influences both encoding and inferences from weather maps; and (5) novice individuals have difficulty integrating different causal principles to make inferences from weather maps. These indicate that recognizing and integrating relevant variables and principles across spatial and temporal dimensions is a comprehension bottleneck. This suggests that an information display which, based on the characteristics of the problem to be solved and the domain, progressively reveals relevant information instead of showing all information at once is likely to improve problem-solving performance. This prediction was tested.

We used the graph in Figure 4. The control display was the animated display from the previous experiment. The experimental condition display showed the same animation, but with the additional property that an initially blank display progressively revealed each component (in fashion similar to the illustration in Figure 3) as the animation progressed.

4.3 Procedure

Fifty five undergraduate students of engineering volunteered to participate in the experiment, in return for a nominal payment. They were recruited from an undergraduate algorithm class in the Computer Science & Software Engineering Department at Auburn University. Subjects were assigned to two matched groups based on their GPA and assigned to a control condition or an experimental condition. The experiment was conducted one subject at a time in an eye tracking laboratory equipped with the SMI head-mounted eye tracker, eye tracking computer, and a stimulus display computer.

First, subjects studied Dijkstra's algorithm for finding the shortest path with a printed tutorial, and then watched an animation of the procedure (similar to Figure 2) as many times as they wanted. No time limit was imposed in this training phase. When a subject indicated that he or she was ready, the person was asked to sit on a high-backed chair, and watched the stimulus display on a 20-inch wall-mounted monitor at eye level at a distance of approximately 3 feet. The experimenter sat behind the subject and controlled the experiment through the eye-tracking computer. The experiment proper began by the subject pressing number key 1. After watching the stimulus display corresponding to the experimental or control condition, the subject depressed number key 3 to indicate that he or she had the answer, and then wrote it on a printed picture of the graph that the experimenter provided. Eye movement data was collected and recorded between these two key presses.

4.4 Process and Outcome Measures

We collected two process measures (eye movements and response times) and one outcome measure (accuracy of answer provided to the question in Figure 4, scored on a scale of 0-6). Response time and accuracy are commonly used metrics of problem solving performance. However, not all problems in visuo-spatial domains have answers that can be unequivocally classified as correct or incorrect. A case in point is developing an action plan for an emergency evacuation from an information display that shows factors such as population distribution, layout of roads, features of the terrain and weather conditions. Here it is as important to ensure that the problem solver has considered all critical elements of the domain as it is to create a feasible plan. Therefore, we developed two derived measures called Coverage and Order, in addition to accuracy and response time, to characterize the quality of problem solving. Coverage and Order are derived from eye movement data, as explained below.

Coverage is defined as the percentage of objects in the display that were attended to for more than a time interval threshold. We set the threshold to 200 milliseconds, approximately equal to two fixations. Coverage is therefore a number between 0 and 100.

A good strategist will not only attend to all relevant objects in the display, but also consider them in the order that best supports reasoning. For example, a crucial feature that separates expert and novice problem solving in meteorological reasoning from weather maps is that novices attend to objects that are perceptually salient whereas experts attend to objects that are thematically relevant [14]. Therefore, we developed a metric called Order that measures how systematically a user attended to causally related elements of the display. This metric is explained next.

Let S be an ordered sequence of display objects that a user attended to during a problem solving session. So S begins with the first display item attended to, and ends with the last item attended to before the solution to the problem is produced. This sequence is generated from eye movement data. In this sequence, if object j appears immediately after object i, and if i can influence j in the event chains of the system, then i-j represents an ordered dependent pair in the sequence S. Consecutive ordered dependent pairs represent ordered subsequences of S. The length of a subsequence is the number of dependent pairs in it. Order of S is defined as the sum of squares of the lengths of subsequences in S. This captures the correctness of the sequential order in which the user visually scanned the display (i.e. each dependent pair indicates that the problem solver considered one pair of display objects in the direction of dependency or causality) weighted by the number of consecutive dependent pairs that have been considered (i.e. if subjects A and B both considered the same number of dependent pairs, but if A looked at longer subsequences than B, the value of Order will be higher for A than B). Order is a number greater than or equal to zero.

From the raw eye movement data we also computed the total fixation duration on each component using a bounding box technique. We additionally determined the total number of fixations of each subject for each problem (excluding fixations on the question and on blank areas of the screen).

5   Results and Conclusions

5.1   Comparing an Animated Display to a Static Display

In this experiment, the experimental condition showed subjects an animation of how to find the shortest path in a graph and the control condition showed a static picture as described in Section 4.1.

We expected that, though subjects in both conditions knew of Dijkstra's procedure, the animated display that showed the operation of the procedure until a certain point would produce better accuracy than a static display that showed only the state of the graph that resulted from partial operation of the procedure. We expected response time for the animated display to be more since the animation was likely to encourage more systematic scans of the display than a static picture. We also expected that animated display would produce more coverage and higher order than the statics display.

Figure 5 shows results of answer accuracy (Ans), response time (RT), Coverage (Co), number of focus shifts (F/S) and Order (Order) in table. We used only 48 subjects' eye movement data (24 in each condition) for the data analysis due to problems with collecting good quality eye movements data from all subjects. In terms of accuracy, there was no significant difference between the two conditions. In terms of response time, the animated increased response times (T-test $= 2.113$, p-value $= 0.04$). It also produced higher Coverage (T-test $= 7.253$, p-value $= 0.0001$). Also, the animated display produced more focus shifts (T-test $= 7.363$, p-value $= 0.0001$) and a higher value of Order (T-test $= 4.257$, p-value$=0.0001$).

		Ans	RT	Co	F/S	Order
Exp 1	Mean	1.42	102.6	74.27	166.7	7.29
N(24)	SD	0.5	30.8	13.07	60.49	5.36
Con 1	Mean	1.27	67.89	46.08	61.2	2
N(24)	SD	0.68	74.35	13.84	35.61	2.89
T-test		0.848	2.113	7.253	7.363	4.257
P-value		0.401	0.04	0.0001	0.00001	0.00001

Fig. 5. Overall results of experiment 1

Even though there was no significant difference in accuracy between the animated and static displays, there were significant differences in response time, Coverage, number of focus shifts and Order. This suggests that animated displays induce viewers to look longer at them, look at and across more display objects, and follow the animations systematically, but these visual behaviors do not necessarily lead to more accurate problem solving. One possible explanation for the animated display not improving accuracy is that subjects did not have control over the animation once it started (except to repeat it if desired). So there could have been a speed mismatch between the external animation and the internal simulation of Dijkstra's procedure. This result therefore adds to the extent literature [15] indicating that animations do not necessarily improve comprehension.

5.2 Comparing a Progressively Revealing Animated Display to an Animated Display That Presents All Information Simultaneously

Here the experimental condition showed an animated graph in which information was progressively revealed and the control condition was an animated display that showed the complete graph.

We expected that the progressively revealing display would produce better accuracy and lower Coverage (because this display followed a just in time approach to revealing information), and higher Order, than the animated display.

Initially, we assigned 28 subjects to the experimental condition and 27 subjects to the control condition but we could use only data from 19 subjects in the experimental condition and data from 24 in the control condition due to bad calibration of the eye tracker. Figure 6 shows results of comparing answer accuracy, response time, Coverage, number of focus shifts and Order in table. The progressively revealing display produced better accuracy (T-test = 2.06, p-value = 0.0458). There was not a significant difference in response time between groups. The control condition had significantly higher Coverage (T-test = 2.067, p-value = 0.0451). Regarding number of focus shifts, there was no significant difference between groups. Regarding Order, experimental condition had a significantly higher value of Order (T-test = 2.238, p-value = 0.0307).

		Ans	RT	Co	F/S	Order
Exp 2	Mean	5	160.6	64.54	209.2	9.26
N(19)	SD	2	69.41	9.9	82.83	6.43
Con 2	Mean	3.71	161.5	71.05	213.4	6.08
N(24)	SD	2.07	86.27	10.52	115.1	2.41
T-test		2.060	0.038	2.067	0.134	2.238
p-value		0.0458	0.97	0.0451	0.894	0.0307

Fig. 6. Overall results of experiment 2

These results suggest that progressively revealing information helps viewers to be more accurate (compared to static and animated displays), scan the display more systematically (compared to static and animated displays), and be more efficient (improve accuracy while viewing a smaller percentage of display objects compared to an animated display), without increasing response time (compared to an animated display).

One can use the notion of visual search [1] to explain the better performance observed with a progressively revealing display, i.e. less information on the display requires less search to find what is relevant. But note that at any stage of solving the shortest path problem, one actually needs to focus only on a few adjacent nodes and their edges regardless of how big the graph is. Therefore, scope of visual search is not affected by whether the rest of the graph is visible or not. Thematic relevance of parts of the graph varies as one executes Dijkstra's procedure, and it is possible that subjects in the animated display condition wasted attentional resources on parts that are irrelevant whereas the progressively revealing display helped them identify relevant display objects at different stages of problem solving. The implication is that animation does not necessarily induce efficient allocation of visual attention, but accuracy and efficiency can be improved through display techniques such as progressive revealing.

5.3   Discussion

This paper describes experiments on diagrammatic reasoning with graphs in which we compared process and outcome measures of problem solving when the display showed local processes through visualization techniques such as animation, to when the display was static; and when the display was initially sparse, with detailed information being progressively revealed, to when the display presented all information simultaneously. One outcome measure (accuracy) and four process measures (response time, coverage, number of focus shifts and order), three of which were derived from eye movements, were analyzed to compare problem solving performance. Results indicate that animations induce viewers to look longer at the display, look at and across more display objects, and follow the animations systematically, but these visual behaviors do not necessarily lead to more accurate problem solving. On the other hand, an information display that, besides being animated, also progressively reveals relevant information improves problem solving performance in terms of accuracy, systematicity and efficiency.

This research has implications for the design of information displays that actively track the viewer's visual attention in order to support reasoning and problem solving. Grant and Spivey [7] report that merely attracting the problem solver's attention to relevant regions of a display through an attention attracting mechanism can dramatically improve accuracy. The problem they studied was Duncker's radiation problem, which was not a problem that required reasoning from initial conditions along pathways of dependencies unlike the shortest path problem. In complex domains where objects participate in spatially distributed events or operations, our results provide guidance on how dynamic displays could be designed to better support diagrammatic reasoning.

However, additional research is required before concrete design recommendations can be made. One limitation of current work is that while the reasoning process that subjects engaged in (a mental simulation of Dijkstra's procedure) is dynamic, the underlying graph does not change, and the procedure itself is well-defined. Real-life problems such as planning the evacuation of a city while viewing traffic maps can be modeled as finding shortest paths in a weighted graph, but such problems are solved under competing constraints and rapid changes in underlying conditions. Therefore, our future work will focus on diagrammatic reasoning with less well-defined, heuristic approaches and more flux in relevant variables.

Acknowledgement. This research was supported by the Office of Naval Research under contract N00014-03-10324. Preparation of this paper was supported by the National Science Foundation (NSF) through an independent research & development contract to the second author, and by the Electronics and Telecommunications Research Institute (ETRI) in case of the other authors. The views and opinions expressed in this paper are those of the authors, not of NSF or ETRI.

References

1. Larkin, J., Simon, H.: Why a diagram is (sometimes) worth ten thousand words. Cognitive Science **11** (1987) 65–100
2. Cheng, P.C.H.: Functional roles for the cognitive analysis of diagrams in problem solving. In: Proceedings of the 18th Annual Conference of the Cognitive Science Society. (1996) 207–212
3. Narayanan, N.H., Hegarty, M.: Multimedia design for communication of dynamic information. International Journal of Human-Computer Studies **57** (2002) 279–315
4. Narayanan, N.H., Suwa, M., Motoda, H.: A study of diagrammatic reasoning from verbal and gestural data. In: Proceedings of the 16th Annual Conference of the Cognitive Science Society. (1994) 652–657
5. Narayanan, N.H., Suwa, M., Motoda, H.: Diagram-based problem solving: The case of an impossible problem. In: Proceedings of the 17th Annual Conference of the Cognitive Science Society. (1995) 206–211

6. Rozenblit, L., Spivey, M., Wojslawowicz, J.: Mechanical reasoning about gear-and-belt diagrams: Do Eye-movements predict performance? In: Proceedings of Mind III: The Annual Conference of the Cognitive Science Society of Ireland. (1998) 158–165
7. Grant, E.R., Spivey, M.J.: Guiding attention produces inferences in Diagram-based problem solving. In: Multidisciplinary studies of diagrammatic representation and inference. (2002)
8. Yoon, D., Narayanan, N.H.: Predictors of success in diagrammatic problem solving. In: Diagrammatic Representation and Inference. (2004) 301–315
9. Hegarty, M.: Mental animation: inferring motion from static diagrams of mechanical systems. Journal of Experimental Psychology: Learning, Memory and Cognition 18 (1992) 1084–1102
10. Schwartz, D.L., Black, J.B.: Analog Imagery in Mental Model Reasoning: Depictive Models. Cognitive Psychology 30 (1996) 154–219
11. Sims, V.K., Hegarty, M.: Mental animation in the visual-spatial sketchpad: evidence from dual-task studies. Memory and Cognition (1997)
12. Hegarty, M., Kozhevnikov, M.: Spatial ability, working memory and mechanical reasoning. Visual and Spatial Reasoning in Design (1999)
13. Hegarty, M., Shimozawa, N.: Interpretation of weather maps by non-experts. Technical report, The ONR METOC Workshop (2001)
14. Lowe, R.K.: Extracting information from an animation during complex visual learning. European Journal of the Psychology of Education 14 (1999) 225–244
15. Tversky, B., Morrison, J.B., Betrancourt, M.: Animation: CAn it facilitate? International Journal of Human-Computer Studies 57 (2002) 247–262

Visual Focus in Computer-Assisted Diagrammatic Reasoning

Sven Bertel*

Department of Mathematics and Informatics, Universität Bremen, Germany
bertel@informatik.uni-bremen.de
http://www.informatik.uni-bremen.de/~bertel

Abstract. Visual focus and mental focus in diagram-based problem solving have been shown to be often interrelated with respect to cognitive and visuo-perceptual representational and procedural properties. A problem solver's visual focus can be used for a rough detection of mental focus in diagrammatic reasoning – sufficient to distinguish certain problem solving modes. An integration of visual focus over time can reveal characteristic patterns of problem solving strategies or solutions models. Such patterns hold formidable potential for computer-assisted and collaborative human-computer reasoning with diagrams.

1 Foci in Vision and Diagrammatic Problem Solving

It is one characteristics of reasoning with diagrams that the human visual system is highly involved in the process. Due to an analogicity between problem domain and pictorial formats, much of the mental apparatus designed for perceiving and making sense of the visual world can be rather directly employed for problem solving. Compared to the same problems in propositional formats, with diagrams, mental higher-level but low-capacity problem solving faculties are freed from much load by transferring some aspects of tasks to earlier, perceptual, high-capacity systems. This gives rise to effects such as the offloading of content to the environment, free inferential rides, or the reading off of information from a chart or picture (e.g. [6]). Mental imagery and spatial reasoning are also involved due to overlaps or interrelations of functional cognitive subsystems (cf. [5]). Such relations have effects on reasoning and perceptual foci, as similar and related patterns of eye movements (i.e. scanpaths) exist under imagery and perception, exhibiting comparable functional roles [4]. Mutual influences have been reported between problem solving qualities and the visual focus [3]. On an abstract level, similiar basic spatial principles have been proposed to exist in control of focus in mental representations, images, and external diagrams [2]. While there clearly are too many abstraction levels of mental processing inbetween as to propose a direct and universal correspondence of visual focus of attention and focus in mental problem solving, there seems to exist enough evidence to, with some caution, use the former as a basis for modeling the latter.

* I gratefully acknowledge support by the German Research Foundation (DFG) for the project R1-[ImageSpace], SFB/TR 8 Spatial Cognition. Also, thanks to Mary Hegarty for fruitful collaboration and to Thomas Barkowsky for comments.

D. Barker-Plummer et al. (Eds.): Diagrams 2006, LNAI 4045, pp. 241–243, 2006.

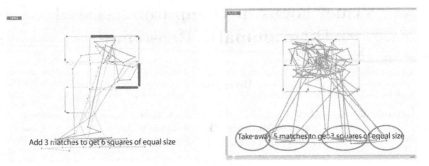

Add 3 matches to get 6 squares of equal size

Take away 5 matches to get 3 squares of equal size

Fig. 1. Fixations in eye movements while solving matchstick configuration problems with multiple solutions. **left:** *Add 3 matches to get 6 squares of equal size.* Activity patterns in fixations can be indicative of mental focus in problem solving and over time reveal the actually constructed solution (indicated by the three bold lines). **right:** *Take away 5 matches to get 3 squares of equal size.* Alternating fixations on different diagrammatic and textual parts of the problem representation can point to a segmentation into partial problems and to a sequential chaining of problem solving steps.

Multiple-Solution Matchstick Problems. In a recent eye tracking study, 18 participants were asked to mentally reconfigure arrangements of matchsticks to fit a given description. All problems had multiple solutions. Eye movements were recorded and after each problem, subjects quickly drew their solutions (cf. Fig. 1). Eye fixation patterns differed across individuals indicating the use of various problem solving strategies. The points that arise on a general level include: (a) the constructed solution was frequently a function of relative fixations and / or time spent over diagram parts; (b) relative frequencies of transitions between diagram parts were often indicative of the solution model; (c) frequent alternating fixations on specific pictorial and textual parts suggest a segmented problem solving process (e.g. pointing to functional problem solving phases or episodes).

2   Collaborative Human-Computer Reasoning

Collaborative human-computer reasoning involves a human and a computational system that jointly reason on shared representations (i.e. a diagram; cf. Fig. 2). Such scenarios are asymmetrical with respect to reasoning strategies and processing capacities. [1] argues for including predictive cognitive processing models within the computational part as to allow for prediction of cognitive parameters (e.g. loads), problem parameters (e.g. complexity), and mental problem solving properties (e.g. preferences in mental model construction). This data could be employed to better adjust the actions of the computational part to the human's cognitive processing needs. Here, the proposed use of the visual focus as a basis for the detection of the current focus in diagrammatic problem solving offers a new method for closer human-computer interaction. E.g., in the matchstick paradigm, good guesses during reasoning as to which solution model the human will construct can be used to generate actions specifically suited for that model

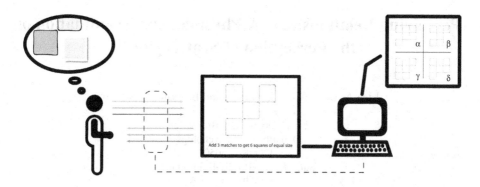

Fig. 2. Coupled human-computer diagrammatic reasoning. A human and a computational reasoner (**left, right**) with individual mental / internal problem solving systems act on a common diagrammatic medium (**center**). Tapping human action channels (gaze & eye fixations, drawing actions) can provide insight into the human's problem solving strategies and status, and can be used for adaptive computational behavior.

(e.g. direct visual attention to diagram parts crucial for model-specific insights: cf. the general approach in [3]). Potential fields of application involve assistive or tutoring scenarios with diagrams as external representations. Individual behavioral measurements can of course not shed full light on higher-level cognitive processes. Given the evidence for a robust connection between foci in vision and problem-solving, however, one can expect to get quite accurate prediction models if other behavioral measurements such as drawing actions are included.

References

1. S. Bertel. Show me how you act on a diagram and I'll tell you what you think. In *Reasoning with Mental and External Diagrams: Computational Modeling and Spatial Assistance. AAAI Spring Symposium.* AAAI; Menlo Park, CA, 2005.
2. D. Engel, S. Bertel, and T. Barkowsky. Spatial principles in control of focus in reasoning with mental representations, images, and diagrams. In C. Freksa et al., editor, *Spatial Cognition IV.*, pages 181–203. Springer; Berlin, 2005.
3. E. R. Grant and M. J. Spivey. Eye movements and problem solving: guiding attention guides thought. *Psychological Science*, 14(5):462–466, 2003.
4. B. Laeng and D. Teodorescu. Eye scanpaths during visual imagery reenact those of perception of the same visual scence. *Cognitive Science*, 26:207–231, 2002.
5. P. Michelon and J. M. Zacks. What is primed in priming from imagery? *Psychological Research*, 67:71–79, 2003.
6. A. Shimojima. Operational constraints in diagrammatic reasoning. In Allwein & Barwise, editor, *Logical reasoning with diagrams*, pages 27–48. Oxford UP, 1996.

Perceiving Relationships: A Physiological Examination of the Perception of Scatterplots*

Lisa A. Best, Aren C. Hunter, and Brandie M. Stewart

University of New Brunswick
Department of Psychology
P.O. Box 5050
Saint John, NB E2L 4L5 Canada
lbest@unbsj.ca

Abstract. Researchers in all areas of science recognize the value of graphical displays and research on graphs has focused on determining which graphical elements enhance readability. To date, no research has examined the physiological processing of graphs. The purpose of this project was to examine the event-related potentials (ERPs) associated with the processing of bivariate scatterplots. Participants viewed scatterplots depicting different linear relationships (positive and negative; strong and weak) and their ERPs were analyzed. Results indicate interesting differences in how scatterplots are processed. Overall, there was differential processing in posterior, medial, and anterior brain sites. Sites on the left and right sides of the brain showed different patterns of activity in response to the scatterplots. In addition, results suggest that different relationships are processed differently in the brain (confirming previous research that has suggested that the perception of covariation is dependent upon the type of relationship depicted on a scatterplot).

1 Introduction

The use of visual aids to represent an idea or relationship dates back thousands of years [29]. Examples of early graphing methods include: a map with coordinates dating from 3800 B.C.E. of Northern Mesopotamia; Egyptian land surveyors around 3200 B.C.E. used coordinate systems around 1000 C.E.; and spatial temporal grids were used by medieval astronomers to show planetary movements [2]. These methods, while certainly graphical in nature do not represent *modern* graphical displays. Descartes is attributed with inventing the coordinate system and it is this system that became synonymous with graphing. In 1765, Joseph Priestly used a rudimentary line chart comparison of human life spans, and this work inspired William Playfair—Playfair popularized the pie chart, bar chart, histogram, and line graph in his publication on the British economy during the 18th century [29]. At this point, graph use in science was fairly common and, by 1830 cumulative frequency distributions and histograms were used to plot distributions of data. By the end of the 19th century, graph

---

* Support for this project was provided by the New Brunswick Innovation Fund.

D. Barker-Plummer et al. (Eds.): Diagrams 2006, LNAI 4045, pp. 244–257, 2006.

use was considered central to science and Funkhouser [7] called the period between 1860 and 1890 the *golden age of graphics.*

Smith, Best, Cylke, and Stubbs [25] surveyed the contents of the three leading psychology journals (*American Journal of Psychology, Pedagogical Seminary,* and *Psychological Review*) and concluded that, at the turn of the 20th century, psychologists analyzed complex data using graphical techniques. Long-standing theories were developed without the use of complex statistical techniques and many of these theories are commonly referred to in present day psychology textbooks. For example, Bartlett's work on memory, Steven's psychophysical power law, Piaget's developmental theory, Freud's psychodynamic theory, and Skinner's theory of operant conditioning are highly cited research programs and each was produced without the use of inferential statistics.

Almost from their advent, graphs have served to clarify patterns and make trends evident. These patterns may not have been evident in textual or tabular presentations, or may only become evident after a careful examination of the material [27]. In 1999, the American Psychological Association made the recommendation that published papers include visual displays because, *Graphical inspection of data offers an excellent possibility for detecting serious compromises to data integrity. The reason is simple: Graphics broadcast; statistics narrowcast.* [30, p. 597].

1.1 Graph Comprehension and Dual-Coding Theories

The field of graphical perception focuses on visualization and involves presenting the same information using different display methods in order to determine which methods of data presentation (e.g., text, tables, types of graphs) lead to more accurate interpretations. Information on a graph is encoded in terms of geometry, texture, and color, etc. and when an observer examines the display, this information is visually decoded. This decoding process has been labeled *graphical perception* [5]. There has been much research on graphical perception and the graphical elements that make it easy (or difficult) to decode the information embedded in a graph. To date this research has focused on optimizing the accuracy with which people make judgments about graphed data.

Pinker's [24] theory of graph comprehension focused on how the brain creates a symbolic representation of visual displays. When a graph is viewed, the initial representation contains the information that allows an observer to decode its underlying meaning. According to this theory, the initial representation includes information about specific elements of the display (e.g., color, shape, size, and location). After these basic graph elements have been deciphered, it is necessary for the cognitive system to perform higher-level analyses. During this process, a graph schema is selected and this schema allows a viewer to recognize the type of graph being viewed and how to interpret the information on the graph. After the correct schema is chosen, it is necessary to translate the basic visual information into conceptual information. Translation allows the viewer to determine the meaning conveyed by the graph and the conclusions that can be made based on this meaning. For example, a viewer may recognize a graph as a scatterplot and interpret the data points as conveying information about the degree of relationship between two variables. Thus, according to this model, visual interpretation involves basic and higher-level perceptual processes, and as Petre and Green [23] point out *Good graphics usually means linking perceptual cues to important information* (p. 57).

Graph perception research has shown that, in many situations, graphs are a powerful means of inducing people to see phenomenon that were previously undetected by them. Because the human visual system is very capable of recognizing and processing visual patterns, a well-constructed visual display can lead to better comprehension than a verbal description of the same phenomenon. Many aspects of human functioning (from simple perception and to higher forms of scientific reasoning) assume an understanding of simple associations and, thus, it follows that research on covariation detection would be of particular interest to graph perception researchers.

According to Paivio [21], the brain has two independent (and interconnected) systems for storing and processing information. The nonverbal system processes images (including graphical displays) and the verbal system processes verbal (including textual) information. Paivio [22] states:

> Human cognition is unique in that it has become specialized for dealing simultaneously with language and with nonverbal objects and events. Moreover, the language system is peculiar in that it deals directly with linguistic input and output (in the form of speech or writing) while at the same time serving a symbolic function with respect to nonverbal objects, events, and behaviors. Any representational theory must accommodate this dual functionality (p 53).

Thus it is assumed that the brain has two independent (and interconnected) systems for storing and processing information. The nonverbal system processes images (including graphs) and the verbal system processes verbal (and textual) information. Paivio explained that as activity spreads associative interconnections are formed between the verbal and visual systems. Referential interconnections allow activity to spread within the system—thus, through these connections related concepts lead to similar activity patterns (for example, the word *pencil* could also activate related words like *eraser* or *pen)*. Mayer and Anderson [15] presented an updated model of this theory and applied it to student learning. These researchers suggest that, as learning occurs, visual and verbal representations form. Their integrated dual-code hypothesis includes connections between the verbal and visual systems. This model contains three major components: a component connecting verbal information and the resulting verbal representation, a component connecting visual information and the resulting visual representation, and a component that includes referential connections between the two representations.

Mayer and his colleagues [15; 16, 18] have examined whether the dual-coding model can be extended to the processing of scientific diagrams. They examined the role of scientific text and diagrams and assessed whether visual information improves recall and problem-solving skills. Overall, they found that illustrations were effective when they included explanatory text. In addition, the length of the text played a role in learning—short textual explanations were more effective than lengthy explanations. If so, it follows that graphical displays accompanied by appropriate figure captions would better convey a message than references in longer paragraphs that refer to the figure.

In an extension of Mayer's research, Holland [8] examined the applicability of dual-coding theory to learning and graphical comprehension. In this study, participants examined a set of time-series data and predicted the correlation coefficient and the next data point in the series. To test the hypothesis that graphical displays enhance decision accuracy, both graphical (scatterplot) and tabular conditions were included. When predicting correlation coefficients, accuracy was similar in the graph and text conditions but when predicting the next point in the data series, accuracy was higher in the graph condition. This extends the applicability of the dual-coding theory to graphical displays and suggests that when making complex predictions accuracy is enhanced when the information is presented on a graph.

There are several implications of this theory. According to Paivio [21, 22], and based on research assessing the dual-coding model, graphical displays are of great use in education [17, 20]. Since the information retrieved from the display is stored and interpreted by two brain systems, ideas presented graphically should be easier to recall and, recall would include enhanced clarity. Associations between the visual and verbal components of the display result in more pathways for retrieving and analyzing the information. Although researchers have not tested the dual-coding theory to determine if graphical displays are a specific image category, it is generally assumed that they are processed in the same areas that other images are processed. If so, one reason why graphical displays are an effective communication device is that the information contained in the graph is processed by both the verbal and nonverbal systems and this *dual coding* makes it easier to assess the information contained in the display.

1.2 Implications of Dual-Coding and Graph Comprehension Theories

Most of the research on dual-coding theory has focused on the functional independence of the verbal and nonverbal systems [28]. Theorists also assume that the two systems are structurally independent and that different areas in the brain are specialized to process specific types of information. Recent advances in neuroscience have allowed for examinations of the structural independence of the systems. Mazoyer, Tzourio-Mazoyer, Mazard, Denis, and Mellet [19] reported activation in the language areas in the frontal and temporal brain areas when participants listened to word definitions. In a mental imagery condition, concrete (easy to form a visual representation) and abstract (difficult to form a visual representation) words were presented to participants. When concrete words were presented, there was activation in the associative visual areas and in the prefrontal areas that correspond to working memory. When abstract words were presented, there was more activation in the language areas of the temporal lobes. Thus, according to dual-coding theory, visualization aids in information processing because, when one visualizes, information processing occurs in separate brain areas. These brain areas communicate and this cross-communication makes the information more memorable and accessible.

Graphical perception researchers assume that graphs are a specific type of image but, to date, there has been little research on how graphs are processed by the brain and, thus to date we do not understand how the brain processes graphical displays. The event-related potential (ERP) is a signal extracted from an electroencephalograph

(EEG) recording and reflects neural activity that is specifically related to sensory, motor, or cognitive events and provides information about how brain activity changes over time. The EEG allows researchers to assess differences in amplitude (cognitive load) and time course (how quickly information is processed). Although *functional* magnetic resonance (*f*MRI) imaging provides optimal spatial resolution and is becoming the standard in cognitive neuroscience research, event related potentials provide better temporal resolution [12] and thus, are more appropriate when processing differences occur over very short time periods.

In an initial study, Adams [2, 3] presented participants with a series of simple images (cat, dog, pen, etc.) and graphs (line graph, bar chart, pie chart) and compared the EEG recordings of these two image types. ERPs from right and left hemisphere sites were analyzed. Data was collected from sites corresponding to early, intermediate, and late perceptual processing. Results indicated that the early processing of graphs and images was similar and there were no evident amplitude or time course differences. Intermediate processing for the two image types was also similar but slight differences in amplitude began to emerge. There were differences in late processing—in frontal lobe sites, the time course of processing was similar but everyday images produced higher amplitudes. This research supports Pinker's [24] theory that basic graph elements are coded before higher (and conceptual) analyses are carried out. It appears that, at the basic perceptual level, graphs and images are processed similarly but, in higher brain centers, processing differs.

1.3 Purpose of the Current Study

Research on the perception of covariation has led to mixed results—some researchers [6, 13] have concluded that people can accurately perceive the degree of relatedness between two variables but others [4, 10] have found lower levels of accuracy. Erlick and Mills [6] presented participants with bivariate scatterplots and asked them to judge the degree of relatedness between the plotted variables. Results suggested a difficulty in predicting correlation coefficients (participants tended to underestimate the degree of correlation). Other researchers [11,26] found that the perception of inverse relationships was particularly difficult for participants. Thus, overall, researchers have concluded that participants are able to make basic covariation judgments but, under some conditions, these decisions are difficult. If it is true that an understanding of simple associations is important to human functioning, research on the physiological correlates of covariation detection could provide interesting information.

The overall goal of this study was to examine how the brain processes statistical information. We were interested in extending the findings of Best and her colleagues [3] and we were specifically interested in examining the physiological processing of information presented on bivariate scatterplots. We were interested in examining whether scatterplots depicting different relationships (positive or negative, strong or weak) lead to similar brain activity. In this experiment, recordings were taken from the left and right temporal and frontal brain sites as participants passively viewed scatterplots depicting different types of relationships.

2 Method

2.1 Participants

Seven undergraduate psychology students were recruited to participate during the Fall 2005 term at the University of New Brunswick Saint John. All participants were enrolled in an introductory psychology course and had instruction on the basic principles of correlation and regression. Participants were awarded 2 bonus points towards their final grade.

2.2 Materials

A grass model 8-10 electroencephalograph (EEG) was used to record electrical activity near the surface of the brain. Using the standard 10-20 electrotrode placement, each electrode site is labelled with a letter and a number. The letter refers to the area of brain underlying the electrode (F – Frontal lobe sites; T– Temporal lobe sites). Even numbers denote sites on the right side of the head and odd numbers denote sites on the left side of the head. In this experiment, electrodes were placed at sites on the posterior (T6, T5), medial (T4, T3), and anterior (F8, F7) of the skull (see Fig. 1). An electro-oculogram placed on the corner of the eyelid indirectly measured the standard potential of the eye during eye blinks. A 12-bit A/D converter sampled continuously at 200Hz, and continuous EEG was parsed, stored, and averaged using a statistical program.

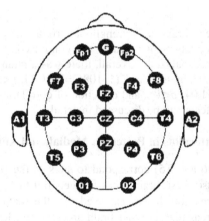

Fig. 1. Standard 10-20 system for EEG electrotrode placement [see 14]

2.3 General Procedure

The participants were led into the lab room, the procedure was explained, and they were asked to fill out consent and demographic forms. ECL electrode gel was spread over the tips of the electrodes to increase conductivity, and an ECL electro cap was placed on the participant's scalp. Mild abrasion of the scalp ensured a proper conductivity level of less than 5 Kohms. Eye blinks were recorded using a pair of electrodes placed just outside the canthus of the right eye correcting for EEG activity due to eye blinks during the procedure. Once prepared, the participants were led into a soundproof box and seated in front of a computer screen. General instructions and the experimental stimuli were presented to them on the screen.

Scatterplots depicting increasing (r = +.9, +.7, +.5, +.3, +.1) and decreasing (r = -.9, -.7, -5, -.3, -.1) linear relationships were presented to participants for 800 msec. Each scatterplot contained 15 data points and contained no textual information. Figure 2 presents samples of the scatterplots. The presentation of the plots was randomized and each of the 10 scatterplots was presented 36 times for a total of 360 experimental trials per participant. Participants were instructed to attend to the scatterplots as they were presented. No response was required of the participants.

Fig. 2. Sample stimuli presented to participants. In Panel A r = +.90. In Panel B r = .70 and in Panel C r = .10.

3 Results

A 3 (brain site) x 2 (side of brain) x 5 (strength of correlation) x 2 (direction of correlation) repeated measures analysis of variance was conducted. In addition, the time of stimulus presentation was analyzed. Overall, there was a main effect for brain site, F $(2, 2172)=13437.04$, $p<.05$, brain side, $F (1, 1086)=2590.11$, $p<.05$, strength of correlation, $F (4, 4344)=533.51.04$, $p<.05$, and direction of correlation, $F (1, 1086)=610.36$, $p<.05$. Each of these effects will be discussed in detail.

3.1 Processing of Scatterplots in Posterior, Medial, and Anterior Brain Regions

Posterior brain sites (T6 and T5) correspond to sites in the right (T6) and left (T5) temporal lobes. Because these brain areas are generally responsible for processing elementary visual information, it was expected that the scatterplots would be processed earlier in these areas than in other brain areas. The ERP voltages differed according to the recording site. Figure 3 shows the initial positive peak (P1) occurring approximately 50 msec after the onset of the stimulus in posterior sites. This waveform (which is actually a combination of two waveforms—C1 and P1) is highly sensitive to basic visual features, such as contrast and spatial frequency [14].

In the medial and anterior sites, P1 occurred later—approximately 75 msec after stimulus onset. In these sites, the amplitude of P1 was smaller, suggesting that these sites are responsible for coding more detailed components of the displays.

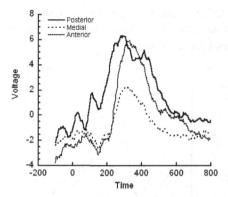

Fig. 3. ERP voltages associated with posterior (T6, T5), medial (T4, T3), and anterior (F8, F7) brain regions

Activity in posterior sites produced an initial negative peak (N1) approximately 75 msec after stimulus onset. The latency of N1 was greater at medial and anterior sites and, in these sites, N1 did not appear until approximately 150 msec after stimulus onset. The N1 wave signals spatial attention and, in posterior sites, is large when participants are faced with a discrimination tasks [14]. In posterior sites the N2 waveform occurred at approximately 150 msec after stimulus onset. The waveform was absent in anterior and medial sites.

In posterior sites, P2 occurred approximately 275 msec after stimulus onset. Luck [14] explained that this wave occurs when a target contains simple stimulus features (the stimuli presented contained only simple stimulus features). Although this wave is often difficult to distinguish in posterior sites, Fig. 3 shows that it is clearly evident. P2 in the medial and anterior sites occurred at approximately 300 msec after stimulus onset. The amplitude of P2 was similar for posterior and anterior sites but was substantially lower for medial sites. The frontal lobe is thought to function as a center of decision making and, it follows that activity in these areas would be high as the information contained on a graph is analyzed.

In posterior sites, an additional positive peak (P3) occurred approximately 450msec after stimulus onset. P3 is generally considered to result as updates to the representation of the environment take place. The amplitude of this wave tends to be higher when participants devote more effort to a task [14]. In this case, P3 might signify that participants were discriminating between the different scatterplots and deciding what types of relationships were represented. Interestingly, P3 was absent in medial and anterior sites.

An examination of voltages on the right and left sides of the brain reveals interesting differences. Figure 4 shows the ERP voltages associated with the left and right brain sites. As can be seen in the Fig., sites on the right side of the brain produced different N1 and P1 components than sites on the left side of the brain. In addition, the N1 and P1 for right hemisphere sites had larger amplitudes than the corresponding components on the left side. Although the P2 for the left side of the brain was slightly earlier (275 msec vs. 325 msec), the P2 components were similar for both sides of the brain. P3 was absent in right hemisphere sites and occurred at approximately 450 msec after stimulus onset for left hemisphere.

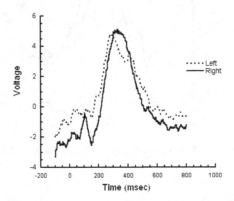

Fig. 4. ERP waveforms for left and right brain regions

The interaction between brain site and brain side was statistically significant, F (2, 2172)=6085.45, $p<.05$. As can be seen in Fig. 5a, in posterior sites, the left and right voltage difference was small. Although the wave amplitude in left posterior sites was slightly higher, the overall pattern of activity was similar.

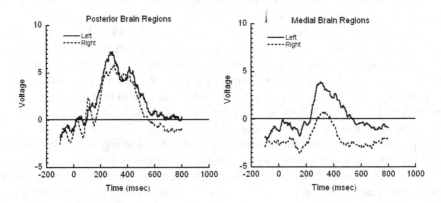

Fig. 5. The ERP waveforms for right and left posterior (T6 and T5) brain regions (Panel A) and right and left medial (T4 and T3) brain regions (Panel B)

In medial sites, the right and left waveforms had similar latencies but the waveform amplitude of the left site was greater than that of the right site (see Fig. 5b). In anterior sites, the pattern of the right and left waveforms was similar but, as can be seen in Fig. 5c, the amplitude of the right waveform was greater than that of the left waveform. It is interesting to note that, although the left posterior site produced more activity than the right posterior site, the right anterior site produced more activity than the left anterior site. This suggests that different areas of the brain are differentially active as the information on the scatterplots was processed. It is possible that this differential processing represents an integration of the visual and verbal pieces of information (as suggested by the dual coding theory).

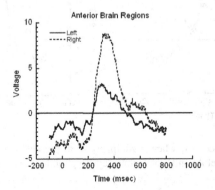

Fig.5c. The ERP waveforms for right and left anterior (F8 and F7) brain regions

3.2 Processing of Different Types of Relationships

To examine whether different types of relationships are processed in a similar manner, the ERPs associated with the different scatterplots were examined. As indicated previously, there was a statistically significant main effect for strength of correlation. Figure 6 presents the ERP voltages for strong (Panel A), moderate (Panel B), and weak (Panel C) relationships. As can be seen in Panel A, the ERP voltages associated with the strong relationships ($r = .7$ and $r = .9$) were quite similar. When the bivariate correlation was very high ($r = .9$), the P3 waveform occurred at approximately 350 msec (it occurred slightly later when the strength of the correlation was $r = .7$). In addition, the amplitude of the waveform was higher when the correlation was $r = .9$. It is interesting to note that the amplitudes associated with the stronger relationships were lower than those associated with moderate or weak relationships. When the correlation was moderate ($r=.5$), the latency of the waveform occurred at approximately 300 msec after stimulus onset and the amplitude of the waveform was higher than that of the stronger relationships. Although the waveforms associated with the strong relationships were similar, the waveforms associated with the weaker correlations were quite different. As can be seen in Fig. 6c, the peak of the waveform was higher when $r = .3$. As with $r = .5$, the peak occurred at approximately 300 msec after stimulus onset. When $r = .1$, the peak occurred at approximately 350 msec after stimulus onset.

The differences between the scatterplots depicting positive and negative relationships are presented in Fig. 7. Although the overall waveforms associated with positive and negative relationships were quite similar, a closer examination illustrates interesting differences. For both sides of the brain, the initial processing (from stimulus

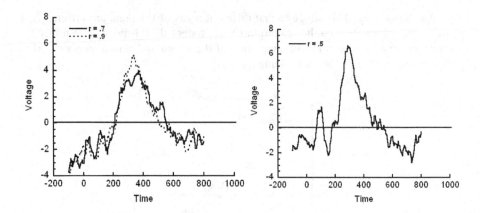

Fig. 6. ERP voltages produced when participants were presented with scatterplots depicting strong relationships (Panel A) and scatterplots depicting moderate relationships (Panel B)

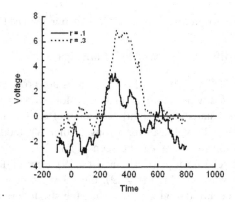

Fig. 6c. ERP voltages produced when participants were presented with scatterplots depicting weak relationships

onset to 200 msec) of positive and negative relationships was quite different. In both cases the early processing of negative relationships produced waveforms that were distinguishable from those of positive relationships. Between 200 and 400 msec after stimulus onset the waveforms produced when participants' viewed positive and negative relationships were quite similar (for both right and left brain regions the amplitude peaked slightly earlier for positive relationships). Between 400 and 800 msec after stimulus onset the waveform associated with negative relationships remained higher than that associated with positive relationships. Given the graph perception research suggesting that people make less accurate judgments when presented with negative relationships, this finding warrants further study.

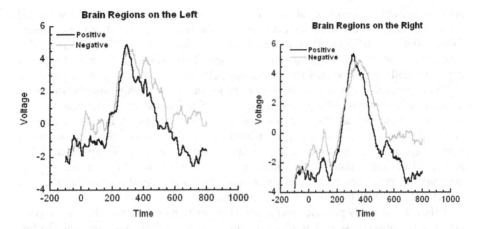

Fig. 7. ERPs of positive and negative relationships as a function of brain side. Panel A presents the waveforms for left side brain regions. Panel B presents the waveforms for right side brain regions.

4 Discussion

One goal of this study was to examine physiological correlates of graph perception. Based on initial results, it appears that different relationships (presented visually on bivariate scatterplots) are processed in different ways. At a basic perceptual level, the processing of the scatterplots was quite similar. Higher level processing was more variable. These results support Pinker's [24] theory of graph comprehension—the waveform (P1) that indicates sensitivity to basic visual features occurred earlier in posterior regions and later in more anterior brain sites. Thus, as Pinker suggested basic visual analyses occur before higher-perceptual analyses can be performed.

Results indicated that the scatterplots were processed earlier in left brain regions. According to dual-coding theory [21, 22] visual and verbal information is processed in different brain systems. In general, the left hemisphere is considered to be responsible for processing language whereas the right hemisphere is crucial for processing spatial information. Differential processing of the right and left brain sites suggest that graph reading involves an integration of language and spatial processing. In posterior regions, processing in left and right sites was similar but medial and anterior regions showed differential processing in right and left brain sites. In medial sites, higher amplitudes were obtained from right hemisphere sites but in anterior regions, higher amplitudes were associated with the left hemisphere recordings. It is possible that after the basic visual features of the graph were analyzed, participants used their visual system to code the specific pattern of data points (spatial task) and their verbal systems to label the relationship (strong vs. weak; positive vs. negative). These findings support both the graphical comprehension and dual-coding theories.

It is interesting to note that the amplitudes associated with the stronger relationships were lower than that of moderate or weak relationships. It is possible that these differences are due to the fact that moderate relationships (r=. 3 and r=. 5) are more difficult to judge than relationships that are very strong or very weak. Jausovec [9] compared the ERPs of gifted and average individuals and concluded that the problem solving skills of gifted individuals involved less mental effort. The *efficiency hypothesis* was developed to explain these findings. This hypothesis states that superior cognitive skills are associated with processing efficiency and proponents suggest that skilled individuals work less hard to solve problems because only task-specific brain areas are active during difficult tasks. In less skilled individuals, brain areas that are irrelevant to the task at hand are also active and this increase in activation leads to poorer performance. The efficiency hypothesis could also explain these results— reading a scatterplot depicting a very strong or very weak relationship requires less mental effort than reading a plot with a moderate relationship. When moderate relationships are presented, participants were not sure what type of relationship they are examining, and thus, their mental effort increased.

Although this study provides preliminary information on the physiological processing of information presented graphically, more research needs to be conducted to further examine the physiological and perceptual processing of graphs. In order to fully examine this relationship, it is necessary to compare physiological and behavioral (i.e., levels of judgmental accuracy) results. A comparison of these types of data would provide information about specific graph characteristics that enhance one's ability to extract information from a graph. In addition, it would then be possible to examine individual differences in graph literacy and determine if these differences are due, at least in part, to how the brain processes graphed information. If the efficiency hypothesis holds, individuals less versed in graphical analyses would show more widespread activity patterns and make less accurate decisions about graphed data.

References

1. Adams, A.: Is a Graph Different than a Picture? An Electrophysiological Examination of Graph Perception. Unpublished honours thesis, University of New Brunswick: Unpublished manuscript (2005)
2. Beniger, J.R., Robyn, D.: Quantitative Graphics in Statistics: A Brief History. The American Statistician 32 (1978) 1-11
3. Best, L.A., Adams, A., Jones, S., Hickman, E., Hunter, A., LeBlanc, A., Stewart, B., Woodland, J.: Is A Picture Worth 1000 Words: An Electrophysiological Examination of Graph Perception. Paper to be presented at the CPA annual meeting (2006)
4. Beyth-Marom, R.: Perception of correlation reexamined. Memory and Cognition 10 (1982) 511-519
5. Cleveland, W. S., McGill, R.: Graphical perception and graphical methods for analyzing scientific data. Science 229 (1982) 828-833
6. Erlick, D. E., Mills, R.: Perceptual quantification of conditional dependency. Journal of Experimental Psychology 73 (1967) 9-14
7. Funkhouser, H.G.: Historical development of the graphical representation of statistical data. Osiris, 3 (1937) 269-404
8. Holland, J. L.: Dual-coding Hypothesis and the Perception of Graphical Displays: Is a Picture Worth a Thousand Words? Unpublished honours thesis, University of New Brunswick, Saint John (2004)

9. Jausovec, N.: Differences in EEG alpha activity related to giftedness. Intelligence 23 (1996) 159-173

10. Jenkins, H. M., Ward, W. C.: Judgment of contingency between responses and outcomes. Washington, DC: APA (1965)

11. Kareev, Y.: Positive Bias in the Perception of Covariation. Psychological Review 102 (1995) 490-502.

12. Kolb, B. Whishaw, I. Q.: Fundamentals of Human Neuropsychology, 5th Ed., New York: W. H. Freeman (2003)

13. Lane, D. M., Anderson, C. A., Kellam, K. L.: Judging the relatedness of variables: The psychophysics of covariation detection. Journal of Experimental Psychology: Human Perception and Performance 11 (1985) 640–649

14. Luck, S. J.: An introduction to the event-related potential technique. Cambridge, MA: MIT Press (2005)

15. Mayer, R. E, Anderson, R. B.: Animations need narrations: An experimental test of a dual-coding hypothesis. Journal of Educational Psychology 83 (1991) 484-490

16. Mayer, R.E., Gallini, J. K.: When is an illustration worth ten thousand words? Journal of Educational Psychology 88 1990 715-726

17. Mayer, R.E., Moreno, R.: Aids to computer-based multimedia learning. Learning and Instruction 12(2002) 107-119

18. Mayer, R. E.: Models for Understanding. Review of Educational Research 59 (1989) 43-64

19. Mazoyer, B., Tzourio-Mazoyer, N., Mazard, A., Denis, M., Mellet, E.: Neural bases of image and language interactions. International Journal of Psychology 37 (2002) 204-208

20. Moreno, R. Mayer, R. E.: Personalized messages that promote science learning in virtual environments. Journal of Educational Psychology 96 (1998) 165-173

21. Paivio, A.: Imagery and verbal processes. New York: Holt Rinehart (1971)

22. Paivio, A.: Mental representations: A dual-coding approach. New York: Oxford University Press (1986)

23. Petre, G., Green, T. R. G.: Learning to read graphics: Some evidence that 'seeing' an information display is an acquired skill. Journal of Visual Languages and Computing 4 (1993) 55-70.

24. Pinker, S.: A theory of graph comprehension. In R. Friedle (Ed.), Artificial intelligence and the future of testing. Hillsdale, NJ: Erlbaum (1990)

25. Smith, L. D., Best, L. A., Cylke, V. A., Stubbs, D. A.: Psychology without p values: Data analysis at the turn of the century. American Psychologist 55 (2000) 260-263

26. Strahan, R. F., Hansen, C. J.: Underestimating correlation from scatterplots. Applied Psychological Measurement 2 (1978) 543–550

27. Tilling, L.: Early experimental graphs. The British Journal for the History of Science 8 (1975) 193-213

28. Vekiri, I.: What is the value of graphical displays in learning? Educational Psychology Review 14 (2002) 261-311

29. Wainer, H.: Visual revelations. Graphical tales of fate and deception from Napoleon Bonaparte to Ross Perot. New York: Copernicus (1997)

30. Wilkinson, L.: Statistical methods in psychology journals: Guidelines and explanations. American Psychologist 54(1999) 594-604.

Using Research Diagrams for
Member Validation in Qualitative Research

Nathan Crilly[1], P. John Clarkson[1], and Alan F. Blackwell[2]

[1] Engineering Design Centre, University of Cambridge, CB2 1PZ, UK
{nc266, pjc10}@cam.ac.uk
[2] Computer Laboratory, University of Cambridge, CB3 0FD, UK
Alan.Blackwell@cl.cam.ac.uk

Abstract. The process of *member validation* requires researchers to present their findings back to the communities that have been studied to gain their appraisal of the work. By depicting subject matter that ranges from the physical to the conceptual, diagrams provide a valuable alternative to the written documents traditionally used in member validation. This paper reports on a study in which diagram-based member validation was used to assess the accuracy and acceptability of the researchers' interpretations. The manner in which the technique was implemented, the benefits that were realized and possible directions for future work are all discussed.

1 Motivation

Qualitative researchers may employ a process of *member validation* to assess the accuracy or acceptability of their interpretations. Such idea-sharing activities have traditionally involved presenting a written report back to the subjects who were studied and recording their responses. Member validation thus offers valuable insights into how the researchers' findings are interpreted by the members of the studied community [1]. However, one particular problem associated with this technique is that members of the studied group do not necessarily have the time, skills or motivation required to analyze and comment on the documents with which they are provided [4]. Such problems prompted Bloor to ask, "how does one frame and present one's analysis in such a form and in such a setting that one can be confident that one's respondents will fully understand it…?" [1].

In comparison with text, diagrams are able to make conceptual relations more visually explicit and this allows researchers to present their ideas, hypotheses or theories in a simple and coherent manner. As such, the preparation and presentation of diagrams is one possible method of alleviating the problems associated with text-based member validation. However, the use of diagrams in qualitative research has traditionally focussed on representations produced by informants rather than researchers [2, 6]. Whilst many qualitative researchers use diagrams to explore and record their own ideas, there is rarely (if ever) any discussion of how these diagrams might be used to communicate with research subjects. This suggests an unnecessary and unhelpful divide between the processes of data analysis and data collection. If researchers are to expend time and analytic effort producing meaningful diagrams, and

D. Barker-Plummer et al. (Eds.): Diagrams 2006, LNAI 4045, pp. 258–262, 2006.
© Springer-Verlag Berlin Heidelberg 2006

these diagrams exhibit good communicative potential, then employing them as a stimulus for member validation is a research technique that promises to be both efficient and effective [3].

2 Field Study

A qualitative research study, undertaken as an exercise in the use of research diagrams, provides the opportunity to explore the benefits offered by diagram-based member validation. The primary objective of the study was to identify the range of factors that influence the aesthetics of products and to define the relationships between those factors. To achieve this, a series of 27 semi-structured field interviews were conducted with members of the design industry.

The interviews were divided into two parts. Firstly, a conventional verbal interview session was conducted focussing on the design process and the determinants of product visual form. Throughout these sessions, the interviewee and interviewer made reference to available visual materials that were relevant to the design projects under discussion (sketches, products, prototypes, photographs, etc.). Secondly, a member validation session was conducted using diagrammatic stimulus materials provided by the researchers. The diagrams represented the researchers' current conceptualization of the domain (Fig. 1) and were presented to the interviewees in a series of images that cumulatively built up the layers of the representation (Fig. 2). This process of sequentially revealing different aspects of the diagram was used to constrain the interviewees' attention to the area under immediate discussion and so that comprehension of each layer could be assessed before presenting the next.

The diagram-based member validation process allowed the interviewees to compare the representation constructed in the stimulus to their own experiences or ideas. This, in turn, allowed the researchers to gain the interviewees' appraisal of the work whilst also eliciting further contributions on the research topic. Such responses can be categorized according to how they related to the diagram. Firstly, showing diagrams to the interviewees provoked comments on the details of the presentation and offered insights into how the diagrams were interpreted. Unanticipated assumptions and misunderstandings noted at this stage were then accommodated in future communications. Secondly, interviewees were encouraged to comment on what the diagram revealed about the researchers' conceptualization of the domain. These comments provided feedback on the underlying structure of the representation, rather than the details of its execution. The researchers thus gained new perspectives on the implications of their assumptions and the limitations of the graphic language used. Thirdly, and in a broader sense, the interviewees were encouraged to discuss the subject in general based on their interpretation of the diagram. The diagrams provided a visual overview of the domain and this prompted the interviewees to comment on the connections between different topics. The process thus elicited information that had been difficult to gain by verbal exchanges or by reference to other visual materials.

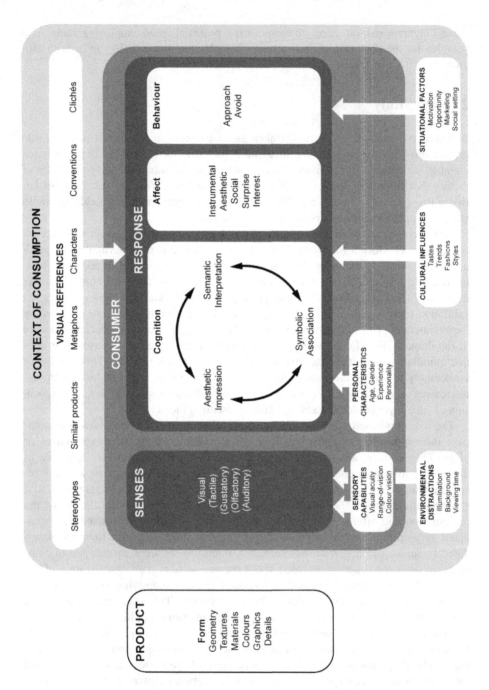

Fig. 1. Example of the kind of diagrammatic structures predominantly used in the study. This diagram corresponds with the right hand side of the illustration provided in Fig. 2.

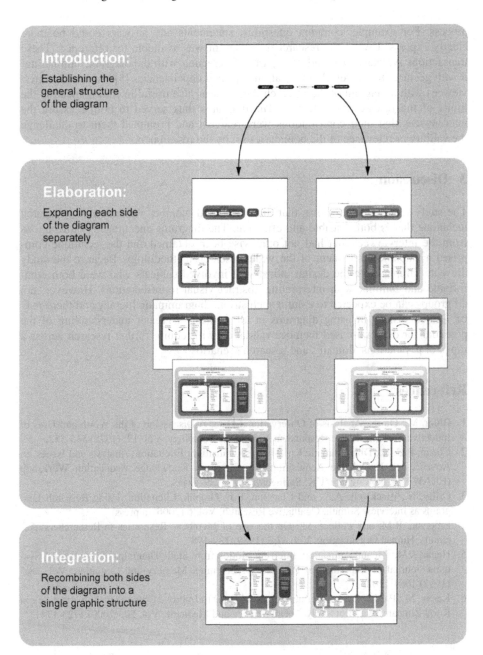

Fig. 2. Illustration of how the diagrams were presented to the interviewees in a progressive sequence of increasing detail

The researchers' interpretation of all three categories of response informed the process of producing new diagrams that were used in later phases of the study. Employing diagrams in this manner also assisted with other aspects of the interviewing process. For example, complex questions, statements and answers could be more clearly expressed when the researchers' diagram was available for reference. These transactions typically involved both parties interacting with the diagram, 'animating' it with gestures to convey flows, relationships and dependencies [5]. Presenting interviewees with a diagram of the research domain was also useful in communicating the limits of the researchers' interests. The diagrams thus served to both reassure the interviewees that their contributions were relevant and prompted them to challenge the validity or relevance of the boundaries set by the researchers.

3 Discussion

The study reported here shows that diagram-based member validation is a research technique that is both feasible and effective. The diagrams encouraged contributions from the interviewees that had not otherwise been obtained and the technique comprised a valuable component of the wider research methodology. Because the study drew its sample from the design industry, it involved subjects who were both comfortable and familiar with interpreting abstract visual representations. However, not all groups can be expected to exhibit such strong diagrammatic literacy and there may be limited benefit in using diagrams in all studies. A greater understanding of the potential for diagram-based member validation will require further research across a range of populations, domains and research scenarios.

References

1. Bloor, M.: On the Analysis of Observational Data: A Discussion of the Worth and Uses of Inductive Techniques and Respondent Validation. Sociology. Vol. 12: (1978) 545-552.
2. Cheng, P.C.-H.: Diagrammatic Knowledge Acquisition: Elicitation, Analysis and Issues. In Advances in Knowledge Acquisition: 9th European Knowledge Acquisition Workshop (EKAW '96). Nottingham, UK: Springer. (1996) 179-194
3. Crilly, N., Blackwell, A.F., and Clarkson, P.J.: Graphic Elicitation: Using Research Diagrams as Interview Stimuli. Qualitative Research. Vol. 6 (2006, in press).
4. Emerson, R.M. and Pollner, M.: On the Use of Members' Responses to Researchers' Accounts. Human Organization. Vol. 47:3 (1988) 189-198.
5. Hegarty, M.: Mental Animation: Inferring Motion from Static Diagrams of Mechanical Systems. Journal of Experimental Psychology: Learning, Memory and Cognition. Vol. 18:5 (1992) 1084-1102.
6. Kesby, M.: Participatory Diagramming as a Means to Improve Communication about Sex in Rural Zimbabwe: a Pilot Study. Social Science and Medicine. Vol. 50: (2000) 1723-1741.

Androcentric Preferences for Visuospatial Representations of Gender Differences

Peter Hegarty, Carmen Buechel, and Simon Ungar

Department of Psychology, University of Surrey, Guildford, GU2 7XH, UK

1 Introduction

Research on explanations shows how category prototypicality can affect the the framing of comparative information about social groups. Spontaneous explanations of group differences focus on the attributes of atypical groups and treat the attributes of prototypical groups as the norm. Thus, gender differences are attributed to women more than men (Miller, Taylor, & Buck, 1991), and sexual orientation differences to lesbians/gay men more than straight women and men (Hegarty & Pratto, 2001). When group differences are explained within an overarching category for which the atypical group is more prototypical (e.g., sexual orientation differences among gay and straight men living with HIV/AIDS), these patterns are not observed (Miller et al., 1991, Experiment 3; Hegarty & Pratto, 2001, Experiment 2).

We tested whether prototypicality similarly affects preferences for the framing of information in visuospatial representations of social group differences. We examined whether people present information about more prototypical entities first and atypical entities second in graphs. As men and women think in androcentric ways that render men more prototypical than women in most social categories (Bem, 1993), this would lead gender differences to be graphed with men's data before women's. As English language users parse spatial information from left-to-right and top-to bottom (Clark & Chase, 1972; Tversky, Kugelman, & Winter, 1991), we defined an androcentric visuospatial representation as one that positioned data about males above or to the left of data about females. We examined preferences for these representations in an experiment (Study 1) and a content analysis (Study 2).

2 Study 1

2.1 Method

Participants. Fifty-four British psychology undergraduates participated in return for course credit.

Materials. Participants were presented with written prompts to draw vertical bar graphs representing differences between two sub-groups within four categories. The categories and sub-groups were fruits (oranges and kumquats), animals (horses and kangaroos), genders (males and females) and nationalities (the British and the French). Each prompt required the participant to graph two properties that varied between the two groups. The prompt to draw a graph representing gender differences

D. Barker-Plummer et al. (Eds.): Diagrams 2006, LNAI 4045, pp. 263–266, 2006.
© Springer-Verlag Berlin Heidelberg 2006

in mathematical and verbal abilities is shown in Figure 1. Participants were randomly assigned to one of two conditions. In one condition, instructions listed the typical group (e.g., *men)* before the atypical group (e.g., *women*) but the atypical group's stereotypical superior trait (e.g., *verbal abilities*) ahead of that of the typical group (e.g., *mathematical abilities*). In the other condition, the order of presentation of both groups and traits were reversed.

Procedure. Participants completed the materials during lab sessions in groups and were thanked and debriefed.

2.2 Results

Each graph that positioned the typical entity first was scored as 1 and each that positioned the atypical entity first was scored as 0. The main hypothesis was supported; participants produced a higher than average proportion of graphs with the typical

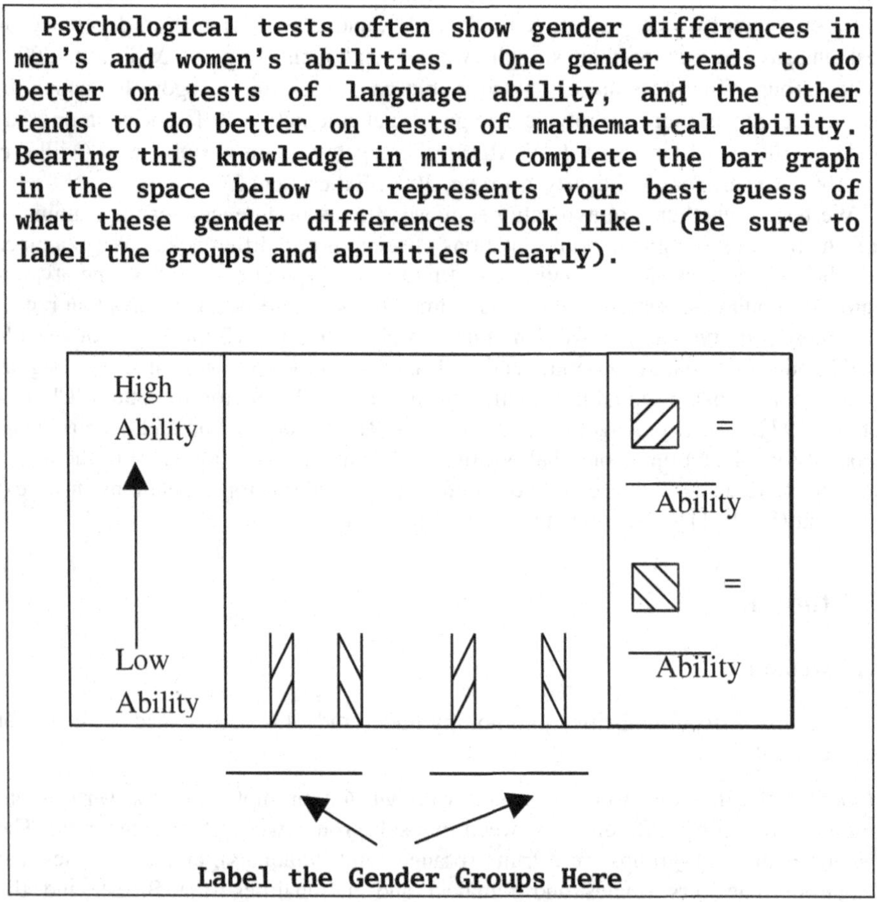
Psychological tests often show gender differences in men's and women's abilities. One gender tends to do better on tests of language ability, and the other tends to do better on tests of mathematical ability. Bearing this knowledge in mind, complete the bar graph in the space below to represents your best guess of what these gender differences look like. (Be sure to label the groups and abilities clearly).
High Ability
Low Ability
=
Ability
=
Ability
Label the Gender Groups Here

Fig. 1. Sample Stimulus Material (Study 1)

entity first ($M = .68$), one-sample t (53) = 5.50, p <.001.   A 2x4 ANOVA was conducted with order as a between groups variable and category as a within-subjects variable.   Two main effects were observed.   As Table 1 shows, participants positioned typical entities first in graphs more often when they were positioned first in text, $F(1, 50) = 8.02$, p <.01, $\eta^2 = .14$. A main effect of category was also observed, $F(1, 50) = 4.26$, p <.05, $\eta^2 = .08$. Tukey's post hoc tests ($\alpha = .05$) showed that the typical entity was positioned first significantly more often in the fruits category than in the other three categories, and significantly more often in the gender category than in the animals category. The interaction was not significant, F <2.6, p >.10.

2.3 Discussion

Participants positioned typical entities ahead of atypical ones on the whole, and males' data before females' data in particular. These results are consistent with the hypothesis that males are positioned first because of an androcentric tendency to see males as more prototypical humans than females. The content analysis (Study 2) explored this hypothesis with scientists' published visuospatial displays.

Table 1. Proportion of Graphs Positioning More Typical Entity First by Order Condition and Category

Category Order	Fruit	Animals	Gender	Nation	All
Atypicals First	.73	.47	.69	.39	.58
Typicals First	.90	.60	.79	.81	.78

3 Study 2

3.1 Method

Articles were sampled from four journals of the American Psychological Association between 1965 to 2004: *Journal of Personality and Social Psychology* (JPSP), *Developmental Psychology* (DP), *Journal of Abnormal Psychology* (JAP), and *Psychology of Women Quarterly* (PWQ). One issue per year of publication was selected at random.   Every article within that issue that reported original empirical data was included in the sample.   Of the 1859 articles sampled, 21% (N = 388) reported one or more empirical gender difference (N = 388). Twenty of these articles (from *DP*) included both parents and children as study participants. For such articles, we treated children as the principal participants of these studies and analyzed reports of gender differences among parents separately.

3.2 Results

On average, each of the 388 articles included 1.71 visuospatial displays (i.e., tables or graphs) representing one or more gender differences. More displays positioned males first than females first (Ms= .74, .26), t (281) = 29.87, p <.001. A 4x2x2 ANOVA with decade of publication, journal of publication, and first author's gender as between-subjects factors, and the proportion of male-first displays as the dependent variable showed only a main effect of publication, F (3, 251) = 5.23, p <.01, η^2= .06. Post hoc tests (α = .05) showed that articles in *PWQ* and *JPSP* had fewer males-first diagrams than articles in *DP*. Proportions of males-first displays in *JAP* articles did not differ significantly from the means for other journals (see Table 2). However, the proportion of androcentric displays within each journal was always significantly greater than .5, all t > 2.14, all p<.005. In contrast, within the 20 articles reporting gender differences between parents, a greater proportion of displays within each article positioned mothers first than fathers first (*Ms*= .81, .19), t(17) = 3.48, p <.01.

Table 2. Females-First and Males-First Displays in APA Publications

Journal Order	DP	JAP	JPSP	PWQ	Parents
Females First	.13	.28	.32	.37	.81
Males First	.87	.72	.68	.63	.19

3.3 Discussion

As in Study 1, males-first visuospatial displays were favored. However, this preference was reversed within an overarching category for which males are less typical. This study is described further in Hegarty and Buechel (in press).

4 General Discussion

These results are consistent with the hypothesis that implicit graph schemas involve a preference for positioning more typical subcategories before less typical ones. Because of androcentism, men's data is usually depicted first. Both relatively naïve undergraduates and relatively sophisticated scientists who are male and female show this preference. These results mirror previous findings that verbal explanation of gender differences refer to females more than males, except when females are more typical of the overarching category (see Hegarty & Buechel, in press; Pratto et al., in press; Miller et al., 1991).

References

Bem, S.L. (1993). The lenses of gender: Transforming the debate on sexual inequality. Yale University Press: New Haven, CT.

Clark, H.H., & Chase, W.G. (1972). On the process of comparing sentences against pictures. Cognitive Psychology, 3, 472-517.

Hegarty, P. & Buechel, C. (in press). Androcentric reporting of gender differences in APA articles, 1965- 2004. Review of General Psychology.

Hegarty, P., & Pratto, F. The effects of category norms and stereotypes on explanations of intergroup differences. Journal of Personality and SocialPsychology, 80, 723-735.

Miller, D. T., Taylor, B., & Buck, M. L. (1991). Gender gaps: Who needs to be explained? Journal of Personality and Social Psychology, 61, 5-12.

Exploring the Notion of 'Clutter' in Euler Diagrams

Chris John*, Andrew Fish, John Howse, and John Taylor

Visual Modelling Group,
School of Computing, Mathematical and Information Sciences,
University of Brighton, Brighton, UK
{C.John, Andrew.Fish, John.Howse, John.Taylor}@brighton.ac.uk
www.cmis.brighton.ac.uk/research/vmg/

Abstract. Euler diagrams are an effective and intuitive way of representing relationships between sets. As the number of sets represented grows, Euler diagrams can become 'cluttered' and lose some of their intuitive appeal. In this paper we consider various measures of 'clutter' for abstract Euler diagrams and show that they compare well with results obtained from an empirical study. We also show that all abstract Euler diagrams can be constructed inductively by inserting a contour at a time and we relate this inductive description to the clutter metrics.

1 Introduction

Euler diagrams consist of contours (closed curves) which divide the plane into zones; they provide an effective and intuitive way of representing relationships between sets. Euler diagrams have numerous applications, including the visualization of statistical data, displaying the results of database queries, representing non-hierarchical computer file systems and for viewing clusters which contain concepts from multiple ontologies; closed curves are a basis for many of the UML notations, including class diagrams and statecharts. However, as the number of contours increases, Euler diagrams tend to become 'cluttered' and lose some of their appeal and effectiveness. An exploration of factors affecting clutter in Euler diagrams was given in [7] and a clutter metric was proposed.

Various additions to the basic Euler diagram notation have been proposed which allow alternative representations of set relationships; for example, the use of shading to denote empty sets [6, 11] and the use of projections [4]. In [7], these are exploited to produce a 'clutter reduction algorithm' for these (extended) Euler diagrams. This can produce semantically equivalent Euler diagrams that are considerably less cluttered.

The clutter measure used in [7] appears to provide an effective measure of clutter. However, there are alternative measures that could be used and intuition varies between individuals on the degree of clutter of various diagrams. In this paper, we present the results of an empirical study designed to test subjects' notion of clutter and consider comparisons between various clutter measures.

---

* The first named author was partially supported by EPSRC studentship 0200109.

D. Barker-Plummer et al. (Eds.): Diagrams 2006, LNAI 4045, pp. 267–282, 2006.
© Springer-Verlag Berlin Heidelberg 2006

We also explore some of the properties of the measures in relation to diagrams comprising more than one connected component, so-called 'nested diagrams' [2] (see section 2.3). We show that any abstract Euler diagram can be constructed inductively by adding a contour at a time. This leads to a deeper understanding of the behaviour of some clutter measures for nested diagrams.

The notion of clutter is an issue in many disciplines both within the area of visual display and beyond. We subscribe to the view that reducing clutter increases the usability of diagrams [10]:

> *If we can better understand clutter, we can create tools to automatically identify it, and even eventually create systems that will advise a designer when the level of clutter is too high, and suggest techniques for reducing visual clutter.*

Metrics for measuring clutter are often specific to a particular domain. For example, [12] measures clutter in aviation displays, [8, 9] for air traffic control and [1] for scatter plots. Our measures are no exception. However, we envisage that aspects of this work will be adaptable to applications further afield; in particular, the notion of separating clutter at the concrete and abstract levels.

In section 2 we review the (concrete and abstract) syntax of Euler diagrams and the notion of 'nesting'. The clutter measure in [7] is introduced in section 3. We then present the results of the empirical study which leads to the definition of two new clutter metrics. In section 4 we show that all Euler diagrams can be built inductively a contour at a time and we relate this to our clutter metrics. We also show how to calculate the clutter measures of nested diagrams. Finally, we consider some directions for further work.

2   Euler Diagram Syntax

Following [5], we define two layers of Euler diagram syntax: an *abstract* or *type* syntax and a *concrete* or *token* syntax. The concrete syntax describes drawn diagrams (on paper or a monitor, for example) and the abstract syntax describes the underlying structure of diagrams.

2.1   Concrete Euler Diagrams

A *concrete Euler diagram* is a subset of the plane bounded by a *boundary rectangle* and properly containing a number of *contours*, each with a distinct label. There are various well-formedness conditions which are usually imposed to ensure diagrams can be interpreted without ambiguity. We require that contours meet transversely (cross one another at a point of intersection) and the diagram has 'connected zones'. Examples of non-well-formed Euler diagrams are given in figure 1. The diagram d_1 contains two contours B and C that do not intersect transversely (they just touch at their point in common) whereas the diagram d_2 has a disconnected zone (shaded).

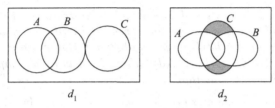

$$d_1 \qquad\qquad\qquad d_2$$

Fig. 1. Non-well-formed Euler diagrams

Definition 1. *A (well-formed) concrete Euler diagram is a quadruple* $\hat{d} = \langle \hat{L}(\hat{d}), \hat{C}(\hat{d}), \hat{Z}(\hat{d}), \hat{\beta} \rangle$ *defined as follows.*

The diagram has a **bounding rectangle**, $\hat{\beta}$, *whose interior contains a finite set,* $\hat{C}(\hat{d})$, *of* **contours** *(simple closed curves). Each contour has a unique label from the set* $\hat{L}(\hat{d})$. *Contours meet transversely and at most two contours meet at any one point.*

The contours divide the interior of $\hat{\beta}$ *into connected 'pieces' called* **zones**; *the set of zones is* $\hat{Z}(\hat{d})$. *Each zone* $\hat{z}$ *is uniquely identified by a subset* $\hat{C}(\hat{z})$ *of the contours: it comprises the region of the plane interior to the contours in the subset* $\hat{C}(\hat{z})$ *and exterior to the contours not in the subset.*

The last part of the definition of 'zone' prevent disconnected zones exhibited in diagram d_2 of figure 1. This also ensures that each zone is described uniquely by a pair of sets $(\hat{C}(\hat{z}), \hat{C}(\hat{d}) - \hat{C}(\hat{z}))$ that partition $\hat{C}(\hat{d})$.

Example 1. The concrete Euler diagram $\hat{d}$ represented in figure 2 has three contours a, b, c with labels A, B, C respectively. There are six zones which are described by the following pairs of sets that each partition $\hat{C}(\hat{d}) = \{a, b, c\}$:

$$(\{a\}, \{b, c\}), (\{a, c\}, \{b\}), (\{a, b, c\}, \varnothing), (\{b, c\}, \{a\}), (\{c\}, \{a, b\}), (\varnothing, \{a, b, c\}).$$

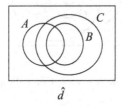

$$\hat{d}$$

Fig. 2. An Euler diagram

2.2   Abstract Euler Diagrams

An abstract Euler diagram encapsulates the *structure* in a concrete Euler diagram but discards some topological details. It includes information about the way the contours interact (as specified by the zones) but 'loses' information about the shape of the contours, for example. To describe abstractly the diagrammatic elements in a concrete diagram, we need to specify its contours and zones. Since only the labels of contours are needed to specify a zone, we can identify contours and labels.

Definition 2. *Let L be a set of contour labels. A* **zone** *(on L) is a pair (x, y) where $x \subseteq L$ and $y = L - x$. The zone (x, y) is* **inside** *the contour labels in x and* **outside** *the contour labels in y. An* **abstract Euler diagram** *d with labels L is a pair $d = \langle L, Z \rangle = \langle L(d), Z(d) \rangle$ where $Z(d)$ is a non-empty set of zones satisfying the following condition.*

(*) *There is a zone inside every contour label:* $\forall l \in L(d) \; \exists (x, y) \in Z(d) : l \in x$.

It is common (see [2, 6] for example) to require the set of zones to satisfy the following condition in addition to (*).

(**) *There is a zone outside all the contour labels:* $(\varnothing, L(d)) \in Z(d)$.

This condition is usually included as a 'drawability condition' since any diagram that fails to satisfy (**) is not well-formed. At the abstract level, however, the condition appears somewhat arbitrary. For example, the diagram $d = \langle \{A, B\}, \{(\{A\}, \{B\}), (\{B\}, \{A\})\} \rangle$ does not satisfy (**) but there is no particular reason why it should not be allowed as an abstract diagram.

2.3   Nesting in Euler Diagrams

The notion of *nesting* of Euler diagrams is defined in [2]. An Euler diagram d is *nested* if it can be constructed by embedding one diagram d_2 into a zone z^* of another diagram d_1. An example is shown in figure 3 where d_2, with contours labelled P and Q, is embedded in the zone $(\{A, C\}, \{B\})$ of the diagram d_1, with contours A, B and C.

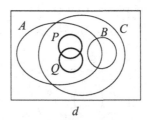

d

Fig. 3. A nested Euler diagram

Definition 3. *An abstract Euler diagram $d = \langle L, Z \rangle$ is* **nested** *if there exist Euler diagrams $d_1 = \langle L_1, Z_1 \rangle$ and $d_2 = \langle L_2, Z_2 \rangle$ and a zone $z^* = (x^*, y^*) \in Z_1$ satisfying the following conditions.*

1. *$\{L_1, L_2\}$ is a partition of L.*
2. *$Z(d) = Z_1^* \cup Z_2^*$ where $Z_1^* = \{(x, y \cup L_2) : (x, y) \in Z_1\}$ and $Z_2^* = \{(x^* \cup x, y^* \cup y) : (x, y) \in Z_2\}$.*
3. *$Z_1^* \cap Z_2^* = \{(x^*, y^* \cup L_2)\}$.*

We say that d_2 is **embedded** *in the zone z^* of d_1, and write $d = d_2 \xrightarrow{z^*} d_1$. A diagram that is not nested is called* **atomic**.

It will be convenient to define the **depth** of z^*, $depth(z^*)$, to be the number of contours enclosing it: $depth(z^*) = |x^*|$. The **components** of a diagram d are its atomic 'pieces' (obtained by successively decomposing as $d_2 \xrightarrow{z^*} d_1$).

Example 2. We shall describe the nested abstract Euler diagram illustrated in figure 3. Let d_1 and d_2 be the diagrams defined as follows:
$d_1 = \langle L_1, Z_1 \rangle$ where $L_1 = \{A, B, C\}$ and

$$Z_1 = \{(\{A\}, \{B, C\}), (\{A, C\}, \{B\}), (\{A, B, C\}, \varnothing),$$
$$(\{B, C\}, \{A\}), (\{C\}, \{A, B\}), (\varnothing, \{A, B, C\})\};$$

$d_2 = \langle L_2, Z_2 \rangle$ where $L_2 = \{P, Q\}$ and

$$Z_2 = \{(\{P\}, \{Q\}), (\{P, Q\}, \varnothing), (\{Q\}, \{P\}), (\varnothing, \{P, Q\})\}.$$

The diagram d is formed by embedding the diagram d_2 into the zone $(\{A, C\}, \{B\})$ of d_1, $d = d_2 \xrightarrow{(\{A,C\},\{B\})} d_1$.

3   Measuring Clutter

Several factors contributing to the clutter of an Euler diagram were identified in [7]. These include the number of zones, the number of contours, the ratio of zones to contours within a component (connected 'piece' of the diagram), and the degree of 'intersection' or 'disjointness' between pairs of contours – diagrams where most pairs of contours intersect tend to appear more cluttered than those where most pairs are disjoint. Various measures of clutter were considered in [7]. A preferred measure, called *contour scoring*, was proposed as a sufficiently sensitive measure of clutter which is straightforward to calculate. In this section, we consider various 'clutter measures' and some of their properties and compare how these measures perform relative to data obtained from an empirical study.

3.1   Contour Scoring

In a concrete Euler diagram, the contour score is calculated simply by counting the number of zones within each contour and summing. In an abstract Euler diagram, the *contour score* is the sum over the zones of the cardinalities of containing sets:

$$CS(d) = \sum_{(x,y) \in Z(d)} |x|.$$

The diagrams d_1, d_2 and d_3 in figure 4 have contour scores of 8, 14 and 28 respectively. Contour scoring is simple to compute and it is able to differentiate between diagrams where some other measures fail. In figure 4, for example, $CS(d_1) \neq CS(d_2)$, even though the number of zones (7), contours (4) and components (2) are the same.

In section 4 we use contour scoring to define bounds on the clutter in a diagram and we consider various patterns of diagrams and their contour scores.

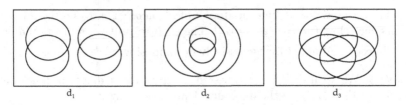

Fig. 4. Illustrating 'contour scoring'

The diagram d_1 in figure 5 can be generalised into a pattern we call a 'chain', displaying a high degree of 'disjointness' between contours; chains have a contour score which is a *linear* function of the number of contours. The diagram d_2 in figure 5 can be generalised into a pattern we call a 'crossed tunnel', displaying a higher degree of 'intersection' between contours; crossed tunnels have a contour score that is a *quadratic* function of the number of contours.

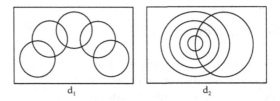

Fig. 5. The 'chain' and 'crossed tunnel' component patterns on 5 contours

3.2   Empirical Testing

To test whether contour scoring gives an accurate measure of clutter, an experiment was conducted to compare people's intuition with the calculated measures. Over 230 experimental subjects, ranging in age from 12 to 40, were asked to score various Euler diagrams on a scale of 1 to 30 purely on the basis of how 'cluttered' they appeared. To ensure an intuitive response, subjects were given an appropriately short time to complete the test. The experiment involved several sets of diagrams to test various properties. Here we show the results of two tests, each involving 16 diagrams. Four (appropriately scaled) measures, introduced in [7], were compared to the experimental data: $\mathcal{Z}ones$ (the number of zones), $\mathcal{Z}:\mathcal{C}$ (the ratio of zones to contours), $\mathcal{Z}\mathcal{S}$ (zone scoring, a measure calculated by summing the vertex degrees of the simple geometric dual graph) and $\mathcal{C}\mathcal{S}$ (contour scoring). The results for one of the sets of 16 diagrams are illustrated in figure 6; the results for the other set was similar. In each graph, the bold line represents the data (the mean score for each diagram) and the dashed line represents the measure.

Each measure appears quite accurate at measuring clutter scored by the subjects. However, closer inspection suggests that contour scoring 'follows' the empirical data most closely. In [7], it was argued that contour scoring was able to detect subtle variations in clutter the other measures were not. For example, figure 7 shows the diagrams in the experiment represented by the points 14 and

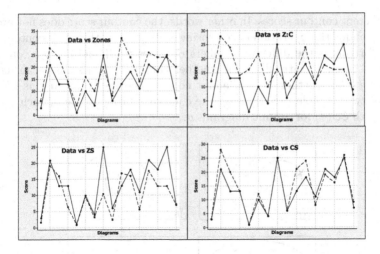

Fig. 6. Empirical data (in bold) compared to four measures (each dashed)

15 (last but two and last but one) on the graphs. The data shows that d_{15} is perceived as being more cluttered than d_{14}. Of the four measures, only $\mathcal{CS}$ scores d_{15} greater than d_{14}; the other measures each score these diagrams equally.

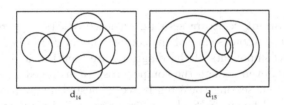

Fig. 7. Diagrams 14 and 15 from the study

Rank correlations support contour scoring as the most accurate of the four measures tested. For both sets of 16 diagrams used in the experiment, the Pearson correlation between the data and $\mathcal{CS}$ was 0.889, compared to 0.721 for $\mathcal{Z}ones$, 0.549 for $\mathcal{ZS}$, and 0.451 for $\mathcal{Z}{:}\mathcal{C}$. These results are not intended as an in-depth statistical analysis, but they do provide evidence of the capability of various measures to score clutter in Euler diagrams. A thorough analysis of results from this large data set is intended for future publication.

3.3   Alternative Measures

The empirical data lends support to contour scoring as an accurate and reliable measure of clutter in Euler diagrams but it can give rise to some potentially undesirable results. Consider the (atomic) diagrams d_1 and d_2 in figure 8, which have equal contour scores (given in the box in the diagram). Inserting a contour into each diagram surrounding the given component produces diagrams d_3 and d_4

with different contour scores. In other words, the contour score does not preserve equality under the operation of 'enclosing a diagram with a new contour'. The reason for this is the different numbers of zones in d_1 and d_2; see corollary 1 (section 4.2). It is natural to ask whether this undermines $\mathcal{CS}$ as an measure for clutter and whether we can adapt contour scoring to avoid this. We now consider two possible variations on the contour scoring measure, each of which agrees with $\mathcal{CS}(d)$ for atomic diagrams d.

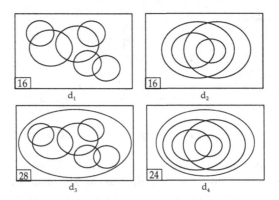

Fig. 8. Illustrating $\mathcal{CS}$ when enclosing a diagram with a new contour

Simple component contour scoring, denoted $\mathcal{SCS}(d)$, ignores nesting and simply sums the contour scores of the components of d. In diagrams d_1 and d_2 in figure 4 the components of the diagrams are identical and hence have the same $\mathcal{SCS}$ score, even though the components are arranged differently. Simple component contour scoring consistently scores nested diagrams lower that $\mathcal{CS}$; see theorem 4. Although $\mathcal{SCS}$ clearly preserves equality under the operation of 'enclosing a diagram with a contour', the diagrams in figure 9 illustrates a weakness in $\mathcal{SCS}$. Here $\mathcal{SCS}(d_1)$ is quite a lot bigger than $\mathcal{SCS}(d_2)$ even though the two diagrams are visually very similar. This example suggests that $\mathcal{SCS}$ may not adequately represent the clutter in nested diagrams where one component is quite 'deeply embedded'.

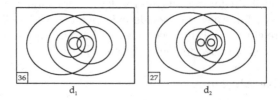

Fig. 9. Two similar diagrams scored differently under $\mathcal{SCS}$

Nested component contour scoring, denoted $\mathcal{NCS}$, takes the sum of the components' contour scores and adds the 'nesting depth' of each component; it represents a natural compromise between $\mathcal{CS}$ and $\mathcal{SCS}$ by taking account of the depth of nesting at the component level. The diagram d_2 in figure 9 has three components with contour scores 25, 1 and 1. Their respective nesting depths are 0, 4 and 5, giving $\mathcal{NCS}(d_2) = 36$. Whilst is is straightforward to compute $\mathcal{NCS}$ for (simple) concrete diagrams, it is more difficult to formulate at the abstract level. Like $\mathcal{CS}$, $\mathcal{NCS}$ fails to preserve equality under the operation of 'enclosing a diagram with a contour' as figure 10 illustrates.

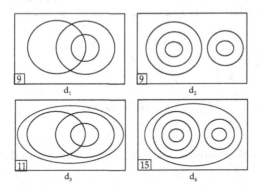

Fig. 10. Illustrating $\mathcal{NCS}$ when enclosing a diagram with a new contour

3.4 Empirical Testing Revisited

Since all three variations of contour scoring exhibit strengths and weaknesses, we consider the empirical study to investigate how accurately the measures correspond to the experimental data. The study was carried out *before* these adaptations to contour scoring were considered. Since the measures differ only on nested diagrams, we consider how they compare on the 14 nested diagrams from the original 32 diagrams used in the study. The results are illustrated in figure 11, where again the experimental data is represented by bold lines and the measures by dashed lines.

All three measures perform well in relation to the data. The performance of simple contour scoring warrants a particular mention; although it consistently scores diagrams lower than the other measures, its rank correlation with the data is 0.893 (as compared to 0.761 for $\mathcal{CS}$ and 0.707 for $\mathcal{NCS}$). This is a surprisingly good match since $\mathcal{SCS}$ ignores *how* components are nested in diagrams. These values may suggest that $\mathcal{SCS}$ more accurately represents the data when scoring nested diagrams. However when all 32 (nested *and* atomic) diagrams are considered, $\mathcal{CS}$ has a slightly higher correlation with the data scores than $\mathcal{SCS}$ (0.889 and 0.857 respectively). As there is little difference in how well the measures represent the experimental data, which is adopted may depend on other factors (e.g., whether equality should be preserved under enclosing with a contour).

4 Generating Euler Diagrams Inductively

We investigate the effect certain diagrammatic transformations have on clutter scores. We define a syntactic rule which can be used to generate any Euler

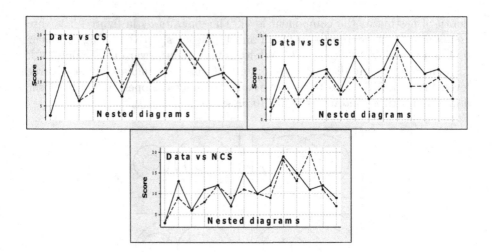

Fig. 11. Empirical data on nested diagrams compared to three measures

diagram inductively. This provides a framework to explore generic diagrammatic patterns and their clutter scores. We concentrate on $\mathcal{CS}$ but the process could be adapted to the other measures.

4.1   Inserting a Contour

We describe how (abstract) Euler diagrams can be constructed inductively by inserting one contour at a time. To define the 'insert contour' rule, we need to specify how the inserted contour relates to each of the existing contours in the diagram. We begin by considering its (one-sided) inverse operation, 'remove contour', which is simpler to describe.

Rule 1 *Remove contour.* Let $d = \langle L(d), Z(d) \rangle$ *be an Euler diagram and let* $l \in L(d)$. *The diagram* d *with* l *deleted,* $d' = d - l$, *is defined by:*

1. $L(d') = L(d) - \{l\}$
2. $Z(d') = \{(x - \{l\}, y - \{l\}) : (x, y) \in Z(d)\}$. □

It is clear that $d - l$ satisfies condition (*) of definition 2 and so is a well-defined Euler diagram.

We now define an operation which adds a new contour label l' to an abstract Euler diagram d. Various systems based on Euler diagrams have included an 'add contour' rule which preserves the semantic content of the diagram (see, for example, [6]). The purpose of our 'insert contour' rule is different. We are not interested here in reasoning with Euler diagrams but rather in building (abstract) Euler diagrams, so our 'insert contour' rule allows a contour to be inserted into a diagram in *any* way that will create a well defined Euler diagram.

Consider inserting a contour into a concrete Euler diagram. For each existing zone z in the diagram, the inserted contour will either entirely enclose z, or will split z in two or will be disjoint from z (strictly, the *interior* of the inserted contour will be disjoint from z). For example, in figure 12, d' is obtained from d by inserting the contour labelled E. The inserted contour entirely encloses the existing zones $(\{B,C\},\{A\})$ and $(\{C\},\{A,B\})$, it splits in two the zones $(\{B\},\{A,C\})$ and $(\varnothing,\{A,B,C\})$ and it is disjoint from the zones $(\{A\},\{B,C\})$ and $(\{A,B\},\{C\})$.

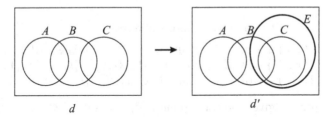

Fig. 12. Inserting a contour

In an abstract diagram, to specify how a new contour label is inserted, we need to 'partition' the set of zones into three subsets:

$Z_1(d)$, the set of zones to be enclosed by the new contour label;

$Z_2(d)$, the set of zones to be split by the new contour label;

$Z_3(d)$, the set of zones that are to be 'disjoint' from the new contour label.

In general, one or more of the sets $Z_1(d), Z_2(d), Z_3(d)$ may be empty so that, strictly, $\{Z_1(d), Z_2(d), Z_3(d)\}$ need not be a partition of $Z(d)$. However, to maintain condition (*) of definition 2, the new contour cannot be added entirely outside all the existing zones, so $Z_1(d) \cup Z_2(d) \neq \varnothing$.

In the example above, represented by figure 12, we have:

$$Z_1(d) = \{(\{B,C\},\{A\}),(\{C\},\{A,B\})\}$$
$$Z_2(d) = \{(\{B\},\{A,C\}),(\varnothing,\{A,B,C\})\}$$
$$Z_3(d) = \{(\{A\},\{B,C\}),(\{A,B\},\{C\})\}.$$

We obtain the zones of d' as follows. Each zone (x,y) in $Z_1(d)$ gives rise to a single zone in d' by adding E to x, producing zones $(\{B,C,E\},\{A\})$ and $(\{C,E\},\{A,B\})$. Each zone (x,y) in $Z_2(d)$ gives rise to a pair of zones in d' by adding E to x and to y separately; this produces zones $(\{B,E\},\{A,C\})$, $(\{B\},\{A,C,E\})$ and $(\{E\},\{A,B,C\})$, $(\varnothing,\{A,B,C,E\})$. Finally, each zone (x,y) in $Z_3(d)$ gives rise to a single zone in d' by adding E to y, producing zones $(\{A\},\{B,C,E\})$ and $(\{A,B\},\{C,E\})$.

Rule 2 *Insert contour*

Let d be a unitary Euler diagram and let $l \notin L(d)$ be a new contour label.

Let $\{Z_1(d), Z_2(d), Z_3(d)\}$ be a 'partition' of $Z(d)$ – one or more of the sets may be empty but $Z_1(d) \cup Z_2(d) \neq \varnothing$. The diagram d with l inserted relative to the partition $\{Z_1(d), Z_2(d), Z_3(d)\}$ is d' where $L(d') = L(d) \cup \{l\}$ and the zones are defined as follows.

1. $(x, y) \in Z_1(d) \Rightarrow (x \cup \{l\}, y) \in Z(d')$.
2. $(x, y) \in Z_2(d) \Rightarrow (x \cup \{l\}, y), (x, y \cup \{l\}) \in Z(d')$.
3. $(x, y) \in Z_3(d) \Rightarrow (x, y \cup \{l\}) \in Z(d')$.   □

Lemma 1. *Suppose the Euler diagram d_2 is obtained from the Euler diagram d_1 by removing contour label l; that is, $d_2 = d_1 - l$. Then d_1 can be obtained from d_2 by inserting l.*

Proof

The zones of d_1 can be partitioned into three sets:

$Z_1(d_1) = \{(x, y) \in Z(d_1) : l \in x$ and $(x - \{l\}, y \cup \{l\}) \notin Z(d_1)\}$;
$Z_2(d_1) = \{(x \cup \{l\}, y) \in Z(d_1) : l \notin x \cup y\} \cup \{(x, y \cup \{l\}) \in Z(d_1) : l \notin x \cup y\}$;
$Z_3(d_1) = \{(x, y) \in Z(d_1) : l \in y$ and $(x \cup \{l\}, y - \{l\}) \notin Z(d_1)\}$.

Removing l from d_1 gives the corresponding 'partition' of $Z(d_2)$:

$Z_1(d_2) = \{(x - \{l\}, y) : (x, y) \in Z_1(d_1)\}$;
$Z_2(d_2) = \{(x, y) : (x \cup \{l\}, y), (x, y \cup \{l\}) \in Z_2(d_1)\}$;
$Z_3(d_2) = \{(x, y - \{l\}) : (x, y) \in Z_3(d_1)\}$.

It is now clear that inserting the contour label l in d_2 relative to this 'partition' $\{Z_1(d_2), Z_2(d_2), Z_3(d_2)\}$ recreates the Euler diagram d_1.   □

The following theorem follows by induction on the number of contour labels.

Theorem 1. *All abstract Euler diagrams can be generated from the trivial diagram $\langle \varnothing, \{(\varnothing, \varnothing)\} \rangle$ by applying rule 2.*   □

Example 3. Let T_n denote the 'crossed tunnel' Euler diagram with n contours ($n \geq 2$); T_5 is shown as diagram d_2 in figure 5. To be precise, let $L(T_n) = \{l^*, l_1, l_2, \ldots, l_{n-1}\}$ where l^* is the label of the contour that intersects all the other contours and suppose that $l_1, l_2, \ldots, l_{n-1}$ label the other contours from 'outside in' so that l_1 contains l_2 which contains l_3 etc. To construct T_{n+1} from T_n, we add the contour label l_n relative to the following partition of $Z(T_n)$:

$Z_1(T_n) = \varnothing$;
$Z_2(T_n) = \{(\{l_1, l_2, \ldots, l_{n-1}\}, \{l^*\}), (\{l^*, l_1, l_2, \ldots, l_{n-1}\}, \varnothing)\}$;
$Z_3(T_n) = Z(T_n) - Z_2(T_n)$.

4.2   Using 'Insert Contour' to Calculate Contour Score

Suppose that the Euler diagram d' is obtained from d by inserting the contour label l relative to the 'partition' $\{Z_1(d), Z_2(d), Z_3(d)\}$ of $Z(d)$. We aim to relate the contour score of d, $CS(d)$ to that for d', $CS(d')$. For $i = 1, 2, 3$, let $CS_i(d)$ denote the contribution to the contour score arising from $Z_i(d)$:

$$CS_i(d) = \sum_{(x,y) \in Z_i(d)} |x|.$$

Then $CS(d) = CS_1 + CS_2 + CS_3$.

The contour label l is added to the containing set x of each zone (x, y) in $Z_1(d)$; since there are $|Z_1(d)|$ such zones, $CS_1(d') = CS_1(d) + |Z_1(d)|$.

Each zone (x, y) in $Z_2(d)$ gives rise to two zones in d', one of which has the additional contour l added to its containing set x; since there are $|Z_2(d)|$ such zones, $CS_2(d') = 2CS_2(d) + |Z_2(d)|$.

Finally, each zone (x, y) in $Z_3(d)$ gives rise to a single zone in d' with the same containing set x, so $CS_3(d') = CS_3(d)$. Combining these equations gives the following theorem.

Theorem 2. *Let d' be obtained from d by inserting the contour label l relative to the 'partition' $\{Z_1(d), Z_2(d), Z_3(d)\}$ of $Z(d)$. With the notation above,*

$$CS(d') = CS(d) + CS_2(d) + |Z_1(d)| + |Z_2(d)|. \qquad \square$$

Corollary 1. *Let d be an abstract Euler diagram and let d' be obtained from d by 'enclosing' it with a new contour (see section 3.3). Then $CS(d') = CS(d) + |Z(d)|$.*
$$\square$$

Figure 13 shows various patterns of Euler diagram that can be generated by inserting one contour at a time. The figure shows a representative diagram for each pattern with 5 contours, together with the contour score for a diagram with n contours. In each case, the formula for $CS(d)$ can be proved by a simple induction argument using theorem 2.

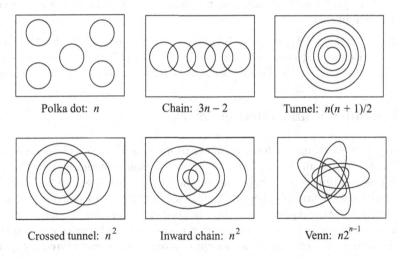

| Polka dot: n | Chain: $3n - 2$ | Tunnel: $n(n + 1)/2$ |
| Crossed tunnel: n^2 | Inward chain: n^2 | Venn: $n2^{n-1}$ |

Fig. 13. Patterns of Euler diagrams and their contour scores

The 'polka dot' and Venn diagrams with n contours give the lower and upper bounds respectively on the contour score for a diagram with n contours.

Theorem 3. *Let d be an abstract Euler diagram with n contour labels. Then $n \leq CS(d) \leq n2^{n-1}$.*
$$\square$$

4.3   Clutter in Nested Diagrams

The next theorem gives the contour score and its variations for a nested diagram; part 1 generalises corollary 1. The theorem relates to some of the properties of the measures discussed in section 3.3.

Theorem 4. *Let* $d = d_2 \xrightarrow{z^*} d_1$ *be an abstract Euler diagram obtained by embedding* d_2 *in the zone* $z^* = (x^*, y^*)$ *of* d_1 *as in definition 3. Then*

1. $\mathcal{CS}(d) = \mathcal{CS}(d_1) + \mathcal{CS}(d_2) + (|Z_2| - 1)depth(z^*)$.
2. $\mathcal{SCS}(d) = \mathcal{SCS}(d_1) + \mathcal{SCS}(d_2)$.
3. $\mathcal{NCS}(d) = \mathcal{NCS}(d_1) + \mathcal{NCS}(d_2) + depth(z^*)$, *provided* d_2 *is atomic.*

Proof. In the notation of definition 3,

$$
\mathcal{CS}(d) = \sum_{(x,y)\in Z(d)} |x|
$$

$$
= \sum_{(x,y\cup L_2)\in Z_1^*} |x| + \sum_{(x^*\cup x,y^*\cup y)\in Z_2^*} |x^* \cup x| - \sum_{(x^*,y^*\cup L_2)\in Z_1^*\cap Z_2^*} |x^*|
$$

$$
= \mathcal{CS}(d_1) + \sum_{(x^*\cup x,y^*\cup y)\in Z_2^*} |x| + \sum_{(x^*\cup x,y^*\cup y)\in Z_2^*} |x^*| - depth(z^*)
$$

$$
= \mathcal{CS}(d_1) + \mathcal{CS}(d_2) + |Z_2|depth(z^*) - depth(z^*).
$$

The result in part 1 now follows. The proofs of parts 2 and 3 follow immediately from the definitions.   □

In example 2 (figure 3), we have $\mathcal{CS}(d_1) = 9$, $\mathcal{CS}(d_2) = 4$, $depth(z^*) = 2$ and $|Z_2| = 4$. Hence $\mathcal{CS}(d) = 9 + 4 + 3 \times 2 = 19$, $\mathcal{SCS}(d) = 9 + 4 = 13$ and $\mathcal{NCS}(d) = 9 + 4 + 2 = 15$. These results are easily verified by direct calculation.

5   Conclusion and Further Work

The empirical work and exploration of the properties of various clutter measures, confirms the claim, made in [7], that contour scoring provides an effective measure. Consideration of the properties of the contour score led to the definition of alternative measures (for nested diagrams). Further and more detailed statistical analysis of the experimental data is needed to confirm the initial findings presented in this paper. The experiment was designed and carried out before the additional measures of simple component contour scoring and nested component contour scoring were defined. A further study to explore subjects' scoring of the clutter of nested diagrams is desirable.

In attempting to understand the properties of clutter measures relating to nested diagrams, it was shown that all abstract Euler diagrams can be built inductively by inserting one contour at a time. This could produce an efficient method of generating a library of Euler diagrams. It is well known that not all concrete diagrams can be built in this way. For example, the Venn-5 diagram in figure 13 has the property that removing any contour results in a non-well-formed Euler diagram with disconnected zones. There are, of course, concrete versions of Venn-5 which can be constructed by adding one contour at a time. It is an open question whether, given any drawable abstract Euler diagram, there exists a concrete instantiation that can be constructed one contour at a time. In considering this question, describing a concrete version of our 'insert contour' rule and exploring its properties will be important. For example, it is not clear under what conditions a contour can be inserted into a concrete diagram to produce a well-formed diagram.

In this paper we consider 'abstract clutter' – clutter arising from the *structure* of the diagram. However, it is clear that various aspects of the layout of concrete diagrams can influence how cluttered it appears. Factors that appear to have an effect include the shape, size and relative positions of contours and whether there is symmetry in the diagram. Further experimental and theoretical work is needed to explore the factors affecting what we might call 'concrete clutter'. This would include investigating various well-formedness conditions which provide different ways of representing the same information content.

Automatic theorem proving algorithms based on Euler diagrams are being adapted to consider minimising the length of a proof [3]. An alternative or additional criterion, designed to improve understandability, might be to use the measures to reduce clutter in proofs.

The ideas presented in this paper could have wider applicability beyond Euler diagram systems. Reducing visual complexity whilst maintaining information content could be seen as a desirable goal for many diagrammatic systems. The idea suggested here – to separate 'structural' (abstract) from 'visual' (concrete) clutter – may prove a useful framework in fulfilling this goal.

Acknowledgement. The authors would like to thank Jane Southern and Helen Medland for help with the empirical work and statistical analysis.

References

1. Bertini, E. and Santucci, G. *Improving 2D Scatterplots Effectiveness through Sampling, Displacement, and User Perception*, Ninth International Conference on Information Visualisation (IV'05), 2005, 826-834.
2. Flower J., Howse J. and Taylor J. *Nesting in Euler diagrams: syntax, semantics and construction.* J. Software Systems Modeling, 3, 2004, 55-67.
3. Flower J, Masthoff J, Stapleton G. *Generating Readable Proofs: A Heuristic Approach to Theorem Proving With Spider Diagrams*, Proc. Diagrams 2004, LNAI 2980, Springer-Verlag, 2004, 166-181.
4. Gil J., Howse J., Kent S. and Taylor J. *Projections in Venn-Euler Diagrams*, Proc. IEEE Symposium on Visual Languages (VL2000), IEEE, 2000, 119-126.
5. Howse J., Molina F., Shin S-J. and Taylor J. *On Diagram Tokens and Types*, Proc. Diagrams 2002, LNAI 2317, Springer-Verlag, 2002, 146-160.
6. Howse J., Stapleton G. and Taylor J. *Spider Diagrams*, LMS J. Comput. Math., 8, 2005,145-194.

7. John C. *Measuring and reducing clutter in Euler diagrams*, Proc. 1st International workshop on Euler diagrams, ENTCS 134, Elsevier, 2005, 103-126.

8. Lloyd N. *Clutter measurement & reduction for enhanced information visualization.* Masters thesis, Worcester Polytechnic Institute. Available at www.wpi.edu/Pubs/ETD/Available/etd-011206-232808/unrestricted/nlloyd.pdf

9. Peng W., Ward M. and Rundensteiner E. *Clutter Reduction in Multi-Dimensional Data Visualization Using Dimension Reordering*, infovis, IEEE Symposium on Information Visualization (INFOVIS'04), 2004, 89-96.

10. Rosenholtz R., Li Y., Mansfield J., and Jin, Z. *Feature Congestion: A Measure of Display Clutter.* Proc. SIGCHI conference on Human Factors in Computing Systems, 2005, 761-770.

11. Shin S.-J. *The Logical Status of Diagrams*, Cambridge University Press, 1994.

12. Wickens, C. *Display Formatting and Situation Awareness Model (DFSAM): An approach to aviation display.* Technical report AHFD-05-14/NASA-05-5, 2005.

Using Channel Theory to Account for Graphical Meaning Generations

Atsushi Shimojima

Faculty of Culture and Information Science
Doshisha University, Japan
ashimoji@mail.doshisha.ac.jp

Abstract. Many graphical representation systems apparently support *derivative meanings*, namely, meaning relations not directly legitimized by the systems' semantic conventions. Practitioners and researchers of graphical communication have long realized the functional importance of the generative nature of graphical meaning, yet no systematic attempts have been made to clarify the informational mechanism behind it. In this project, we develop a mathematical framework in which meaning derivation properties of graphical systems are explicitly modeled and accounted for.

The two charts in Figure 2 are the results of presenting the two sets of data in Figure 1 in the form of scatter plots. The example is borrowed from Tufte [1].

A		B	
10.0	8.04	10.0	7.46
8.0	6.95	8.0	6.77
13.0	7.58	13.0	12.74
9.0	8.81	9.0	7.11
11.0	8.33	11.0	7.81
14.0	9.96	14.0	8.84
6.0	7.24	6.0	6.08
4.0	4.26	4.0	5.39
12.0	10.84	12.0	8.15
7.0	4.82	7.0	6.42
5.0	5.68	5.0	5.73

Fig. 1. Two sets of data presented in the scatter plots in Figure 2

On the one hand, each of these scatter plots is "intertranslatable" with the corresponding table: one provides sufficient information to reproduce the other, and vice versa. In this limited sense, each plot has the same informational content as the corresponding table. On the other hand, there is a definite sense in which each plot reveals more information than the table does. The particular shape formed by the dots on a scatter plot seems to indicate some general fact about the data, or more precisely, about the situation that the data are about. Thus, the particular shape appearing in plot A indicates that Y-values and X-values are positively correlated, and the shape in plot B indicates that Y-values and X-values are mostly proportional.

It is clear, and almost trivial, that the particular shape and the density of the cloud formed in some scatter plot carries information that the corresponding table of data would not. In fact, this type of additional informational relations are quite prevalent in graphical representation systems, and as the above example illustrates, their existence is

D. Barker-Plummer et al. (Eds.): Diagrams 2006, LNAI 4045, pp. 283–285, 2006.

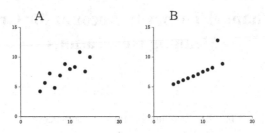

Fig. 2. Scatter plots presenting the data in Figure 1

often the very reason why a given system is preferred to others as a method of displaying certain sets of information. Nevertheless, when it comes to the question how such additional informational relations arise in a given graphical system, things are much less clear. The aim of this project is to give an detailed answer to this general question.

Beside being prevalent, graphical meaning generation is a *functionally important* phenomenon. Some researchers hypothesize that the utility of a graphical system depends on what repertoire of derivative meanings it allows the viewer to extract, and how easily the viewer can do so [2, 3]. A related hypothesis is that the proficiency or expertise of reading graphics depends on the ability to appreciate derivative meanings in the graphics [4, 5, 2, 6] It has been also hypothesized that evaluations of derivative meanings in a graphic forms a class of mental operations relatively independent from evaluations of basic meanings, whose occurrences depend on the given purposes of reading the graphic [5, 7, 8].

Despite its apparent prevalence and functional importance, the phenomenon of graphical meaning generation has received little explicit attention in the literature of graphics semantics. No semantic theories of graphics, either grammatical, model-theoretic, or algebraic, have ever attempted to track its logical origin. Also, the phenomenon is apparently independent of any informational relations, such as "secondary notations" [9] and "graphical implicatures" [10], that have been studied in pragmatic accounts of graphics. Pinker's theory [2] offers a systematic account of the conditions for a cognitive system to comprehend derivative meanings, yet as a psychological theory, it is silent about the logical relationship between derivative meanings and basic meanings.

In the full paper, we try to build a framework of graphics semantics in which meaning derivation properties of graphical systems are explicitly modeled and accounted for. In our account, derivative meaning relations in a graphical system are results of the logical interactions involving (i) constraints installed by the semantic conventions, (ii) constraints originated in the domain of representations, and (iii) constraints originated in the domain of represented objects. The account is based on channel theory, developed by Barwise and Seligman [11].

Beside being the first comprehensive treatment of the phenomenon of graphical meaning generation, this project is of deeper philosophical interest. In a nutshell, channel theory is a machinery to combine logics. Its basic strategy is to model an environment as a system ("distributed system" as they call it) consisting of smaller component domains coming with their own "local logics". These component domains are often overlapped and interconnected, meaning that their local logics interact in interesting ways, giving rise to new "theorems" governing the entire system.

On its mathematical side, channel theory provides an exact prescription of what system-wide logic results from its component local logics, given the particular overlapping and connectedness of their respective domains. Underlying this endeavor is the philosophical view that regularities governing a complex environment, of which we cognitive agents make heavy use for inference, for problem-solving, and even for survival, can be captured as such theorems in the combined system.

With this background, the full paper reports the rather interesting discovery that this philosophy and mathematics of channel theory has its natural application in accounting for the meaning generation phenomenon in graphical representation systems. Thus, on the one hand, our finding is a partial confirmation of the philosophical thesis underlying channel theory. On the other hand, our finding demonstrates that meaning generation in graphical systems is not an isolated phenomenon based on weird capabilities of certain graphical representation systems, but an instance of the general capability for multiple local logics to combine and generate new theorems in a larger system. Channel theory thus offers a broader perspective from which to view graphical meaning, hinting at a new way of building semantic theories of graphical representations.

References

1. Tufte, E.R.: The Visual Display of Quantitative Information. Graphics Press, Cheshire, CN (1983)
2. Pinker, S.: A theory of graph comprehension. In: Aritificial Intelligence and the Future of Testing. L. Erlbaum Associates, Hilsdale, NJ (1990) 73–126
3. Lohse, G.L.: A cognitive model for understanding graphical perception. Human-Computer Interaction **8** (1993) 353–388
4. Lowe, R.K.: Diagram prediction and higher order structures in mental representation. Research in Science Education **24** (1994) 208–216
5. Guthrie, J.T., Weber, S., Kimmerly, N.: Searching documents: Cognitive processes and deficits in understanding graphs, tables, and illustrations. Contemporary Educational Psychology **18** (1993) 186–221
6. Gilhooly, K.J., Wood, M., Kinnear, P.R., Green, C.: Skill in map reading and memory for maps. Quarterly Journal of Experimental Psychology **40A** (1988) 87–107
7. Kinnear, P.R., Wood, M.: Memory for topographic vontour maps. British Journal of Psychology **78** (1987) 395–402
8. Ratwani, R.M., Trafton, J.G., Boehm-Davis, D.A.: Thinking graphically: Extracting local and global information. In: Proceedings of the Twenty-fifth Annual Conference of the Cognitive Science Society. (2003) 958–963
9. Petre, M., Green, T.R.: Requirements of graphical notations for professional users: Electronics cad systems as a case study. Le Travail humain **55** (1992) 47–70
10. Marks, J., Reiter, E.: Avoiding unwanted conversational implicatures in text and graphics. In: AAAI-90: Proceedings of Eighth National Conference on Artificial Intelligence. Volume 1., Menlo Park, CA/Cambridge, MA, AAAI Press/MIT Press (1990) 450–455
11. Barwise, J., Seligman, J.: Information Flow: the Logic of Distributed Systems. Cambridge University Press, Cambridge, UK (1997)

Toward a Comprehensive Model of Graph Comprehension: Making the Case for Spatial Cognition

Susan Bell Trickett and J. Gregory Trafton

Naval Research Laboratory, 4555 Overlook Drive,
Washington DC 20375
trickett@itd.nrl.navy.mil, trafton@itd.nrl.navy.mil

Abstract. We argue that a comprehensive model of graph comprehension must include spatial cognition. We propose that current models of graph comprehension have not needed to incorporate spatial processes, because most of the task/graph combinations used in the psychology laboratory are very simple and can be addressed using perceptual processes. However, data from our own research in complex domains that use complex graphs shows extensive use of spatial processing. We propose an extension to current models of graph comprehension in which spatial processing occurs a) when information is not explicitly represented in the graph and b) when simple perceptual processes are inadequate to extract that implicit information. We apply this model extension to some previously published research on graph comprehension from different labs, and find that it is able to account for the results.

1 Introduction

Until recently, models of graph comprehension have mostly focused on simple graphs and tasks, for which information is explicitly represented in the graph. Examples of this type of graph/task combination include extracting trends from bar graphs, reading off values from bar and line graphs, comparing values in bar graphs, and the like (e.g, (Cleveland, 1985). Recently, however, researchers have begun to question the extent to which these simple, context-lean graphs and tasks represent the true nature of graph use beyond the psychology laboratory. In reality, graphs may be used to make predictions, and thus require information that is not explicitly represented (DeSanctis & Jarvenpaa, 1989) and graphs may be used as problem-solving tools (Cheng *et al.*, 2001; Scaife & Rogers, 1996; Tabachneck-Schijf *et al.*, 1997; Trafton *et al.*, 2000; Trafton & Trickett, 2001). Researchers have thus been attempting to understand more complex graph/task interactions, as well as the use of more complex graphs themselves. These newer models have begun to account for such factors as the iterative nature of graph interpretation (Carpenter & Shah, 1998; Peebles & Cheng, 2003), the familiarity of the graph (Peebles & Cheng, 2003), the role of graph and domain knowledge (Freedman & Shah, 2002), and the importance of expertise (Roth & Bowen, 2003; Tabachneck-Schijf et al., 1997).

In their recent review of the graph comprehension literature, Shah, Freedman, and Vekiri draw a distinction between perceptual and conceptual processes (Shah *et al.*, 2005). In their interpretation, perceptual processes are "bottom-up encoding

D. Barker-Plummer et al. (Eds.): Diagrams 2006, LNAI 4045, pp. 286–300, 2006.
© Springer-Verlag Berlin Heidelberg 2006

mechanisms," which focus on the visual features of the display, whereas conceptual processes equate to "top-down encoding processes," which influence interpretation. They propose that perceptual processes account for performance on "simple, fact-retrieval tasks," but they further argue that "If visual features do not automatically evoke a relationship, either because the relationships are not visually integrated in a graph or because the graph viewer does not have the prior knowledge required to make an interpretation, information must be retrieved by complex inferential processes."

Although several models agree that these "complex inferential processes" are an essential part of the graph comprehension process under some circumstances, they remain largely unspecified. Indeed, in taking into account both perceptual and conceptual processes, Shah et al. identify five factors that play a role in predicting performance on graph comprehension: display characteristics, data complexity, the viewer's task, the viewer's prior domain knowledge, and the viewer's knowledge about graphs. What current models lack is a means to specify precisely how these factors will influence the type of complex inferential processes that will be engaged.

We investigated what happens in complex, problem-solving domains when scientists are unable to extract the information they need from the visualization they are using (Trafton & Trickett, 2001). Based on an in-depth analysis of several hours of verbal protocols, we found not only that it was extremely common for the scientists to confront situations where they were unable to directly extract the information they needed, but also that in these cases, they used spatial transformations more frequently than any other strategy to generate this information. We concluded that models of graph comprehension should be expanded to include spatial processing, particularly in complex domains for which complex visualizations are required.

Our most recent research has focused on complex graphs in another complex domain (meteorology). Common tasks in this domain certainly include fact retrieval (e.g., temperature, wind-speed, etc., at a specific location), in which information is explicitly represented and little specialized domain knowledge is required to extract it. However, equally commonly, forecasters use graphs to draw inferences (e.g., finding the pressure at location C when it is given for locations A and B), for which information is not explicitly represented and some domain knowledge is required. They must also make complex weather predictions, for which information is not explicitly represented and a great deal of domain knowledge is needed. Our approach has been to observe and record experts and journeymen using weather graphs as part of their regular work, and to interview them about their strategies. Our results have been consistent: the verbal protocols show that in this domain, at least, people use a great deal of spatial processing to extract and use information from data visualizations (Trafton et al., 2000; Trafton & Trickett, 2001). Further evidence of spatial processing is found in meteorologists' gestures when they talk about how they performed the task (Trafton *et al.*, in press). Additionally, in keeping with the important role of domain knowledge in graph comprehension (Freedman & Shah, 2002; Roth & Bowen, 2003; Tabachneck-Schijf et al., 1997), experts use far more spatial processing than journeymen (Trafton et al., in press). This general result—that spatial processing is prevalent in complex graph comprehension—has been replicated in studies of fMRI research, in addition to the original work in astronomy and computational fluid dynamics. All these domains share some important characteristics with the meteorology domain (in terms of the complexity of the visualizations, the task, and the domain) (Trafton & Trickett, 2001).

Based on these data, we have become convinced that spatial processing is an important component of a comprehensive model of graph comprehension, and specifically, that it plays a crucial role in guiding the "complex inferential processes" discussed above that are involved when information cannot be directly extracted from the graph. Yet, curiously, spatial processing is not explicitly included in any of the current models of graph comprehension. This is something of a puzzle, because these models are highly successful in accounting for graph comprehension behavior in many graphs and tasks, including some in which information must be inferred rather than extracted directly. Indeed, some models (Lohse, 1993; Peebles & Cheng, 2003; Pinker, 1990) are explicitly and exclusively propositional; others are simply non-committal (Freedman & Shah, 2002; Roth & Bowen, 2003). The goal of this paper is to investigate why this should be so—that is, why current theories of graph comprehension, so successful in analyzing many of the behaviors associated with graph interpretation, do not account for our data. We propose a refinement of current theories that enables them to predict when spatial processing will occur, and how it will guide graph comprehension in these complex situations.

First, we provide a brief definition of spatial processing. Though this definition is a simplification, it is nonetheless useful because it provides operational means by which we can identify spatial processing in our verbal protocol data and in our analyses of the requirements of graph tasks in the graph comprehension literature. Second, we briefly describe a generic model of graph comprehension, based on recent analyses by Shah, Freedman, and Vekiri (2005). We analyze two situations, one in which needed information can be directly extracted from the graph, and a second in which it must be inferred using perceptual processes. The purpose of this description is to establish a terminology that applies to graph comprehension tasks specifically addressed by this model, which we can then use to describe the tasks and actions involved in the more complex meteorological task. Third, we summarize the results of several studies in complex domains that show spatial processing is used in these tasks. We also analyze several specific instances of forecasting tasks, to show *when* and *how* spatial processing is used. We show that when the desired information was not explicitly represented in the graph *and* when perceptual processing could not generate the type of information needed, spatial processing was used. Finally, we apply our model to an analysis of a graph comprehension tasks from the graph comprehension literature, and show how it provides a viable interpretation of performance on these tasks.

2 Spatial Processes

Baddeley was instrumental in establishing the distinction between verbal and spatial processing (Baddeley, 1999) and in further distinguishing between spatial and visual processing (Baddeley & Liebeman, 1980). Spatial processing involves "the internalized reflection and reconstruction of space in thought" (Hart & Moore, 1973).

Operationally, we define spatial processing in two ways. Spatial processing involves maintaining spatial information (e.g., the relative locations of objects) in working memory (so-called spatial working memory). Instances of spatial processing can therefore be identified by means of task analysis (Gray *et al.*, 1993). Spatial processing can also be identified via the use of mental spatial transformations, which occur when a spatial object is transformed from one mental state or location into another mental state or location. Mental spatial transformations—which we refer to simply as spatial transformations—occur in a mental representation that is an analog of physical space and are frequently part of a problem-solving process. There are many types of spatial transformations: creating a mental image, modifying that mental image by adding or deleting features, mental rotation (Shepard & Metzler, 1971), mentally moving an object, animating a static image (Bogacz & Trafton, 2005; Hegarty, 1992), making comparisons between different views (Kosslyn *et al.*, 1999; Trafton *et al.*, 2005), and any other mental operation which transforms a spatial object from one state or location into another.

Note that we distinguish between spatial processing (i.e., the use of spatial transformations) and purely perceptual processing, in which graph users are able to make *direct* or *explicit* comparisons from the graph itself, without the need to hold spatial information in working memory. Thus, comparing two adjacent bar heights on a graph requires only perceptual processing, whereas comparing a bar height on a displayed graph with one on a remembered graph requires spatial processing, because the remembered bar height would have to be projected onto the displayed graph for the comparison to occur (assuming that specific values had not been remembered).

Most graph comprehension research is not designed to specifically identify the type of processing—verbal or spatial—used. However, while people are doing graph-related tasks, it is possible to elicit verbal protocols, that "dump" the contents of working memory during problem-solving (Ericsson & Simon, 1993). These verbal protocols can then be coded, and instances of spatial processing can be identified. We have conducted several studies in which meteorologists (and scientists) give verbal protocols while making their forecasts, and we have coded the spatial transformations in those protocols. Our IRR for this coding has been consistently good.

3 A Generic Model of Graph Comprehension

Several general models of graph comprehension are based on one proposed by Pinker (Pinker, 1990), in which the visual features of the display, gestalt processes, and the graph schema all interact to allow the user to extract the conceptual message of the graph. To summarize this model: 1) the user has a goal (which is provided) to extract a specific piece of information 2) the user looks at the graph, and the graph schema and gestalt processes are activated 3) the salient features of the graph are encoded, based on gestalt principles 4) the user now knows which cognitive/interpretive strategies to use, because the graph is familiar (graph knowledge)—that is, the user knows where to look at what steps to take 5) the user extracts necessary goal-directed visual chunks 6) the user may compare two or more visual chunks and 7) the user extracts the relevant information to satisfy the goal (answer the question).

Figure 1 shows a simple bar graph, depicting information from the US Census Bureau (www.census.gov). Suppose the goal is to extract the amount of lifetime earnings for woman with a high school diploma. This information is explicitly represented in the graph, and can be directly extracted when the user executes steps 1 through 5, and step 7. Suppose, however, the goal is to extract how much more women earn if they complete a bachelor's degree. In this case, the information is not explicitly represented; however, it can be easily extracted by repeating steps 1 through 5 for the bachelor's degree bar, and calculating the difference. Other information that is not explicitly represented can also be extracted in a straightforward manner—for example, the trend of earnings as education increases can be extracted by using the perceptual process of noticing that each successive bar is a bit taller than the bar to its left (i.e., by comparing visual chunks). Similarly, the answer to the question of who earns the most can be extracted by locating the tallest bar (again, comparing the bar heights, or visual chunks). None of these questions, which are typical of questions posed for this type of graph, requires the use of spatial processes; all can be answered by using the perceptual processes built in to the generic model.

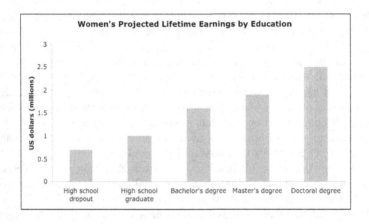

Fig. 1. Simple bar graph

Building on this model, Shah and Carpenter have shown that the processes involved are iterative, rather than serial (Carpenter & Shah, 1998), and Freedman and Shah have further adapted the model to account for the role of domain knowledge, which may influence the last stage, in accordance with the user's expectations (Freedman & Shah, 2002). Nonetheless, these basic perceptual processes have been sufficient to account for many tasks in the graph comprehension literature, such as fact-retrieval from line or bar graphs, trends for line and bar graphs, making proportion judgments from pie charts, making comparison judgments, determining the slope of a regression line, and so on. Variability in performance on these tasks is likely due to other factors, such as graph or domain knowledge (Freedman & Shah, 2002).

In this generic model, although graphs depend on spatial arrays, the processes by which information is extracted are largely perceptual (whether the information is represented explicitly or implicitly). Graphs depict relationships by means of the strategic arrangement of spatial elements, and those relationships can be easily extracted because spatial attributes are automatically encoded relationally, e.g., a higher line is encoded as meaning a greater value (Pinker, 1990). What makes some graphs better than others for particular tasks is precisely this characteristic of graphs—e.g., trend information is more easily extracted from line graphs than bar graphs (Tversky, 1995) because a line with an increasing slope is encoded as going from less to more. The trend can thus be interpreted by means of virtually effortless perceptual processing.

4 Tasks in Complex Domains

In this section, we examine how well the generic model outlined above accounts for performance in the meteorological and scientific domains. The simplest type of task in meteorology is straightforward fact retrieval. Figure 2 shows a typical meteorological graph. Note that this graph is significantly more complex than the bar graph in Figure 1: it shows four variables (wind speed, wind direction, temperature, and sea level pressure) in addition to latitude and longitude lines overlaid onto a map of the Southeastern U.S. Despite the graph complexity, however, in such cases, when the given goal is to extract a specific piece of information, the generic model is quite adequate. For example, if a forecaster wanted to know the current temperature at Pittsburgh, he would take the same steps as those outlined above: first, he would look at the weather graph, activating the graph schema, then he would find Pittsburgh and encode the color, thereby extracting the required visual chunk; the graph schema would guide him to translate the color into a temperature value, by looking at the legend, and he would "read off" the appropriate value from the legend. The forecaster might make several iterations between looking at Pittsburgh and the legend (Trafton *et al.*, 2002),, but the basic mechanisms from the generic model can easily account for performance on this task. The model also supports some tasks in which information is not directly represented, such as the comparison "Which is hotter, Pittsburgh or Washington?" As in the comparison task in the bar graph, the forecaster could answer this question by using perceptual processes, locating both Pittsburgh and Washington and their associated colors, and either looking up the values on the legend or (a more likely expert strategy) comparing the colors to see which represents the hotter temperature.

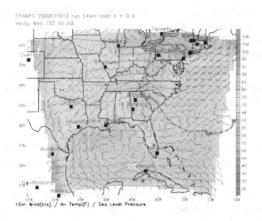

Fig. 2. Meteorological graph (note that the original graph is in color)

In many forecasting tasks, however, information cannot be directly extracted from the graph, and in these cases, it may not be possible to extract the information by perceptual processes, such as directly comparing visual chunks. Trafton, Marshall, Mintz, and Trickett (Trafton et al., 2002) conducted an eyetracking study in which forecasters were asked to perform a number of routine forecasting tasks. The tasks were designed to have certain characteristics: asking for quantitative information, where the answers were explicitly represented in the graph; asking for quantitative information that was imprecisely represented (i.e., the values were represented by a symbology that the user must know in order to extract the needed information); and asking for quantitative information that was entirely *implicit* in the graph (e.g., what is the pressure at location C, when it was given for locations A and B, so that values must be inferred).

In another study involving scientists rather than forecasters, Trafton, Trickett, and Mintz (2005) investigated the use of mental imagery in scientific visualization. Two astronomers and a physicist were observed using complex visualizations to analyze data. Trafton et al. compared the number of spatial transformations in the verbal protocols with the number of physical transformations they performed on the visualizations (i.e., creating a new visualization or adjusting a current one). There were significantly more spatial transformations than physical transformations, suggesting that the scientists frequently used spatial processing in preference to the computer's visualization capabilities. More interestingly, Trafton et al. found that comparisons were extensively used to tie the internal and external visualizations together. They divided these comparisons into two types: comparing two external visualizations and comparing an external and an internal visualization. They also coded the type of comparison made: comparing features (such as color or size) or aligning (i.e., making an estimation of "fit" between the two images). They found that the type of comparison was related to the type of visualization being compared. When two external visualizations were compared, the scientists most frequently compared features of the visualization; however, when an internal and an external visualization were compared, the scientists most frequently made alignments. In terms of our model, when the information was explicitly represented (in the external visualizations), the scientists used a perceptual strategy of comparing visible features; when the information was not explicitly represented, they used a spatial strategy of aligning one visualization with another, in order to estimate the overall "fit" between the two.

In addition to these studies, further data from our study of forecasters shows a large number of situations in which data must be inferred, e.g. resolving discrepancies between models or between a representation of certain phenomena and the forecaster's conflicting belief about the phenomena, integrating large amounts of disparate information into a comprehensive mental model, comparing visual chunks that are no longer on the visible display, but must be recalled in memory and mentally juxtaposed, as well as projecting the changes in the visualization that will likely occur over time, given current conditions. We analyze detailed examples of some of these situations below, in order to show where the information came from and how it was generated. We consistently find that when forecasters cannot directly extract the information from the display, they do one of two things: either they recall it, and given the nature of the domain, much of this recalled information is spatial, or they generate it by means of spatial transformations, a form of spatial processing.

Table 1. Resolving a discrepancy between two weather models

Utterance	Information Explicitly Available?	Action
You also have a 12 max 14	Yes	Extract information
winds are not supporting that	Yes	Note discrepancy
The next chart has it moving down further to the south	No	Recall from spatial memory (previous visualization, no longer on screen)
there is a low coming off the coast that is propagating around	No	Spatial transformation (accompanied by hand gesture tracing location of imagined low)
so I would move it further to the south	No	Spatial transformation
and that just supports what I said about ours, OK	N/A	Resolve discrepancy

Table 1 shows an instance of a forecaster trying to resolve a discrepancy between two visualizations. Her goal was to determine whether or not to maintain a high-seas warning, and the chart on display showed the projected sea heights in different locations of a particular model for the period of interest. She begins by reading off the projected sea height in the location she is interested in, "12 max 14" (i.e., high enough to be of concern to her). However, information about wind speed conflicts with this information. She then recalls another visualization that showed the high seas area in transition further to the south (at this stage, she is using her memory of relative location, i.e., spatial processing). Her next utterance is a spatial transformation, in which she mentally creates and moves a low pressure system (it is not represented on the current visualization, but is recalled from memory and projected onto the current visualization). This is followed by another spatial transformation, in which she mentally moves the area of high seas further to the south. Implicitly, she performs a mental comparison of these transformed mental representations, and finally indicates that in this new location, the high seas makes sense and the discrepancy is resolved. It is important to note that, just as in the use of more simple graphs, there is an interaction between the demands of the task and the type of visualization. In the current case, however, this interaction is complex and requires several different types of processing: information must be extracted directly from the current visualization, both spatial and non-spatial information must be recalled from previously accessed visualizations, and spatial information must be both superimposed and manipulated on the current visualization.

Table 2. Discrepancy between weather model and forecaster's mental model

Utterance	Information Explicitly Available?	Action
I can't buy an 82 out of the weather bureau at all...	N/A	Note discrepancy
and having a hard time understanding why they're coming up with what they got.	N/A	Reiterate discrepancy (no easy resolution)
They have um Brunswick a max temperature of 78 for Friday,	Yes	Extract information
we push a front through	No	Spatial transformation
and we go to 82 degrees	No	Non-spatial projection
That's just no way you would think anything like that would happen....	No	Spatial transformation (mental comparison)
I'm not buying that	N/A	Reiterates disbelief
So again what I'm gonna do	N/A	Conclusion
I'm gonna more or less stay with what I had yesterday,	N/A	Conclusion
I'm going 77	N/A	Conclusion

Table 2 illustrates a similar case of discrepancy resolution, only in this instance the discrepancy is between the National Weather Service prediction and the forecaster's own mental model. The forecaster begins by doubting the NWS temperature prediction and puzzles over how it was constructed. He then attempts to reconstruct the process, recalling the prediction for the day prior to the disputed forecast, and performs a spatial transformation on that model, mentally moving a front through the relevant area (the front is not represented on the current visualization). He updates the mental representation after the front has hypothetically moved through, and projects the disputed 82 degrees on that update. He performs another spatial transformation by mentally comparing the two relevant chunks—the updated mental representation and the representation containing the 82 degrees, and finds them still discrepant. As a result, he cannot believe the NWS forecast is valid, and resolves the discrepancy by sticking with his own mental model. As in the previous example, spatial transformations are a crucial part of how he uses the visualization to resolve the problem.

Similarly, Table 3 shows an instance of a forecaster needing to compare visual chunks that are not visible on any current display. In this situation, the forecaster is attempting to update a paper chart, by integrating information from all the previous visualizations (from disparate models) she has viewed. She first notes a discrepancy between one of these models and the others, by performing a mental comparison of the two representations (a spatial transformation). Her second utterance indicates that one of the remembered models did display the lows that are absent from the Canadian model. In each of the next three utterances, she performs some form of mental comparison between her memory of the Canadian model and her memory of the ENSAP model. In the end, she determines a placement for the low on the third representation, a paper chart that she is attempting to update.

Table 3. Compare visual chunks not on the visible display

Utterance	Information Explicitly Available?	Action
Also, one thing I'm noting is that the Canadian model is having a problem picking up the two lows	No	Spatial transformation: comparison (2 mental representations)
that are circulating around this cut-off low off of the coast of Greenland	No	Recall from memory
They do have something there	No	Spatial transformation: Comparison (alignment)
But they're not putting a central pressure on it, as ENSAP is	No	Spatial transformation: Comparison (alignment)
And it...they're definitely there	No	Spatial transformation (projection)
So I'll put an X where I think that low should be	N/A	Resolution

From these examples, it appears that spatial transformations serve a particular purpose in this domain, namely, they provide a means whereby the forecasters can *generate* needed information that is not explicitly represented in the visualization. This information is constructed by performing spatial transformations on the information that is explicitly represented, and developing new mental representations based on those (mentally) transformed visualizations. These new mental representations can be further manipulated or used as the basis for comparisons.

We thus propose that current models of graph comprehension do not include spatial processing because for the tasks and graphs used in most studies of graph comprehension, either the information can be directly extracted from the display, or if not, it can be inferred using direct perceptual processing of the available visual chunks. We propose that models of graph comprehension should account for these conditions, in that when the information can neither be extracted directly nor inferred from perceptual processes, spatial processing will be used. We now test this aspect of this model by applying it to a graph task in the graph comprehension literature.

5 Tasks in the Graph Comprehension Literature

In this section, we turn to a re-analysis of a previously published graph comprehension study, in order to test our model of spatial processing. We have chosen to focus our analyses on one of the tasks investigated by Simkin and Hastie (Simkin & Hastie, 1987) because their stated aim was to establish elementary codes that can account for the processes that operate "when people decode the information represented in a graph," that is, that can apply across different graph/task combinations and account for differences in performance across different graphical representations.

Simkin and Hastie used three different graph types—simple bar, divided bar, and pie charts—and three tasks—discrimination, comparison judgment, and proportion judgment. The graphs were similar to those in Figure 3. We focus only on the comparison judgment task, because according to Simkin and Hastie's task analysis, all their different elementary codes are involved in this task across the different graph types. For the comparison task, participants were asked to assess the percentage the smaller visual chunk was of the larger. This information is not explicitly represented in the graph; the question of interest to us is whether it can be extracted by purely perceptual means, or whether spatial processing must be used.

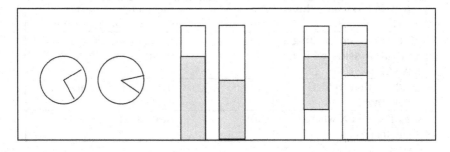

Fig. 3. Left to right: pie, bar, and divided bar graphs, of the type used by Simkin and Hastie

Simkin and Hastie developed four elementary processes by which people extract information from graphs: anchoring, scanning, projection, and superimposition. Anchoring involves selecting a portion of the graph as a baseline, or anchor, against which other judgments can be made (e.g., 50% of a bar). Scanning involves moving the eye from the anchor to the edge of the distance to be judged (e.g., from the midpoint of a shaded area to its edge). Projection involves mentally drawing a line from a point in one image to a point in anther. Superimposition involves moving elements of the graph to a new position, to create overlap with other elements in the graph.

In terms of our model, anchoring, scanning, and projection can be considered perceptual processes, at least in these very simple tasks. Although projection involves mentally drawing a line, this extension most likely does not make much, if any, demand on spatial working memory, because of the direct juxtaposition of the start and end points. Superimposition, however, does involve spatial processing, because a spatial element, in which size and/or angle are crucial, must be mentally moved from one spatial location to another. Superimposition is thus a spatial transformation.

Simkin and Hastie's task analysis showed that for the simple bar graph, projection, anchoring, and scanning are required, whereas for the divided bar and pie graphs, superimposition, anchoring, and scanning are needed. In this case, our model predicts that the task would take longer and be less accurate for divided bar and pie graphs than for simple bar graphs, because it involves at least one spatial transformation, and spatial processing is more effortful than purely perceptual processing. This is indeed what Simkin and Hastie found: reaction time was significantly shorter for simple bar than divided bar and pie graphs, though these did not differ significantly from each other, and accuracy for simple bar graphs was greater than for divided bar or pie graphs, though again, these two were not significantly different from each other.

This analysis suggests some validation of our model, in that we are able to match Simkin and Hastie's independently established elementary processes to our distinction between perceptual and spatial processing, and show that graph/task combinations that involve spatial processing were found to be more difficult than one that did not. The analysis also shows that there are simple graph tasks that can be—and have been—performed in the psychology laboratory for which perceptual processing alone cannot account for performance on the task. Although Simkin and Hastie did not analyze elementary graphing processes according to their perceptual or spatial nature, it is clear that at least one of their processes—superimposition—involves spatial cognition. Since Simkin and Hastie's goal in establishing these four elementary codes was to "develop a vocabulary of elementary mental processes that can be combined to build information-processing models of performance in common graph-perception tasks," it would be a simple matter to perform task analyses on other tasks to determine which of the elementary codes (perceptual or spatial) is involved.

6 General Discussion

We have proposed that a comprehensive model of graph comprehension needs to include spatial processing as an important component of the graph comprehension process, and we have proposed a model that predicts when spatial processing is required. In accordance with prior models of graph comprehension, we have argued that when information can be extracted directly from the graph, the task can be accomplished by perceptual processing alone. However, we have proposed an extension to those models, such that when information is not explicitly represented and the information cannot be extracted by perceptual processes, spatial processing will likely be used. Whether perceptual processing is sufficient in such cases depends on the interaction of the task and graph. We have suggested that in more complex domains, in which the graphs and tasks are more complex, it is more likely that spatial processing will be required. However, we have also shown that spatial processing can be needed for simple tasks performed on simple graphs, depending on the graph type.

We believe that the reason spatial processing has not been part of graph comprehension models is that the focus on simple tasks and graphs has made it unnecessary, since in general, simple perceptual processes are sufficient to account for performance. Differences in performance on different graph types have been attributed to a better or worse match between the task and the graph. In addition, the strength of graphs as a form of representation is that they can make implicit things explicit (at least, good graphs do), so graphs are designed and selected so that it *is* possible to make direct comparisons between visual chunks. However, as tasks, domains, and visualizations become more complex, this transparency may not always be possible or even desirable (forecasters, for example, don't want simpler graphs, they want many variables represented). As graph comprehension research moves out of the psychology laboratory into the "real world" of practice, it will be more important for graph comprehension models to include spatial processing.

Although it may be rare for simple graph/task combinations to require spatial processing, such cases do exist. In addition to Simkin and Hastie's comparison judgment task, consider the case of a simple two-by-two psychology experiment, represented by a simple bar graph as in Figure 4. In this instance, the interaction is explicit, but in order to determine whether there is a main effect, the spatial strategy of mentally averaging the relevant bar heights is most likely used. Other problem-solving uses of graphs (e.g., (Cheng et al., 2001; Scaife & Rogers, 1996; Tabachneck-Schijf et al., 1997) are likely to involve spatial processing as well.

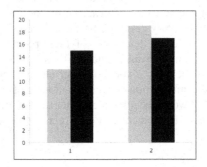

Fig. 4. Simple bar graph: Depending on the task, spatial processing may be used

In general, many spatial cognition tasks can be solved by a non-spatial strategy, for example, mathematically. If this strategy is chosen, there is obviously no spatial cognition at work. However, we believe that people will prefer spatial strategies in graph comprehension tasks, in part because the graph is a spatial array, and using spatial cognition does not require "translation" of the spatial code. Our model makes testable predictions as to when spatial processing will be used, and the type of strategy used in these instances where multiple strategies are possible is, in fact, an empirical question.

Our model extension embraces more recent refinements of the generic model discussed earlier. For example, eyetracking data from our studies (Trafton et al., 2002) confirms the iterative nature of perceptual processing (Carpenter & Shah, 1998). Furthermore, expertise appears to be an important factor in spatial processing of graphical information (c.f.,(Freedman & Shah, 2002)—both domain and graph knowledge. Without these two important types of knowledge, it is unlikely that a graph user would be able to generate the necessary spatial transformations to extract the needed information.

Without incorporating spatial processing, we believe current models of graph comprehension will be incomplete. Including spatial processing in our models will help us to understand why some representations might be better than others at a cognitive level, by shedding light on processes that underlie different graph/task interactions. It can help identify situations in which spatial processing is unavoidable and can help us make predictions about performance using these graphs. In some situations, it can help us design better graphs, by developing creative ways to reduce the number of spatial transformations required (for example, by facilitating direct comparisons that can be performed perceptually). In sum, we propose that including spatial processing is an important step in building a comprehensive model of graph comprehension.

References

Baddeley, A. D. (1999). *Essentials of human memory*. Hove, UK: Psychology Press, Ltd.

Baddeley, A. D., & Liebeman, K. (1980). Spatial working memory. In R. S. Nickerson (Ed.), *Attention and performance VIII* (pp. 521-539). Hillsdale, NJ: Lawrence Erlbaum Associates.

Bogacz, S., & Trafton, J. G. (2005). Understanding dynamic and static displays: Using images to reason dynamically. *Cognitive Systems Research 6*(4), 312-319.

Carpenter, P. A., & Shah, P. (1998). A model of the perceptual and conceptual processes in graph comprehension. *Journal of Experimental Psychology:Applied, 4*(2), 75-100.

Cheng, P. C.-H., Cupit, J., & Shadbolt, N. R. (2001). Supporting diagrammatic knowledge acquisition: An ontological analysis of cartesian graphs. *International Journal of Human-Computer Studies, 54*(457-494).

Cleveland, W. S. (1985). *The elements of graphing data*. Monterey, CA: Wadsworth.

DeSanctis, G., & Jarvenpaa, S. L. (1989). Graphical presentation of accounting data for financial forecasting: An experimental investigation. *Accounting, Organizations adn Society, 14*, 509-525.

Ericsson, K. A., & Simon, H. A. (1993). *Protocol analysis: Verbal reports as data*. (2nd ed.). Cambridge, MA: MIT Press.

Freedman, E. G., & Shah, P. (2002). *Toward a model of knowledge-based graph comprehension*. Paper presented at the Diagrammatic Representation and Inference, Second International Conference, Diagrams 2002, Callaway Gardens, GA, USA.

Gray, W. D., John, B. E., & Atwood, M. E. (1993). Project ernestine: Validating a goms analysis for predicting and explaining real-world performance. *Human-Computer Interaction, 8*(3), 237-309.

Hart, R. A., & Moore, G. T. (1973). The development of spatial cognition: A review. In R. M. Downs & D. Stea (Eds.), *Image and environment: Cognitive mapping and spatial behavior* (pp. 246-295). Chicago: Aldine Publishing Company.

Hegarty, M. (1992). Mental animation: Inferring motion from static displays of mechanical systems. *Journal of Experimental Psychology: Learning, Memory and Cognition, 18*(5), 1084-1102.

Kosslyn, S. M., Sukel, K. E., & Bly, B. M. (1999). Squinting with the mind's eye: Effects of stimulus resolution on imaginal and perceptual comparisons. *Memory and Cognition, 27*(2), 276-287.

Lohse, G. L. (1993). A cognitive model for understanding graphical perception. *Human Computer Interaction, 8*, 353-388.

Peebles, D., & Cheng, P. C.-H. (2003). Modeling the effect of task and graphical representation on response latency in a graph reading task. *Human Factors, 45*(1), 28-46.

Pinker, S. (1990). A theory of graph comprehension. In R. Freedle (Ed.), *Artificial intelligence and the future of testing* (pp. 73-126). Hillsdale, NJ: Lawrence Erlbaum Associates, Inc.

Roth, W.-M., & Bowen, G. M. (2003). When are graphs worth ten thousand words? An expert-expert study. *Journal of the Learning Sciences, 21*(4), 429-473.

Scaife, M., & Rogers, Y. (1996). External cognition: How do graphical representations work? *International Journal of Human-Computer Studies, 45*(185-213).

Shah, P., Freedman, E. G., & Vekiri, I. (2005). The comprehension of quantiative information in graphical displays. In P. Shah & A. Miyake (Eds.), *Cambridge handbook of visuospatial thinking* (pp. 426-476). New York, NY: Cambridge University Press.

Shepard, R., & Metzler, J. (1971). Mental rotation of three-dimensional objects. *Science, 171*, 701-703.

Simkin, D., & Hastie, R. (1987). An information-processing analysis of graph perception. *Journal of the American Statistical Association, 82*(398), 454-465.

Tabachneck-Schijf, H. J. M., Leonardo, A. M., & Simon, H. A. (1997). Camera: A computational model of multiple representations. *Cognitive Science, 21*(3), 305-350.

Trafton, J. G., Kirschenbaum, S. S., Tsui, T. L., Miyamoto, R. T., Ballas, J. A., & Raymond, P. D. (2000). Turning pictures into numbers: Extracting and generating information from complex visualizations. *International Journal of Human Computer Studies, 53*(5), 827-850.

Trafton, J. G., Marshall, S., Mintz, F. E., & Trickett, S. B. (2002). Extracting explicit and implicit information from complex visualizations. In M. Hegarty, B. Meyer & H. Narayanan (Eds.), *Diagramatic representation and inference* (pp. 206-220). Berlin Heidelberg: Springer-Verlag.

Trafton, J. G., Trickett, S., Stitzlein, C. A., Saner, L., Schunn, C. D., & Kirschenbaum, S. S. (in press). The relationship between spatial transformations and iconic gestures. *Spatial cogntion and computation.*

Trafton, J. G., & Trickett, S. B. (2001). A new model of graph and visualization usage. In J. D. Moore & K. Stenning (Eds.), *The proceedings of the twenty third annual conference of the cognitive science society* (pp. 1048-1053). Mahwah, NJ: Erlbaum.

Trafton, J. G., Trickett, S. B., & Mintz, F. E. (2005). Connecting internal and external representations: Spatial transformations of scientific visualizations. *Foundations of Science, 10*, 89-106.

Tversky, B. (1995). Cognitive origins of graphic productions. In F. T. Marchese (Ed.), *Understanding images: Finding meaning in digital imagery.* (pp. 29-53). New York: Springer-Verlag.

Author Index

Lecture Notes in Artificial Intelligence (LNAI)

Vol. 3835: G. Sutcliffe, A. Voronkov (Eds.), Logic for Programming, Artificial Intelligence, and Reasoning. XIV, 744 pages. 2005.

Vol. 3830: D. Weyns, H. V.D. Parunak, F. Michel (Eds.), Environments for Multi-Agent Systems II. VIII, 291 pages. 2006.

Vol. 3817: M. Faundez-Zanuy, L. Janer, A. Esposito, A. Satue-Villar, J. Roure, V. Espinosa-Duro (Eds.), Nonlinear Analyses and Algorithms for Speech Processing. XII, 380 pages. 2006.

Vol. 3814: M. Maybury, O. Stock, W. Wahlster (Eds.), Intelligent Technologies for Interactive Entertainment. XV, 342 pages. 2005.

Vol. 3809: S. Zhang, R. Jarvis (Eds.), AI 2005: Advances in Artificial Intelligence. XXVII, 1344 pages. 2005.

Vol. 3808: C. Bento, A. Cardoso, G. Dias (Eds.), Progress in Artificial Intelligence. XVIII, 704 pages. 2005.

Vol. 3802: Y. Hao, J. Liu, Y.-P. Wang, Y.-m. Cheung, H. Yin, L. Jiao, J. Ma, Y.-C. Jiao (Eds.), Computational Intelligence and Security, Part II. XLII, 1166 pages. 2005.

Vol. 3801: Y. Hao, J. Liu, Y.-P. Wang, Y.-m. Cheung, H. Yin, L. Jiao, J. Ma, Y.-C. Jiao (Eds.), Computational Intelligence and Security, Part I. XLI, 1122 pages. 2005.

Vol. 3789: A. Gelbukh, Á. de Albornoz, H. Terashima-Marín (Eds.), MICAI 2005: Advances in Artificial Intelligence. XXVI, 1198 pages. 2005.

Vol. 3782: K.-D. Althoff, A. Dengel, R. Bergmann, M. Nick, T.R. Roth-Berghofer (Eds.), Professional Knowledge Management. XXIII, 739 pages. 2005.

Vol. 3763: H. Hong, D. Wang (Eds.), Automated Deduction in Geometry. X, 213 pages. 2006.

Vol. 3755: G.J. Williams, S.J. Simoff (Eds.), Data Mining. XI, 331 pages. 2006.

Vol. 3735: A. Hoffmann, H. Motoda, T. Scheffer (Eds.), Discovery Science. XVI, 400 pages. 2005.

Vol. 3734: S. Jain, H.U. Simon, E. Tomita (Eds.), Algorithmic Learning Theory. XII, 490 pages. 2005.

Vol. 3721: A.M. Jorge, L. Torgo, P.B. Brazdil, R. Camacho, J. Gama (Eds.), Knowledge Discovery in Databases: PKDD 2005. XXIII, 719 pages. 2005.

Vol. 3720: J. Gama, R. Camacho, P.B. Brazdil, A.M. Jorge, L. Torgo (Eds.), Machine Learning: ECML 2005. XXIII, 769 pages. 2005.

Vol. 3717: B. Gramlich (Ed.), Frontiers of Combining Systems. X, 321 pages. 2005.

Vol. 3702: B. Beckert (Ed.), Automated Reasoning with Analytic Tableaux and Related Methods. XIII, 343 pages. 2005.

Vol. 3698: U. Furbach (Ed.), KI 2005: Advances in Artificial Intelligence. XIII, 409 pages. 2005.

Vol. 3690: M. Pěchouček, P. Petta, L.Z. Varga (Eds.), Multi-Agent Systems and Applications IV. XVII, 667 pages. 2005.

Vol. 3684: R. Khosla, R.J. Howlett, L.C. Jain (Eds.), Knowledge-Based Intelligent Information and Engineering Systems, Part IV. LXXIX, 933 pages. 2005.

Vol. 3683: R. Khosla, R.J. Howlett, L.C. Jain (Eds.), Knowledge-Based Intelligent Information and Engineering Systems, Part III. LXXX, 1397 pages. 2005.

Vol. 3682: R. Khosla, R.J. Howlett, L.C. Jain (Eds.), Knowledge-Based Intelligent Information and Engineering Systems, Part II. LXXIX, 1371 pages. 2005.

Vol. 3681: R. Khosla, R.J. Howlett, L.C. Jain (Eds.), Knowledge-Based Intelligent Information and Engineering Systems, Part I. LXXX, 1319 pages. 2005.

Vol. 3673: S. Bandini, S. Manzoni (Eds.), AI*IA 2005: Advances in Artificial Intelligence. XIV, 614 pages. 2005.

Vol. 3662: C. Baral, G. Greco, N. Leone, G. Terracina (Eds.), Logic Programming and Nonmonotonic Reasoning. XIII, 454 pages. 2005.

Vol. 3661: T. Panayiotopoulos, J. Gratch, R. Aylett, D. Ballin, P. Olivier, T. Rist (Eds.), Intelligent Virtual Agents. XIII, 506 pages. 2005.

Vol. 3658: V. Matoušek, P. Mautner, T. Pavelka (Eds.), Text, Speech and Dialogue. XV, 460 pages. 2005.

Vol. 3651: R. Dale, K.-F. Wong, J. Su, O.Y. Kwong (Eds.), Natural Language Processing – IJCNLP 2005. XXI, 1031 pages. 2005.

Vol. 3642: D. Ślęzak, J. Yao, J.F. Peters, W. Ziarko, X. Hu (Eds.), Rough Sets, Fuzzy Sets, Data Mining, and Granular Computing, Part II. XXIII, 738 pages. 2005.

Vol. 3641: D. Ślęzak, G. Wang, M. Szczuka, I. Düntsch, Y. Yao (Eds.), Rough Sets, Fuzzy Sets, Data Mining, and Granular Computing, Part I. XXIV, 742 pages. 2005.

Vol. 3635: J.R. Winkler, M. Niranjan, N.D. Lawrence (Eds.), Deterministic and Statistical Methods in Machine Learning. VIII, 341 pages. 2005.

Vol. 3632: R. Nieuwenhuis (Ed.), Automated Deduction – CADE-20. XIII, 459 pages. 2005.

Vol. 3630: M.S. Capcarrère, A.A. Freitas, P.J. Bentley, C.G. Johnson, J. Timmis (Eds.), Advances in Artificial Life. XIX, 949 pages. 2005.

Vol. 3626: B. Ganter, G. Stumme, R. Wille (Eds.), Formal Concept Analysis. X, 349 pages. 2005.

Vol. 3625: S. Kramer, B. Pfahringer (Eds.), Inductive Logic Programming. XIII, 427 pages. 2005.

Vol. 3620: H. Muñoz-Ávila, F. Ricci (Eds.), Case-Based Reasoning Research and Development. XV, 654 pages. 2005.

Vol. 3614: L. Wang, Y. Jin (Eds.), Fuzzy Systems and Knowledge Discovery, Part II. XLI, 1314 pages. 2005.

Vol. 3613: L. Wang, Y. Jin (Eds.), Fuzzy Systems and Knowledge Discovery, Part I. XLI, 1334 pages. 2005.

Vol. 3607: J.-D. Zucker, L. Saitta (Eds.), Abstraction, Reformulation and Approximation. XII, 376 pages. 2005.

Vol. 3601: G. Moro, S. Bergamaschi, K. Aberer (Eds.), Agents and Peer-to-Peer Computing. XII, 245 pages. 2005.

Vol. 3600: F. Wiedijk (Ed.), The Seventeen Provers of the World. XVI, 159 pages. 2006.

Vol. 3596: F. Dau, M.-L. Mugnier, G. Stumme (Eds.), Conceptual Structures: Common Semantics for Sharing Knowledge. XI, 467 pages. 2005.